Protein Formulation
and Delivery

DRUGS AND THE PHARMACEUTICAL SCIENCES

A Series of Textbooks and Monographs

Executive Editor

James Swarbrick

PharmaceuTech, Inc.
Pinehurst, North Carolina

Advisory Board

Larry L. Augsburger
University of Maryland
Baltimore, Maryland

Harry G. Brittain
Center for Pharmaceutical Physics
Milford, New Jersey

Jennifer B. Dressman
Johann Wolfgang Goethe University
Frankfurt, Germany

Anthony J. Hickey
University of North Carolina School of
Pharmacy
Chapel Hill, North Carolina

Jeffrey A. Hughes
University of Florida College of
Pharmacy
Gainesville, Florida

Ajaz Hussain
Sandoz
Princeton, New Jersey

Trevor M. Jones
The Association of the
British Pharmaceutical Industry
London, United Kingdom

Stephen G. Schulman
University of Florida
Gainesville, Florida

Vincent H. L. Lee
University of Southern California
Los Angeles, California

Elizabeth M. Topp
University of Kansas School of
Pharmacy
Lawrence, Kansas

Jerome P. Skelly
Alexandria, Virginia

Peter York
University of Bradford School of
Pharmacy
Bradford, United Kingdom

Geoffrey T. Tucker
University of Sheffield
Royal Hallamshire Hospital
Sheffield, United Kingdom

Protein Formulation and Delivery

Second Edition

Edited by

Eugene J. McNally

Gala Biotech, a Catalent Pharma Solutions Company
Middleton, Wisconsin, USA

Jayne E. Hastedt

ALZA Corporation
Mountain View, California, USA

informa

healthcare

New York   London

Informa Healthcare USA, Inc.
52 Vanderbilt Avenue
New York, NY 10017

© 2008 by Informa Healthcare USA, Inc.
Informa Healthcare is an Informa business

No claim to original U.S. Government works
Printed in the United States of America on acid-free paper
10 9 8 7 6 5 4 3 2

International Standard Book Number-10: 0-8493-7949-0 (Hardcover)
International Standard Book Number-13: 978-0-8493-7949-9 (Hardcover)

Library of Congress Cataloging-in-Publication Data

Protein formulation and delivery / edited by Eugene J. McNally, Jayne E. Hastedt. -- 2nd ed.
   p. ; cm. --   (Drugs and the pharmaceutical sciences ; 175)
   Includes bibliographical references and index.
   ISBN-13: 978-0-8493-7949-9 (hardcover : alk. paper)
   ISBN-10: 0-8493-7949-0 (hardcover : alk. paper)
   1. Protein drugs--Dosage forms. I. McNally, Eugene J., 1961- II. Hastedt, Jayne E.
III. Series: Drugs and the pharmaceutical sciences ; v.175.
   [DNLM: 1. Protein Conformation. 2. Drug Delivery Systems. 3. Drug Design.
4. Drug Stability. 5. Proteins--administration & dosage.   W1 DR893B v.175 2007
/ QU 55.9 P9667 2007]

RS431.P75P77 2007
615'.19--dc22   2007023435

Visit the Informa Web site at
www.informa.com

and the Informa Healthcare Web site at
www.informahealthcare.com

Preface

While working with the contributors on this second edition of *Protein Formulation and Delivery* it has been interesting to note the dramatic changes and advances that have taken place in the biotechnology field. One of the most significant has been the improvement in product titers in cell culture where progress has taken protein levels from tens of milligrams per liter to multi-grams per liter. The availability of larger amounts of protein and the resulting decrease in cost of goods has created new opportunities for products and delivery routes, and has made extensions to higher dose levels possible. All of these advances have had a cascade effect, placing increased expectations on formulation scientists to produce products with longer shelf-lives, at higher concentrations, that are stable at ambient storage conditions. This progressive period in the field of biotechnology presents significant challenges to the protein formulator.

In this edition we have strived to provide a basic understanding of areas of formulation and delivery for first-time formulators as well as analysts working in the field, while providing examples of new and novel areas that are of interest to seasoned protein formulators. We have added more drug delivery chapters, expanded the characterization and formulation sections, and included a new chapter on specifications in an effort to prepare professionals in the field to address these current challenges.

We would like to thank all of the contributors to the first edition of *Protein Formulation and Delivery* and those who have returned to update their original work, as well as all of the new contributors, who have enabled us to offer readers a new perspective on this stimulating area of formulation and drug delivery.

Eugene J. McNally
Jayne E. Hastedt

Contents

Contributors

Ehud Arbit Emisphere Technologies, Inc., Tarrytown, New York, U.S.A.

Tony Auffret Sandwich Laboratories, Pfizer Ltd, Kent, U.K.

Frank K. Bedu-Addo Integritas Drug Development Consulting, LLC, Cincinnati, Ohio, U.S.A.

Brooks Boyd Zogenix, Inc., Emeryville, California, U.S.A.

Paul Bridges Genentech, Inc., San Francisco, California, U.S.A.

Paul M. Bummer University of Kentucky, Lexington, Kentucky, U.S.A.

Andrew R. Clark Nektar Therapeutics, San Carlos, California, U.S.A.

Stephen J. Farr Zogenix, Inc., Emeryville, California, U.S.A.

Larry A. Gatlin Parenteral Center of Emphasis, Groton Laboratories, Pfizer Inc., Groton, Connecticut, U.S.A.

Isabel Gomez-Orellana Emisphere Technologies, Inc., Tarrytown, New York, U.S.A.

Carol Hasselbacher Stability, Corporate Product Quality, Amgen Inc., Longmont, Colorado, U.S.A.

Jayne E. Hastedt ALZA Corporation, Mountain View, California, U.S.A.

Helmut Hoffmann Boehringer Ingelheim Pharma GmbH & Co. KG, Biberach an der Riss, Germany

Sandy Koppenol   Nastech Pharmaceutical Company, Inc., Bothell, Washington, U.S.A.

Lawrence S. Linn   Zogenix, Inc., Emeryville, California, U.S.A.

Shingai Majuru   Emisphere Technologies, Inc., Tarrytown, New York, U.S.A.

Paul McGoff   ZymoGenetics, Seattle, Washington, U.S.A.

Eugene J. McNally   Gala Biotech, a Catalent Pharma Solutions Company, Middleton, Wisconsin, U.S.A.

Anthony Mire-Sluis   Corporate Product Quality, Amgen Inc., Thousand Oaks, California, U.S.A.

Wassim Nashabeh   Regulatory Affairs, Genentech Inc., San Francisco, California, U.S.A.

Sandra Pisch-Heberle   Boehringer Ingelheim Pharma GmbH & Co. KG, Biberach an der Riss, Germany

David S. Scher   Alkermes, Inc., Cambridge, Massachusetts, U.S.A.

Evgenyi Y. Shalaev   Parenteral Center of Emphasis, Groton Laboratories, Pfizer Inc., Groton, Connecticut, U.S.A.

Steven J. Shire   Genentech, Inc., San Francisco, California, U.S.A.

Stanley M. Speaker   Pharmaceutical Sciences–Global Biologics, Pfizer Inc., Chesterfield, Missouri, U.S.A.

Cynthia L. Stevenson   Nektar Therapeutics, San Carlos, California, U.S.A.

Dirk L. Teagarden   Pharmaceutical Sciences–Global Biologics, Pfizer Inc., Chesterfield, Missouri, U.S.A.

Wei Wang   Pharmaceutical Sciences–Global Biologics, Pfizer Inc., Chesterfield, Missouri, U.S.A.

James Wright   Alkermes, Inc., Cambridge, Massachusetts, U.S.A.

1

Overview of Protein Formulation and Delivery

James Wright

Alkermes, Inc., Cambridge, Massachusetts, U.S.A.

PROTEIN AND PEPTIDE REACTIONS

The experience of developing a number of protein and peptide formulations has taught me to look for three principal degradation pathways in formulations of proteins: deamidiation, oxidation, and aggregation. One of the most important aspects of these degradation pathways for proteins is the potential to diminish the potency and increase the immunogenicity of a biopharmaceutical product. A significant example of a biologic product that has shown formulation-dependent immunogenicity issues was observed in Exprex® (erythropoietin) between 1998 and 2003. The effect of a formulation change was a ten-fold increase in the incidence of pure red-cell aplasia in patients. Formulation changes, extraction of organics from stoppers, and the route of administration may have all played a role in the observed increase in incidence of red-cell aplasia for this product. This complex and important story emphasizes the need for having a broad awareness of the potential interactions of formulation, container, analytical methods, and route of delivery with a biological. This book is an attempt to present the broad range of knowledge required to formulate protein drug products.

Bummer and Koppenol cover two critical degradation pathways in depth in Chapters 2 and 3. The problems associated with these protein degradation processes are extensively discussed in these chapters for good reason: they are the constant unwanted companions of development scientists in the protein delivery field. The deamidation reaction is primarily the result of the hydrolysis of asparagine to aspartic acid. Thus, the reaction is most likely to proceed in an aqueous environment. If the protein is formulated in the solid state, the reaction rate is minimized. As discussed, this reaction has been extensively studied, and the effects of pH, temperature, and buffers are well known and well documented. This extensive literature gives development scientists the opportunity to systematically optimize solution stability with respect to deamidation. In addition, the reactivity of asparagine is largely determined by neighboring residues, and thus, the reactivity of the protein or peptide toward deamidation can be predicted from structure.

Oxidation is in many ways a more complex process than deamidation in that the reactants and catalysts of the reaction are numerous and complex. There are several matters of significant concern; these include the ability to cause oxidation of specific residues through metal ion catalysis, by vapor phase hydrogen peroxide, photochemically, or through organic solvent, or even to diminish long-term stability in the presence of pharmaceutical excipients. Thus, each unit operation in the production and storage of a protein product is a potential source of oxidation.

The conformational stability of the protein is also of fundamental concern in formulating proteins. Problems with conformational stability are often expressed as soluble aggregates and/or insoluble particulates (large aggregates). Aggregated proteins are a significant problem in that they are associated with decreased bioactivity and increased immunogenicity. The potential to produce aggregated forms is often enhanced by exposure of the protein to shear, liquid–air, liquid–solid, and even liquid–liquid interfaces. This means that there is the potential to denature and aggregate a protein in almost any unit operation in the downstream processing of a protein, including the formulation of it. However, a protein can often be stabilized against aggregation by optimizing pH, temperature, and ionic strength, and through the use of surfactants.

PREFORMULATION AND ANALYTICAL DEVELOPMENT

The development of stability-indicating analytical methods (see Chapter 4) for the drug product usually starts with the preformulation studies. During the preformulation studies, methods are selected and the first evaluation of the protein on short-term stability is conducted. Since one of the most significant purposes of the preformulation work is to define the processes that destroy protein integrity and activity, it is critical that the analytical methods chosen for the task indeed be indicative of stability. The three basic protein degradation processes (deamidation, oxidation, and aggregation) must be followed by accurate, precise, and simple analytical methods.

Bedu-Addo has prepared a very complete review of analytical mythologies and preformulation studies designed to support the development of proteins and peptides. The review starts with protein charge, size, and aggregation state. The chemical reaction of deamidation of a protein or a peptide is a common and important chemical decomposition reaction. The deamidation of a protein can be followed using a number of techniques: isoelectric focusing and ion exchange chromatography are two of the most common. Each of these analytical techniques allows the development scientist to follow the disappearance of the starting isoforms and the generation of reaction products. Often, these reaction products (the more deamidated state of the molecule) are easy to isolate and test in bioactivity assays and in pharmacokinetic studies. These activity studies are important steps to take when there is concern that the molecule is prone to deamidation, since the practical consequences of deamidation can be judged scientifically.

Oxidation is often determined using reversed-phase high-performance liquid chromatography or by peptide analysis. Aggregation is usually determined by size exclusion chromatography (SEC) or by sodium dodecyl sulfate-polyacrylamide gel electrophoresis (SDS-PAGE). These two techniques give different answers because they address different questions. SDS-PAGE detects and quantifies irreversible aggregates, whereas SEC will detect noncovalent aggregates as well. The distinction is critical, since the stability of the dosage form may be judged by either or both methods. Bedu-Addo also reviews biophysical characterization techniques and interpretation for proteins.

During the preformulation and analytical development phase, there is opportunity to determine the correlation between the different analytical techniques and the biological activity assay. Often, an activity assay is developed that can judge the suitability and stability of the dosage form. It is imperative to know early in the development process as to which analytical techniques predict and which degradation processes correlate with losses in activity. Understanding this relationship between biological activity and independent analytical characterization is an important aspect of the preformulation package.

SOLUTION FORMULATION STABILITY

McGoff and Scher review the issues of protein solubility, aggregation, and adsorption. The authors provide a systemic framework for evaluating the physical stability of proteins. The problems associated with physical instability are serious in that they lead to aggregation, adsorption to surfaces, and loss of biological activity. Thus, formulation studies to determine optimal stabilization conditions should include variables such as temperature, solution pH, buffer ion, salt concentration, protein concentration, and the effect of surfactants. In addition, the protein should be characterized and stabilized against adsorption and/or denaturation at interfaces.

The effect of concentration of the protein on the stability of the formulation is also a critical concern. Proteins undergo concentration-dependent aggregation and adsorption, and thus the effect of protein concentration on physical stability must be accounted for. The testing of the adsorption potential of a protein with the range of materials that might be used in packaging or processing is an important part of the formulation effort.

SOLID-STATE FORMULATION STABILITY

The pharmaceutical solid-state product of a protein or peptide is usually produced by freeze-drying. A freeze-dried product is produced by a process that, if done correctly, is friendly to protein structure and activity and can prolong the shelf life of the product. The reason that proteins have enhanced stability in the solid state is that molecular mobility is drastically reduced.

A protein is usually present as an amorphous solid in a lyophilized product, with a glass transition temperature associated with the material. In very rough terms, at temperatures above the glass transition temperature, there is significant molecular mobility, whereas below the glass transition temperature, reduced mobility, and reaction rates are achieved. Thus, the stabilization of a protein is dependent on the use of a freeze-drying process and formulation that enhances protein conformation and reduces molecular mobility. Reduced mobility and enhanced protein stability is achieved through temperature, moisture, and excipient control during and after the freeze-drying process. A very systematic review of freeze-drying and the selection excipients are presented in Chapters 8 and 9.

PROTEIN DELIVERY

Approaches to the challenging problems of protein drug delivery are broadly covered by authors who have practical development experience. Solution formulation development is covered thoroughly by McGoff and Scher with high-concentration formulation development reviewed by Stevenson. In addition, needle-free injectors are reviewed by Farr, Boyd, Bridges, and Linn. The injectable (subcutaneous, intramuscular, or intravenous) product is usually the first dosage form developed for delivery of a protein and becomes the standard by which other routes of delivery are judged by. As was seen with Eprex, seemingly simple changes in protein formulations can lead to very significant changes in product performance.

Pulmonary delivery of proteins presents important advantages over injectable formulations and the formulation of proteins for pulmonary delivery is reviewed by Clark. Pulmonary delivery makes administration of the protein much more patient friendly. However, the stability of drug product, consistency of delivery, complexity of the delivery device, and the concerns over safety are all significant development challenges that must be overcome for application of pulmonary delivery of proteins.

CONCLUSION

The development of a formulation for a protein is an amazingly complex task. First of all, proteins have complex structures and often contain structural heterogeneity. This is why a strong preformulation program is necessary. Also, because of this complexity and heterogeneity, a wide array of chemical, physical, spectroscopic, and biological assays are required to adequately characterize a protein and to determine its stability in the formulation.

The complexity in protein structure is matched by the complexity of the ways this fragile structure can undergo degradation. As we know, the manufacturing process can have a profound impact on protein structure through pH, heat, oxidation, interfacial degradation, and shear, just to name a few of the potential variables to control. Excipients, containers, delivery systems, and container closure systems may also interact with the formulated protein and are thus critical aspects of the development process. The complexity of protein structure, the wide array of sophisticated analytical techniques required to fully characterize protein formulations, and the many different paths of protein degradation make it clear that development of a stable well-characterized protein product is a demanding challenge. It is also clear that even small changes in the formulation or process may lead to significant changes in product performance as reviewed in the chapter presented by Hastedt and McNally.

2

Chemical Considerations in Protein and Peptide Stability

Paul M. Bummer

University of Kentucky, Lexington, Kentucky, U.S.A.

DEAMIDATION

Introduction

The deamidation reactions of asparagine (Asn) and glutamine (Gln) side-chains are among the most widely studied nonenzymatic covalent modifications to proteins and peptides (1–7). Considerable research efforts have been extended to elucidate the details of the deamidation reaction in both in vitro and in vivo systems, and a number of well-written, in-depth reviews are available (1–5,8,9). This work touches only on some of the highlights of the reaction and on the roles played by pH, temperature, buffer, and other formulation components. Possible deamidation-associated changes in the protein structure and state of aggregation also are examined. The emphasis is on Asn deamidation, since Gln is significantly less reactive.

Reaction Mechanism

The primary reaction mechanism for the deamidation of Asn in water-accessible regions of peptides and proteins at basic or neutral conditions is shown in Figure 1. For the present, discussion is confined to the intramolecular mechanism, uncomplicated by adjacent amino acids at other points in the primary sequence. Under alkaline conditions, the key step in the reaction is the formation of a deprotonated amide nitrogen, which carries out the rate-determining nucleophilic attack on the side-chain carbonyl, resulting in a tetrahedral intermediate and finally the formation of the five-member succinimide ring. For such a reaction, the leaving group must be

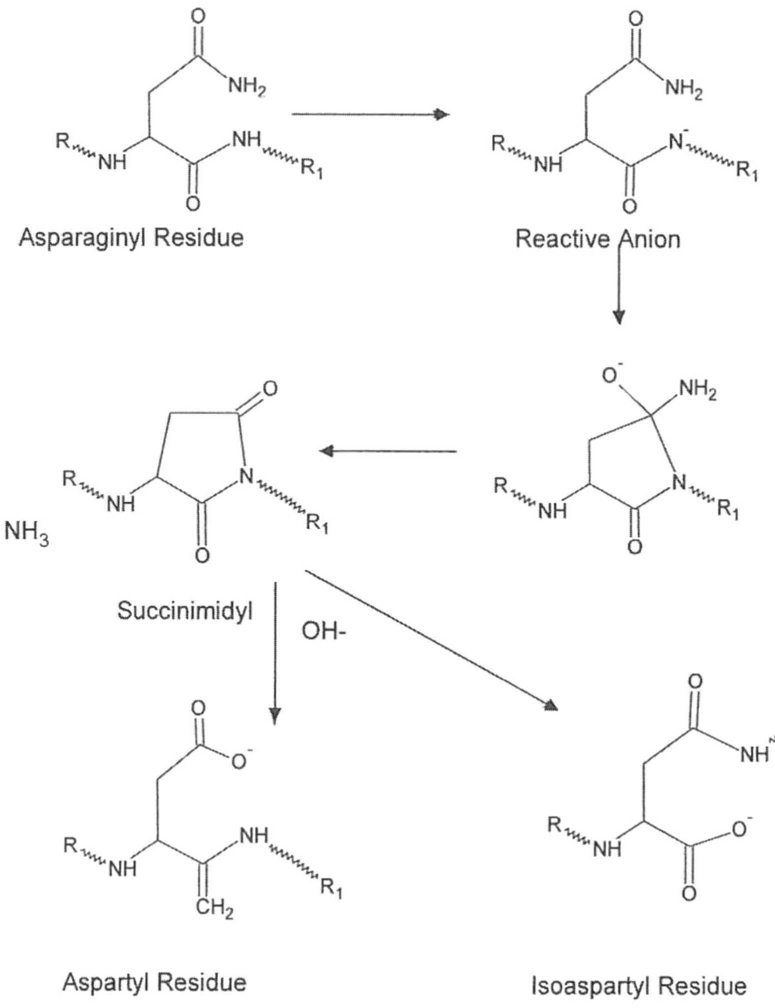

Figure 1 Proposed reaction mechanism for deamidation of asparaginyl residue. Note the formation of the succinimidyl intermediate and the two possible final products.

easily protonated, and in this case, it is responsible for the characteristic formation of ammonia (NH_3). The succinimide ring intermediate is subject to hydrolysis, resulting in either the corresponding aspartic acid or the isoaspartic acid (β-aspartate). Often, the ratio of the products is 3:1, isoaspartate to aspartate (10–12). In the case of acid catalysis (pH < 3), a tetrahedral intermediate is also formed, but breaks down with the loss of NH_3 without going through the succinimide (13–17). The reaction also appears to be sensitive to racemization at the α-carbon, resulting in mixtures of D- and L-isomers (10,13–15). The rate of degradation of the parent peptide in aqueous media often follows pseudo-first-order kinetics (16,17).

A number of other alternative reactions are possible. The most prevalent reaction appears to be a nucleophilic attack of the Asn side-chain amide nitrogen on the peptide carbonyl, resulting in main-chain cleavage (10,16,18). This reaction (Fig. 2) is slower than that of cyclic imide formation and is most frequently observed when Asn is followed by proline, a residue incapable of forming an ionized peptide-bond nitrogen.

pH Dependence

Under conditions of strong acid (pH 1–2), deamidation by direct hydrolysis of the amide side-chain becomes more favorable than formation of cyclic imide (16,19). Under these extreme conditions, the reaction is often complicated by main-chain cleavage and denaturation. Deamidation by this mechanism is not likely to produce isoaspartate or significant racemization (16).

Under more moderate conditions, the effect of pH is the result of two opposing reactions: (*i*) deprotonation of the peptide-bond nitrogen, promoting

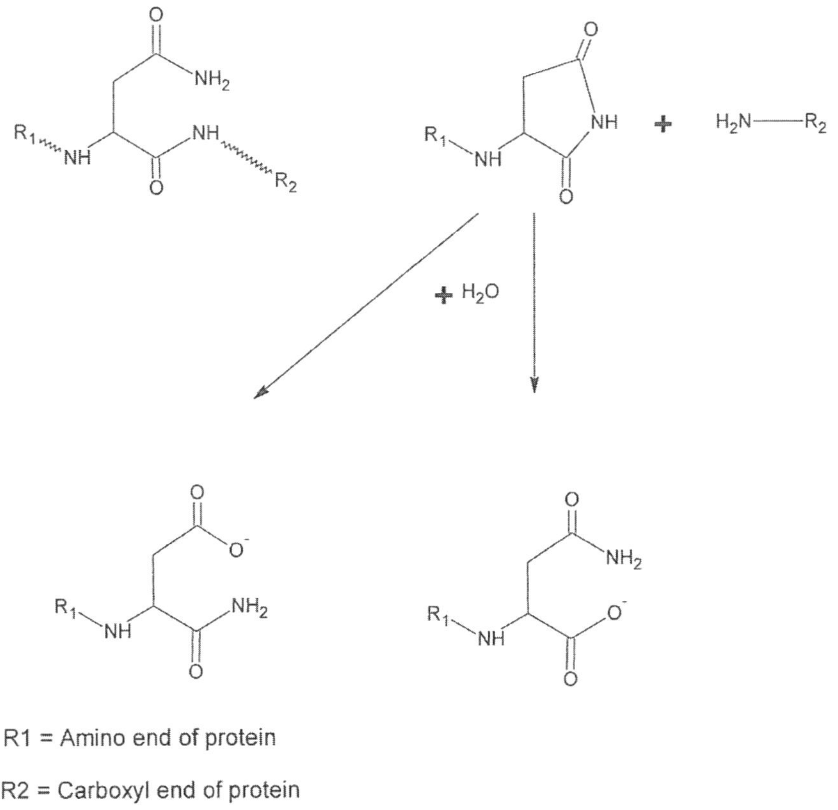

R1 = Amino end of protein

R2 = Carboxyl end of protein

Figure 2 Proposed reaction mechanism for main-chain cleavage by asparaginyl residues.

the reaction and (*ii*) protonation of the side-chain–leaving group, inhibiting the reaction. In deamidation reactions of short chain peptides uncomplicated by structural alterations or covalent dimerization (20), the pH-rate profiles exhibit the expected "V" shape, with a minimum occurring in the pH range of 3 to 4 (16). Computation studies by Peters and Trout (21) have been helpful in shedding light on the effect of pH. These authors have suggested that under mildly acidic conditions (3 < pH, 4), the rate-limiting step is the attack of the deprotonated nitrogen on the side-chain. The rate-limiting step at neutral pH is the hydrogen transfer reaction, while under basic conditions (pH > 7), it is the elimination of NH_2^- from the tetrahedral intermediate. Experimental studies have shown that the increase in rate on the alkaline side of the minimum does not strictly correlate with the increase in deprotonation of the amide nitrogen, indicating that the rate of reaction is not solely dependent on the degree of the peptide-bond nitrogen deprotonation (16,19). The pH minimum in the deamidation reaction measured in vitro for proteins may (22) or may not (23) fall in the same range as that of simple peptides. Overall pH-dependent effects may be modified by structure-dependent factors, such as dihedral angle flexibility, water accessibility, and proximity of neighboring amino acid side-chains (see section Peptide and Protein Structure).

Effect of Temperature

The temperature dependence of the deamidation rate has been studied in a variety of simple peptides in solution (16,24,25). Small peptides are easily designed to avoid competing reactions, such as oxidation and main-chain cleavage, and are thus useful to isolate attention directly on the deamidation rate. In solution, deamidation of small peptides tends to follow an Arrhenius relationship. Activation energies of the reaction do tend to show pH dependence, and a discontinuity in the Arrhenius plot is expected when the mechanism changes from direct hydrolysis (acid pH) to one of cyclic imide (mildly acidic to alkaline pH).

The deamidation rate of proteins also shows temperature dependence (23,26,27) under neutral pH. For deamidation reactions alone, temperature-associated rate acceleration in proteins may be due to enhanced flexibility of the molecule, allowing more rapid formation of the cyclic imide (28), or it may occur by catalysis by side-chains brought into the vicinity of the deamidation site (5).

The availability of water appears to be an important determinant in temperature-associated effects. In studies of lyophilized formulation of Val-Tyr-Pro-Asn-Gly-Ala, the deamidation rate constant was observed to increase about an order of magnitude between 40°C and 70°C (29). In contrast, in the solid state, the Arrhenius relationship was not observed. Further, the deamidation in the solid state showed a marked dependence upon the temperature when the peptide was lyophilized from a solution of pH 8, while little temperature dependence was observed when lyophilization proceeded from solutions at either pH 3.5 or pH 5. The authors related this temperature difference to changes in the reaction mechanism that may occur as a function of pH.

Adjuvants and Excipients

The influence on deamidation by a variety of buffer ions and solvents has been examined. As pointed out by Cleland et al. (4) and reinforced by Tomizawa et al. (13), many of these additives are unlikely to be employed as pharmaceutical excipients for formulation, but they may be employed in protein isolation and purification procedures (30). Important clues to stabilization strategies can be gained from these studies. In the following, it is fruitful to keep in mind the importance of the attack of the ionized peptide-bond nitrogen on the side-chain carbonyl and the hydrolysis of the cyclic imide (Fig. 1).

Buffers

Buffer catalysis appears to occur in some but not all peptides and proteins studied (5). Bicarbonate (16) and glycine (12) buffers appear to accelerate deamidation. On one hand, the phosphate ion has been shown to catalyze deamidation, both in peptides and in proteins (12,13,16,31–34), generally in the concentration range of 0 to 20 mM. Capasso et al. (35,36) observed the acceleration of deamidation by acetate, carbonate, Tris, morpholine, and phosphate buffers only in the neutral to basic pH ranges. On the other hand, Lura and Schrich (37) found no influence on the rate of deamidation of Val-Asn-Gly-Ala when buffer components (phosphate, carbonate, or imidazole) were varied from 0 to 50 mM. A general acid–base mechanism by which the phosphate ion catalyzes deamidation was challenged in 1995 by Tomizawa et al. (13), who found that the rate of lysozyme at 100°C did not exhibit the expected linear relationship of deamidation rate on phosphate concentration. Although not linked to deamidation, it is worthwhile to note that at pH = 8 and 70°C, tris(hydroxymethyl) aminomethane buffer (Tris) has been shown to degrade to liberate highly reactive formaldehyde in forced stability studies of peptides (38).

Ionic Strength

The effects of ionic strength appear to be complicated and not open to easy generalizations. Buffer and ionic strength effects on deamidation are evident in proteins at neutral to alkaline pH (5). In selected peptides and proteins, the catalytic activity of phosphate has been shown to be reduced moderately in the presence of salts NaCl, LiCl, and Tris HC1 (12,13). Of these salts, NaCl showed the least protective effect against deamidation (13).

In the peptide Gly-Arg-Asn-Gly at pH 10, 37°C, the half-life $t_{1/2}$ of deamidation dropped from 60 hours to 20 hours when the ionic strength was increased from 0.1 to 1.2 (22). However, in the case of Val-Ser-Asn-Gly-Val at pH 8, 60°C, there was no observable difference in the $t_{1/2}$ of deamidation when solutions without salt were compared to those containing 1 M NaCl or LiCl (12). Interestingly, for lysozyme at pH 4 and 100°C, added salt showed a protective effect against deamidation, but only in the presence of the phosphate ion (13).

In reviewing the data above, Brennan and Clarke (17) tentatively attributed the promotion of deamidation by elevated levels of ions to enhanced stabilization of the ionized peptide-bond nitrogen, promoting attack on the side-chain amide carbonyl. Other mechanisms would include disruption of tertiary structure in proteins that may have stabilized Asn residues, in some as-yet unknown fashion. That promotion of deamidation is observed in some cases of peptides, and inhibition in others does suggest rather complex and competing effects. Clearly, the stabilizing effects, when observed at all, are often at levels of salt too concentrated for most pharmaceutical formulations.

Solvents

The effect of various organic solvents on the rate of deamidation has not received much attention; it would be expected, however, that in the presence of a reduced dielectric medium, the peptide-bond nitrogen would be less likely to ionize. Since the anionic peptide-bond nitrogen is necessary in the formation of the cyclic imide, a low dielectric medium would retard the progress of the reaction and be reflected in the free energy difference for ionization of the peptide-bond nitrogen (17). Following this hypothesis, Brennan and Clarke (39) analyzed succinimide formation of the peptide Val-Tyr-Pro-Asn-Gly-Ala [the same peptide employed by Patel and Borchardt (16) in studies of pH effects in aqueous solution] as a function of organic cosolvent (ethanol, glycerol, and dioxin) at constant pH and ionic strength. The lower dielectric constant media resulted in a significantly lower rate of deamidation, in agreement with the hypothesis. It was argued that the similar rates of deamidation for different cosolvent systems of the same effective dielectric constant indicated that changes in viscosity and water content of the medium did not play a significant role.

The effect of organic cosolvents on deamidation in proteins is even less well characterized than that of peptides. Trifluoroethanol (TFE) inhibits deamidation of lysozyme at pH 6 and 100°C (13), and of the dipeptide Asn-Gly, but does not inhibit the deamidation of free amino acids. The mechanism of protection is not clear; direct interaction of the TFE with the peptide bond was postulated, but not demonstrated. An alternative hypothesis is that TFE induces greater structural rigidity in the protein, producing a structure somewhat resistant to the formation of the cyclic imide intermediate. Other, pharmaceutically acceptable solvents, ethanol and glycerin, did not exhibit the same protective effects as TFE on lysozyme.

Of course, in dosage form design, organic solvents such as TFE are not useful as pharmaceutical adjuvants. The effects of low dielectric may still supply a rationale for the solubilization of peptides in aqueous surfactant systems, where the hydrophobic region of a micelle or liposome could potentially enhance the stabilization of the Asn residues from deamidation. As pointed out by Brennan and Clarke (17), the results of experiments in organic solvents can have implications on the prediction of points of deamidation in proteins as well. For Asn residues near the surface of the protein, where the dielectric constant is expected to approach that of water, the deamidation rate would be expected to be high. For Asn residues buried in more hydrophobic regions of the protein, where polarities are thought to be more in line with that of ethanol or dioxane (40), reaction rates would be expected to be considerably slower.

Computational studies on the effects of solvent on the reaction were carried out recently by Catak et al. (41). They report that, in the absence of water, the overall activation energy barrier is on the order of 50 kcal/mol, and that this drops to a value of about 30 kcal/mol in the presence of water. In all, about three water molecules participate directly in the reaction, assisting in hydrogen transfer and in the cyclization outlined in Figure 1.

Polymers and Sugars

Considerable interest has developed in the stabilization of proteins and peptides in solid matrices, either polymeric or sugar based. In most solid polymer matrices, the primary role is to improve pharmacokinetic and pharmcodynamic properties of the active by modifying release characteristics and most studies are designed with this intention in mind (42). Sugars are usually employed as an aid to lyophilization of proteins, with the intent of maintaining the tertiary structure and preventing aggregation (43).

The state of the polymer and the activity of water appear to be critical factors in the stabilization of the peptide against deamidation. In general, the observed degradation rate constants exhibit the following rank order: solution > rubbery polymer > glassy polymer (38,44–46). However, this observation does not appear to be valid in every case (44). It has been proposed that up to 30% of a peptide may bind to polyvinylpyrrolidone (PVP) in the solution state, complicating the kinetic analysis (47).

Peptide stability in polymer matrices that are themselves also undergoing degradation provides a unique challenge. For example, it has been observed that PVP may form adducts with the *N*-terminus of peptides (48). Systematic studies of the deamidation of a model peptide in films of the copolymer polylactic–glycolic acid (PLGA) have shown that the reaction is the primary route of degradation only after longer storage times at higher water content (49). The delay in the onset of deamidation of peptide in PLGA may be related to the time necessary to establish an "acidic microclimate" that arises from the hydrolysis of the polymer (50). In support of this acid-catalyzed deamidation hypothesis in PLGA films, the reaction product, isoaspartate was not found.

Computational studies may supply additional insight. Computer simulations of the mobility of peptide, water, NH_3, and polymer in PVP matrix have been carried out by Xiang and Anderson (51). They observed that the diffusivity of water, NH_3, and peptide were between two and three orders of magnitude slower in PVP compared to aqueous solution. Importantly, the conformational dynamics of the peptide in the glassy polymer exhibited a higher energy barrier between states than seen for the peptide in water. Thus, two of the critical events in the process of deamidation, the conformational changes necessary to form the cyclic intermediate in the glassy polymer and the diffusion away of the NH_3 after release, are both slowed considerably in the solid state.

The effect of sugars on the deamidation of a model peptide has been examined. At pH = 7, a solution of peptide in a 5% sucrose or mannitol reduced the deamidation rate to about 16% of that found in the absence of sugars (52). When stored in the solid state, the rate of reaction was even slower, although sucrose appeared to stabilize the peptide to a greater extent than did mannitol. It was observed that sucrose remained amorphous during the test period while mannitol crystallized, complicating the interpretation of the data (53). Cleland et al. (43) determined that 360:1 was the optimal sugar–antibody molar ratio necessary to inhibit aggregation and deamidation over a three-month period. Sugars sucrose, trehalose, and mannitol were able to stabilize the protein so long as less than 8.4% moisture was present.

Our understanding of the stabilization of peptides and proteins in polymer and sugar matrices is far from complete, and additional insight into the molecular mechanism might benefit from the bounty of studies carried out with small molecules in similar systems. Experiments must be designed carefully and interpreted with caution so as to clearly separate the solvent effects of water and perhaps even NH_3 on the reaction from the plastisizing effects on the matrix.

Peptide and Protein Structure

The ability to identify which Asn or Gln residues in a therapeutic protein or peptide may be vulnerable to deamidation would have great practical application in preformulation and formulation studies. The effects of various levels of structure—primary, secondary, and tertiary—are believed to be complex and varied. At present, only primary structure effects have been characterized in a systematic manner.

Primary Sequence

The primary sequence of amino acids in a peptide or protein is often the first piece of structural data presented to the formulation scientist. Considerable effort has been spent to elucidate the influence of flanking amino acids on the rates of deamidation of Asn and Gln residues. The potential effects of flanking amino acids are best elucidated in simple peptides, uncomplicated by side reactions or secondary and tertiary structure effects.

Effect of amino acids preceding Asn or Gln: In an extended series of early studies, Robinson and Rudd (24) examined the influence of primary sequence on the deamidation of Asn or Gln in the middle of a variety of pentapeptides. Mild physiologic conditions (pH 7.5 phosphate buffer at 37°C) were employed. A few general rules can be extracted from this work:

1.   In practically every combination tested, Gln residues were less prone to deamidation than Asn. For the two residues placed in the middle of otherwise identical host peptides, the half-life of the reactions differed by a factor ranging from two- to threefold.

2. In peptides Gly-X-Asn-Ala-Gly, steric hindrance by unionized X side-chains inhibits deamidation. The rank order of deamidation rate found was Gly > Ala > Val > Leu > Ile, with the $t_{1/2}$ ranging from 87 to 507 days. It remains unclear why bulky residues inhibit the reaction, but reduced flexibility of the sequence may be a factor. A similar effect was noted when Gln replaced Asn. In this case, $t_{1/2}$ ranged from 418 to 3278 days, in accordance with the diminished reactivity of Gln.

3. For the same host peptide, when the X side chain was charged, the deamidation rate of Asn followed the rank order of Asp > Glu > Lys > Arg.

Effect of amino acids following Asn or Gln: Early experiments on dipeptides under extreme conditions indicated a particular vulnerability of the Asn-Gly sequence to deamidation (54). More recent studies of adrenocorticotropic hormone (ACTH)-like sequence hexapeptide Val-Tyr-Pro-Asn-Gly-Ala under physiologic conditions (55) have verified that deamidation is extremely rapid ($t_{1/2}$ of 1.4 days at 37°C). The formation of the succinimide intermediate is thought to be the basis for the sequence dependence (10) of deamidation. It is generally believed that bulky residues following Asn may inhibit sterically the formation of the succinimide intermediate in the deamidation reaction.

Steric hindrance of the cyclic imide formation is not the only possible genesis of sequence-dependent deamidation. The resistance to cyclic imide formation in the presence of a carboxyl-flanking proline peptide may be related to the inability of the prolyl amide nitrogen to attack the Asn side-chain (10). The computational studies of Radkiewicz et al. (56) suggest that the effect of the adjacent residue may largely be attributed to electrostatic/inductive effects influencing the ability of the peptide nitrogen atom to ionize (as seen in Fig. 1). In the case of glycine, the inductive effect is insufficient to explain the results, and the authors argue that the ability of glycine to sample more conformational space compared to other amino acids may help stabilize the nitrogen anion. Experimentally, the replacement of the glycyl residue with the more bulky leucyl or prolyl residues resulted in a 33- to 50-fold (respectively) decrease in the rate of deamidation (10). Owing to the highly flexible nature of the dipeptide, the deamidation rate observed in Asn-Gly is thought to represent a lower limit.

In more recent studies, deamidation of Val-Tyr-X-Asn-Y-Ala, a peptide sequence derived from ACTH, was examined with different residues in both flanking positions (57). When X was histidine (and Y is glycine), no acceleration of deamidation was found relative to a peptide where X is proline. Placing a His following the Asn was found to result in similar rates of deamidation when X was phenylalanine, leucine, or valine. The rate when X was histidine was slower than that of alanine, cysteine, serine, or glycine. These results indicate that histidine does not have unique properties in facilitating succinimide formation. Of interest was the observation that histidine on the carboxyl side of the Asn did seem to accelerate main-chain cleavage products.

Some of the general rules for peptides may also show higher levels of dependence on primary sequence. Tyler-Cross and Schrich (12) studied the influence of different amino acids on the adjacent amino end of the pentapeptide Val-X-Asn-Ser-Val at pH 7.3. For X = His, Ser, Ala, Arg, and Leu, deamidation rates were essentially constant and approximately seven times slower than the Val-Ser-Asn-Gly-Val standard peptide. Of special interest to the investigators was the observation of no difference in deamidation rates between those amino acids with and without β-branching (such as valine for glycine). This is in direct contrast to the findings of Robinson and Rudd (24) of 10-fold differences in deamidation for valine substitution for glycine in Gly-X-Asn-Ala-Gly, shown earlier. Under the mild alkaline conditions of Patel and Borchardt (16), Val-Tyr-X-Asn-Y-Ala, no difference in the deamidation rate constants was observed when proline was substituted for glycine in the X position.

Data mining: Data-mining approaches have been employed to formulate a semiquantitative means of predicting the effect of primary structure on rate of deamidation. Capasso (58) proposed the extrathermodynamic relationship shown in Equation 1:

$$\text{Log } k_1 = X_p + Z_{Asn} + Y_p \tag{1}$$

Here k_1 is the observed rate constant for deamidation, X_p is the average contribution of the specific amino acid that precedes Asn, Y_p is the average contribution due to the amino acid that follows Asn, and Z_{Asn} is the value when both the preceding and following amino acids are glycine. Over 60 peptides were included in the database. As expected, the greatest influence on the deamidation rate in peptides was found to arise from the identity of the following amino acid. Some of the values for Y_p are listed in Table 1. As suggested previously, relative to the effect of glycine, bulky hydrophobic amino acids such as valine, leucine, and isoleucine

Table 1   Rate Constants Reported for the Reaction[a] of OH[•] with the Side Chains of Selected Amino Acids (101)

Amino acid	k (L/mole-s)[b]
Cysteine	4.7×10^{10}
Tyrosine	1.3×10^{10}
Tryptophan	1.3×10^{10}
Histidine	5×10^{9}
Methionine	8.3×10^{9}
Phenylalanine	6.5×10^{9}
Arginine	3.5×10^{9}
Cystine	2.1×10^{9}
Serine	3.2×10^{8}
Alanine	7.7×10^{7}

[a]Most values determined via radiolysis.
[b]The pH values of many of these studies have not been listed.

appear to show the slowest deamidation rate, while the smaller, more polar histidine and serine show a rate closer to that of glycine. There do appear to be some
discrepancies between these results and those mentioned earlier (12,16,57), in
particular with respect to the experimentally observed effect of histidine. Clearly,
different databases may give different results. At best, Equation 1 may be viewed
as a first approximation for estimating deamidation rates in formulation studies.

Robinson et al. (59–61) have taken a different approach to mining by
including means to account for the three dimensional structure of the side chains
and by avoiding the use of data gathered in the presence of the known catalyst,
phosphate buffer. A method has been proposed to estimate the deamidation reaction half-life at 37°C and pH = 7.4 based on the primary sequence (61). The
extent to which this method may accurately predict the deamidation rate of peptides in pharmaceutical systems has not yet been rigorously tested experimentally, but if proven valid, it would supply a rather useful tool in guiding early
formulation studies.

Secondary and Tertiary Structure

X-ray or nuclear magnetic resonance (NMR) data can provide a detailed map of
the three-dimensional structure of the protein or peptide. The role of secondary
and tertiary structures in intramolecular deamidation of proteins has been discussed by Chazin and Kossiakoff (62). It is beyond the scope of this work to
present a comprehensive review of the details of deamidation reactions in specific proteins. Excellent reviews of a variety of specific proteins exist (6). For the
most part, detailed mechanisms relating the secondary and tertiary structures of
proteins to enhancement of rates of deamidation are not yet available.

Clear differences in the deamidation rates of some proteins are evident when
native and denatured states are compared (13,63). Denaturation is thought to enhance
main-chain flexibility and water accessibility (62). Sufficient conformational flexibility is required for the Asn peptide to assume the dihedral angles of $\Phi = -120$°C
and $\Psi = +120$°C necessary for succinimide formation. In as much as such angles
tend to be energetically unfavorable (64) in native proteins, it may be expected that
Asn residues in the midst of rigid secondary structures, such as helices, may be
resistant to deamidation. Other reactions, such as cross-linking might also give rise
to rigid regions of the protein and enhanced resistance toward deamidation (65).

The direct influence of secondary structure on deamidation may be best understood in terms of hydrogen-bonding patterns that give rise to defined structures.
The α-helix is characterized by the hydrogen-bonding of the main-chain carbonyl
oxygen of each residue to the backbone nitrogen-hydrogen of the fourth residue
along the chain. The resulting bond is close to the optimal geometry, and therefore
maximal energy, for such an interaction (464). Hydrogen bonds in β-sheets are not
of fixed periodicity as in the helix, but can exhibit comparable bond energies. Citing
structural data for trypsin (66), Chazin and Kossiakoff (62) argue that strong main-
chain hydrogen-bonding of the peptide nitrogen following Asn is an important factor in modulating deamidation. Since formation of the succinimide intermediate

requires the peptide nitrogen to be free to attack the side-chain carbonyl, participation in a strong hydrogen bond by that nitrogen would inhibit the reaction. X-ray crystallography or NMR data may be helpful in identifying Asn residues in native structures likely to be protected by such a mechanism (62). Perhaps studies modeled along the lines of guest–host relationships would be helpful in elucidating further the influence of secondary structure on deamidation (67).

Effects of Deamidation on Secondary and Tertiary Structure

The effects of deamidation on the secondary and tertiary structure of the reaction-product protein have been difficult to generalize (7). In 1994, in an extensive and detailed series of studies, Darrington and Anderson showed that deamidation strongly influences the noncovalent self-association (68) and covalent dimer formation (68,69) of human insulin. The noncovalent dimer formation of triosephosphatase (70) is inhibited by deamidation, probably by charge repulsion arising from the resulting additional anionic charges present in the hydrophobic faces of the monomers.

Deamidation in concentrated solutions of food proteins tends to show increased viscosity, possibly due to enhanced charge interactions between formerly uncharged portions of the protein molecule (63). The isoelectric point of the deamidated molecule is shifted toward lower values, possibly resulting in the modified potential for adsorption to solid surfaces (71). Foamability of protein solutions subject to deamidation is greatly enhanced, probably because of partial unfolding (63).

Deamidation can destabilize a protein, making thermal (70) or chemical (13) denaturation more likely. Folding patterns may be influenced (72,73), and changes in secondary structure can result (70). Other proteins appear to be resistant to structure alterations secondary to deamidation (74,75).

OXIDATION

Introduction

Oxidation has been identified as another of the major degradation pathways in proteins and peptides and can occur during all steps of processing, from protein isolation to purification and storage (76,77). A change in the biological activity of a therapeutic agent potentially can arise from an altered enzymatic activity, inhibited receptor binding properties, enhanced antigenicity, or increased sensitivity to in vivo proteases. In some instances, biological activity is completely or partially lost upon oxidation, while in other instances, no effect on bioactivity is observed. The molecular mechanism of altered bioactivity often comes about either by oxidation of a critical residue at or near the enzyme active site or receptor binding site, or by a dramatic change in the structure of the protein upon oxidation. At present, no general rules are evident to predict with certainty all the effects of oxidation on the biological activity of a particular protein.

The chemistry of autoxidation (i.e., oxidation, not enzyme- or radiation-catalyzed) in nonprotein drug molecules has been reviewed (78,79). There are three main steps that make up any free radical chain reaction oxidation mechanism, namely: initiation, propagation, and termination. In the initiation step, free radical generation is catalyzed by transition metal ions, light energy, or thermal energy. Once initiated, oxidation reactions propagate by chain reactions of organic substances with reactive oxygen species such as singlet oxygen, hydroxyl, and peroxyl radicals. The propagation steps are either hydrogen atom abstraction or addition to olefin. In the termination step, free radicals, both alkyl and reactive oxygen, are consumed without producing further radicals among the products For the purposes of pharmaceuticals, it is important to emphasize the role of both trace metal ions and dissolved oxygen in accelerating oxidation (76,80).

Oxidation in Pharmaceutical Proteins and Peptides

In living systems, a variety of well-characterized reactive oxygen species are produced (81,82). In pharmaceutical formulations, identifying a single oxidation initiator is often difficult, since a variety of initiation possibilities exist, such as photochemical (83,84), metal ion catalyzed (85,86), and high energy γ-radiation (85). Even something as seemingly simple as sonication may promote the generation of reactive oxygen species (87). It has been convincingly shown that the extent of protein oxidation, and subsequent loss of biological activity, exhibits strong dependence upon the oxidation system employed (84,88–90).

In pharmaceutical proteins, transition metal ion catalysis of oxidation has received the lion's share of attention (89,91,92), while much less attention has been devoted to light energy and thermal energy (83,84).

Metal Ion Catalysis of Oxidation

Because of their importance in biological systems, a variety of metal ion–catalyzed oxidation systems have been identified and cataloged (85). Since the metal ion–catalyzed systems tend to be amenable to laboratory manipulations, they have been employed in stability studies (89,91,92). More importantly, trace levels of metal ions known to initiate oxidation are often present as contaminants in pharmaceutical systems (76), making an understanding of metal ion catalysis highly relevant to the job of formulation stabilization.

Iron(II) and copper(II) salts, in the presence of molecular oxygen and water, will slowly oxidize to form O_2^- (superoxide radical) by Equation 2.

$$Fe(II) + O_2 \Leftrightarrow Fe(III) + O_2^- \tag{2}$$

The superoxide radical is not stable at neutral pH and undergoes dismutation to form hydrogen peroxide by Equation 3.

$$2O_2^- + 2H + \Leftrightarrow H_2O_2 + O_2 \tag{3}$$

Hydrogen peroxide reacts further to produce hydroxyl radicals ($OH^•$) by Equation 4.

$$H_2O_2 + Fe(II) \Leftrightarrow Fe(III) + OH^• + OH^- \tag{4}$$

Hydroxyl radicals are capable of abstracting hydrogen atoms with bond energies less than 89 kcal/mol (93–95), producing a carbon-centered radical by Equation 5.

$$OH^\bullet + RH \Longleftrightarrow H_2O + R^\bullet \tag{5}$$

In the presence of oxygen, the carbon-centered radical forms the organic radical ROO$^\bullet$. ROO$^\bullet$ is capable of entering a variety of chain-reaction propagation and termination reactions (91). Overall, at least four different reactive oxygen species, each able to oxidize pharmaceutical proteins, may be produced. In solutions of free amino acids, oxidation by OH$^\bullet$ shows a strong dependence on bicarbonate ion concentration (85,96). It has been suggested that the bicarbonate ion may be required to interact with the amino acid and Fe(II) to form a hybrid complex.

Site-Specific Metal Ion–Catalyzed Oxidation

Radiolysis studies have shown that all amino acid side-chains are vulnerable to oxidation by reactive oxygen species. The same oxygen radicals, when produced by metals (Eqs. 2–5), tend to attack preferentially only a few amino acid residues, most notably His, Met, Cys, and Trp. In addition, metal ion–catalyzed oxidation of proteins can show relative insensitivity to inhibition by free radical scavenger agents (91,92). These observations have led to the hypothesis that metal ion–catalyzed oxidation reactions are "caged" processes in which amino acid residues in the immediate vicinity of a metal ion binding site are specific targets of the locally produced reactive oxygen. Schoneich and Borchardt have discussed the following reaction (92):

$$D-Fe(II) + O_2 + H^+ \rightarrow D-Fe(III)-OOH \tag{6}$$

D is some binding ligand, such as a buffer species, peptide, or protein. By this mechanism, any amino acid residues capable of forming a metal ion binding site are potential sources of reactive oxygen species. Since reaction of the oxygen radical usually occurs in the immediate region of its production before escape into the bulk solution by diffusion, free radical scavengers are unlikely to be efficient formulation protective agents (92). It has been suggested that the terminal hydroxyl group of serine, the free carboxyl groups of aspartic and glutamic acids, the imidazole ring of histidine, and the free amino or free carboxyl groups of *N*-terminal and *C*-terminal (respectively) residues participate in binding metal ions to proteins (96). Further, since a metal ion binding site may be formed by appropriate residues upon folding of the protein molecule, these amino acids need not be adjacent to each other in the primary sequence.

Oxidation by Hydrogen Peroxide Addition

Addition of hydrogen peroxide has been employed as a means to study oxidation of proteins (88,97,98), the advantage being that the concentration and identity of the initiating oxidant is known. In some instances, hydrogen peroxide has been shown to be an oxidant specific for methionine (99), while in other instances, oxidation of cysteine and tryptophan residues also occurs (100). Hydrogen peroxide is thought to oxidize only residues easily accessible on the surface of the folded protein, but more recent evidence suggests oxidation of both surface and buried residues (99). It has been proposed that *t*-butyl hydroperoxide may be a highly specific oxidizer of surface-localized methionine residues (99).

A highly detailed mechanism of the oxidation by hydrogen peroxide of the amino acid cysteine to the disulfide has been recently published (98), showing the formation of cysteine sulfenic acid as an intermediate. These authors proposed a two-step nucleophilic reaction where the thiolate anion attack on the neutral hydrogen peroxide is the rate-determining step. Effects of buffer pH, temperature, and ionic strength were all included in the model. Well-controlled studies such as this will go a long way to elucidating details of the molecular mechanism(s) of oxidation.

Specific Amino Acid Side-Chains

Overview of Amino Acid Oxidation

First-order rate constants for the reaction of OH with the side-chains of selected amino acids have been listed (101) and representative values are shown in Table 1. As a free amino acid, cysteine is the most sensitive to reaction followed closely by the aromatic side-chains. Overall, the reaction rate constants vary less than three orders of magnitude, with the higher values approaching the diffusion limit. Although these values are specific for reactions with free amino acids and thus do not take into account effects due to accessibility, they may be useful as early estimates of the sensitivity of a peptide to OH. Not listed in Table 1 are the reported values for the reaction of OH with selected small peptides (101). In general, the value for the reaction rate constant of a peptide appears to be close to that of the most reactive individual amino acid (101).

One of the most damaging reactions of oxygen radicals is that of hydrogen abstraction from the peptide backbone, in particular at the α-carbon (102,103) As shown in Figure 3, in the presence of O_2, a peroxyl radical is formed that can convert to an imine, followed by hydrolysis of the backbone (102,103).

Methionine

Methionine has been identified as an easily oxidized amino acids in proteins (Table 1), and oxidation of this residue has received considerable attention. Oxidation deprives methionine of its ability to act as a methyl donor, which will influence the bioactivity of proteins dependent on that function (104). The reaction product of methionine oxidation is the corresponding sulfoxide and, under more strenuous oxidation conditions, the sulfone (Fig. 4). These are not the only possible reaction products, but they are usually the first to appear.

Not surprisingly, mechanisms of oxidation of methionine appear to be highly dependent on the reactive oxygen species under consideration (84). Peroxide (105), peroxyl radicals (106), singlet oxygen (105), and hydroxyl radical (91) have all been shown to oxidize methionine residues to sulfoxides and other products. The identity of major oxidizing species present in these solutions remains a matter of

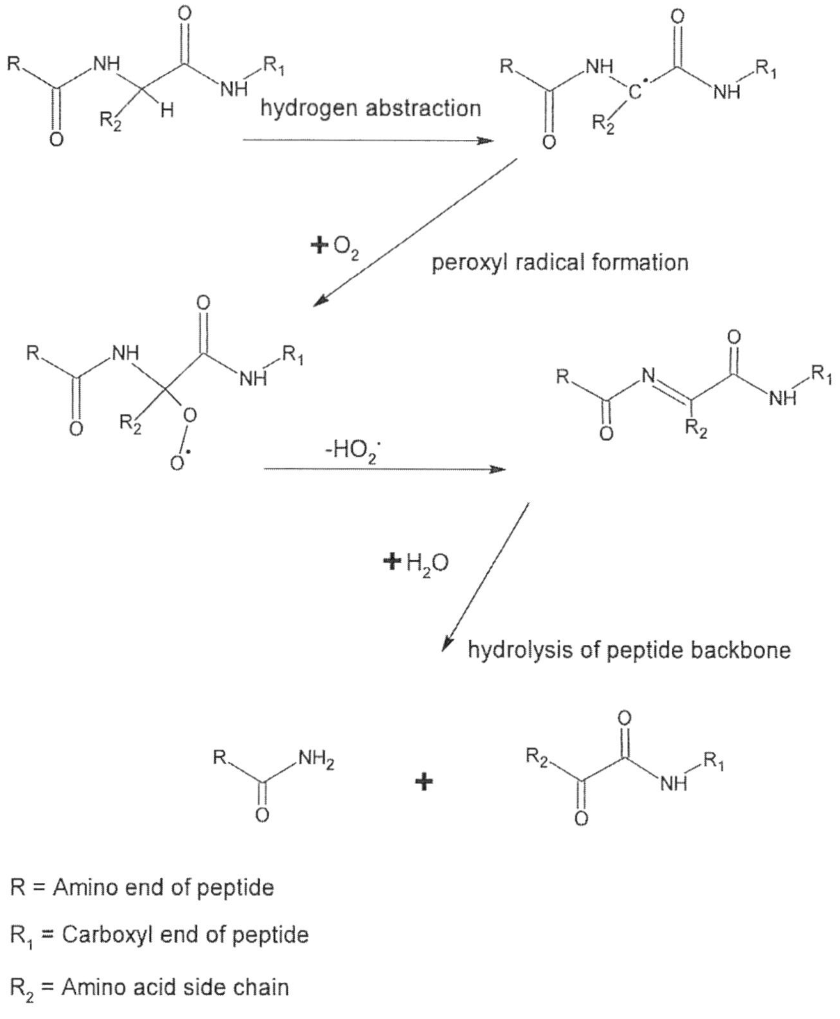

R = Amino end of peptide

R_1 = Carboxyl end of peptide

R_2 = Amino acid side chain

Figure 3   Involvement of peroxyl radical in the hydrolysis of peptide backbone.

controversy (107,108). The reaction mechanisms for proteins in pharmaceutical systems are incomplete, because not all products and intermediates are known.

   Oxidation of methionine in recombinant human relaxin: *Photocata-lyzed oxidation.* A series of papers spanning the 1990s studied methionine oxidation initiated by light (84), hydrogen peroxide (88), and ascorbic acid–Cu(II) (89) in recombinant human relaxin. Upon exposure to light of an intensity of 3600 candles for 5 to 17 days, both methionine residues, Met-B4 and Met-B25, located on the surface region of the B-chain were oxidized to the sulfoxide derivative (84). The identity of the reactive oxygen species formed upon exposure to light was not reported, but peptide mapping results suggest a wide variety of reaction products.

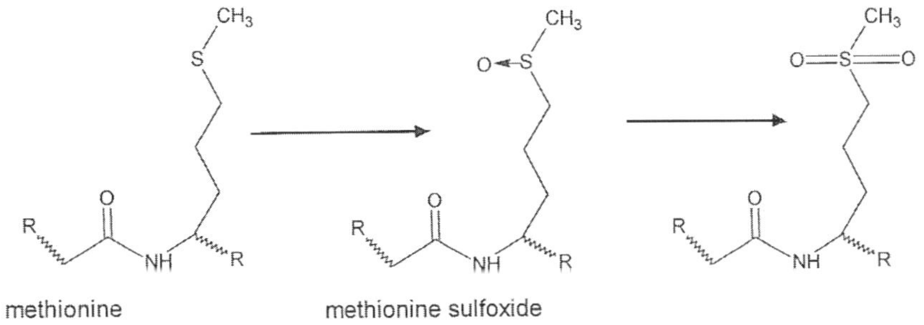

Figure 4 Oxidation of methionine first to the sulfoxide and then to sulfone derivatives.

Hydrogen peroxide. In the presence of added hydrogen peroxide, the methionines (Met-B4 and Met-B25) were the only residues of relaxin to be oxidized (88). Three products were isolated, the monosulfoxide for each methionine and the corresponding disulfoxide. The reaction rate was independent of pH (range 3–8), ionic strength (0.007–0.21 M NaCl), or buffer species (lactate, acetate, Tris). Interestingly, the rate of reaction of the two methionine groups differed, with oxidation at Met-B25 being more rapid than at Met-B4. The oxidation rate of Met-B25 was equivalent to that observed for free methionine and for methionine in a model peptide of the relaxin B-chain (B23–B27). The reduced rate of oxidation at the solvent-exposed residue Met-B4 relative to Met-B25 suggests that accessibility of the residues to H_2O_2 may play a role in the reaction.

Pro-oxidant system ascorbic acid–Cu(II). Contrary to the results observed in the presence of hydrogen peroxide, in the presence of the pro-oxidant system ascorbic acid–Cu(II), a pH-dependent precipitation of relaxin was observed (89). Approximately 80% of protein was lost from solution within 25 minutes at pH 7 to 8. Chromatographic results indicated that the aggregate was not held together by covalent forces. In a second significant aspect of the study (88), in the presence of ascorbic acid–Cu(II), investigators observed oxidation of histidine and methionine (89). One final important difference is that Met-B4 was oxidized preferentially over Met-B25. All these differences are consistent with the conclusion that the oxidant system employed for in vitro studies can have a major impact on the results. Clearly, the issue of identifying the radical species responsible for oxidation of methionine, or any other residue, is of primary importance in setting down complete reaction mechanisms.

Methionine oxidation studies with model peptides: As has been pointed out, development of a molecular-level understanding of oxidation in protein drug delivery systems has been hampered by a lack of characterization of the reaction mechanism and the products. A trail-blazing work by Li and coworkers (91,92,109) has begun for addressing the much-needed mechanistic description of the effects of pH and primary sequence on oxidation pathways of methionine in simple model peptides. These authors have primarily employed the metal ion–catalyzed pro-oxidant system and a series of simple methionine-containing peptides. Considerable efforts have been expended with specific radical scavengers to identify the reactive oxygen species responsible for oxidation.

Buffers and pH: Using the pro-oxidant systems of dithiothreitol/Fe(III) to generate reactive oxygen species, oxidation of methionine in Gly-Gly-Met, Gly-Met-Gly, and Met-Gly-Gly was studied as a function of pH. The degradation rate followed first-order kinetics with respect to peptide, while mass balance comparisons showed that sulfoxide was not the terminal degradation product. The rate of loss of the parent peptide did not vary with pH in the range 6 to 8.1. The rate of loss was observed to accelerate with pH beyond this range.

Li et al. (109) found that the second-order rate constants for the degradation of His-Met in the ascorbic acid–Fe(III) pro-oxidant system show a maximum at pH 6.4. The appearance of a maximal pH was attributed to competing effects of pH on ascorbic acid. Deprotonation of ascorbate at a higher pH ($pK_1 = 4.1$) facilitates electron donation to Fe(III) and accelerates the initiation reaction, while at the same time, ascorbate becomes a better oxygen radical scavenger, inhibiting the reaction. The buffer species also seems to play a role in the kinetics of degradation. In buffers of equal ionic strength, methionine oxidation was faster in the presence of phosphate than in the presence of Tris or HEPES. Phosphate buffers may facilitate the electron transfer from Fe(II) to oxygen, promoting the reaction (109). Buffer species such as Tris or HEPES have a weak affinity for metal ions (110) and result in methionine oxidation reaction rates that are somewhat less than that of phosphate. Tris and HEPES have also been reported to be scavengers of hydroxyl radicals (111), which would be expected to further inhibit reactions in which the hydroxyl radical is the primary reactive oxygen species. In temperature studies, the energy of activation was found to be 23.9 ± 2 kJ/mol, but it is unknown whether this characterizes the formation of the oxidizing species, the oxidation of methionine, or both.

Primary sequence: Li et al. (109) also studied the effect of primary structure on methionine oxidation. When in a terminal position, Met-Gly-Gly or Gly-Met-Met, the first-order degradation rate constants are greater than that of the mid-position Gly-Met-Gly. The inclusion of histidine in His-Met greatly accelerates the degradation of methionine. The greatest degradation rates are observed in His-Gly-Met and His-Pro-Met, where methionine is separated by one residue from histidine. Even His-Gly-Gly-Gly-Gly-Met shows enhanced degradation rates (by a factor of 5) compared to Gly-Gly-Met. Whether this is related to the metal ion binding and localized oxidation is not yet known with certainty. The authors do note that the degradation products of these reactions have not been characterized, and complete reaction mechanisms are not yet available.

Histidine

Histidine is also highly susceptible to oxidation, either by photocatalyzed or by metal ion–catalyzed mechanisms. Photooxidation of proteins in vivo has been extensively studied (112,113). Photosensitizing agents such as methylene blue

(114) or rose bengal (115) are required for photooxidation to take place via the production of singlet oxygen ($^1O_2$). A cycloperoxide ring is produced by the addition of $^1O_2$ to the imidazole ring of histidine (116). The kinetics of photoreactions are often very complex (117), being further complicated by issues of histidine accessibility to solvent and $^1O_2$ (118) as well as simultaneous metal-catalyzed oxidation (119). A variety of products are produced, including the amino acids aspartic acid and Asn (76,120).

From the standpoint of pharmaceutical formulations, recent evidence suggests that the oxidation products of ascorbic acid (a frequently employed antioxidant) may be potent photosensitizing agents, enhancing histidine oxidation in proteins. Ortwerth et al. (119) reported $^1O_2$ concentrations in the millimolar range and H_2O_2 in the micromolar range after one hour of irradiation with ultraviolet light in the presence of dehydroascorbate and diketogulonic acid (by-products of ascorbic acid oxidation). Complete protection against photooxidation can be attained by protection from light or removal of all dissolved oxygen gas (112).

There has been some study of the effect of primary structure on the photooxidation of histidine. Miskoski and Garcia (112) found little difference in the rate of photooxidation of histidine as the free amino acid and in dipeptides His-Gly and Gly-His. Changing the solvent to acetonitrile/water (1:1) resulted in an order-of-magnitude decrease in the rate of oxidation in all three substrates, which is suggestive of polarity effects on the rate of reaction.

Histidine appears to be particularly sensitive to transition-metal-catalyzed oxidation, presumably because it often forms a metal binding site in proteins (121). Fenton chemistry at the bound metal ion (such as that as in Eq. 6) could result in high localized concentrations of reactive oxygen species. Histidine residues at the N-terminus appear to be especially susceptible to site-directed metal ion–catalyzed oxidation (122). Metal-catalyzed oxidation of histidine results in the production of 2-oxo-imidazoline (Fig. 5) (123). By-products of the proposed reaction include aspartic acid. In the metal ion–catalyzed oxidation of polyhistidine, the production of aspartic acid is accompanied by scission of the histidyl peptide bond (120), but it remains unclear whether the scission is a part of the reaction mechanism or merely reflects the instability of the 2-oxo-imidazoline ring to the conditions of isolation and analysis (120,124). Chain scission is not frequently observed in proteins upon histidine oxidation.

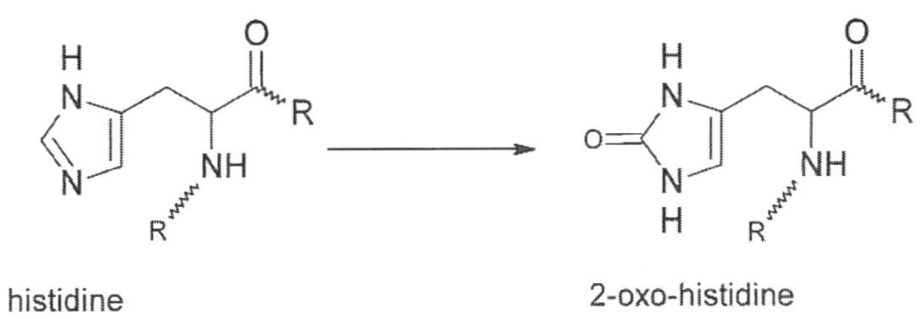

Figure 5 The first product of histidine oxidation, 2-oxo-histidine.

Cysteine

Metal ion–catalyzed oxidation of cysteine residues usually results in the formation of both intra- and intermolecular disulfide bonds (125,126). Further oxidation of the disulfide results in sulfenic acid.

The mechanism may be summarized as follows (107):

Formation of thiyl radical:

$$RS^- + M^{+n} \Rightarrow RS^{\bullet} + M^{+(n-1)} \tag{7}$$

Formation of disulfide radical anion:

$$RS^- + RS^{\bullet} \Rightarrow RS^{\bullet -}SR \tag{8}$$

Formation of superoxide:

$$RS^{\bullet -}SR + O_2 \Rightarrow RSSR + O_2^{\bullet -} \tag{9}$$

Generation of peroxide:

$$RSH + O_2^{\bullet -} + H^+ \Rightarrow RS^{\bullet} + H_2O_2 \tag{10}$$

Regeneration of M^{+n}:

$$M^{+(n-1)} + O_2 \Rightarrow M^{+n} + O_2^{\bullet -} \tag{11}$$

This mechanism can result in the production of reactive oxygen species capable of further oxidative damage to the disulfide as well as to other residues in the vicinity. When metal ions are made unavailable by chelation with ethylenediaminetetraacetic acid (EDTA), cysteine oxidation is greatly reduced (127). In general, a pH of 6 appears to be optimal for the oxidation of cysteine in proteins (128). At low pH, the protonation of the sulfhydryl (pK_a 8.5) inhibits reaction with the metal ion (Eq. 7). In the alkaline region, electrostatic repulsion of two ionized cysteines is thought to result in an increased separation of the two residues and a reduced reaction rate (Eq. 8). Oxidation in the absence of a nearby thiol has also been observed (129).

Tryptophan

Tryptophan is well known to be a target of reactive oxygen species superoxide (130), singlet oxygen (131), hydroxyl radical (132), and peroxide (133). The most prominent reaction products of tryptophan oxidation appear to be N'-formylkynurenine and 3-hydroxykynurenine (134). Monohydroxyl derivatives of tryptophan at the 2, 4, 5, 6, and 7 positions have also been observed (Fig. 6). N'-formylkynurenine may be also formed by photooxidation (135). Metal ion catalysis of oxidation appears to play a role in the photolytic mechanism (134).

Very little work has been directed toward an understanding of the influence of primary sequence upon tryptophan photooxidation (136,137). It is known that inclusion of Trp in a peptide bond significantly reduces photocatalyzed radical yield (138). At neutral pH and in the presence of dissolved oxygen,

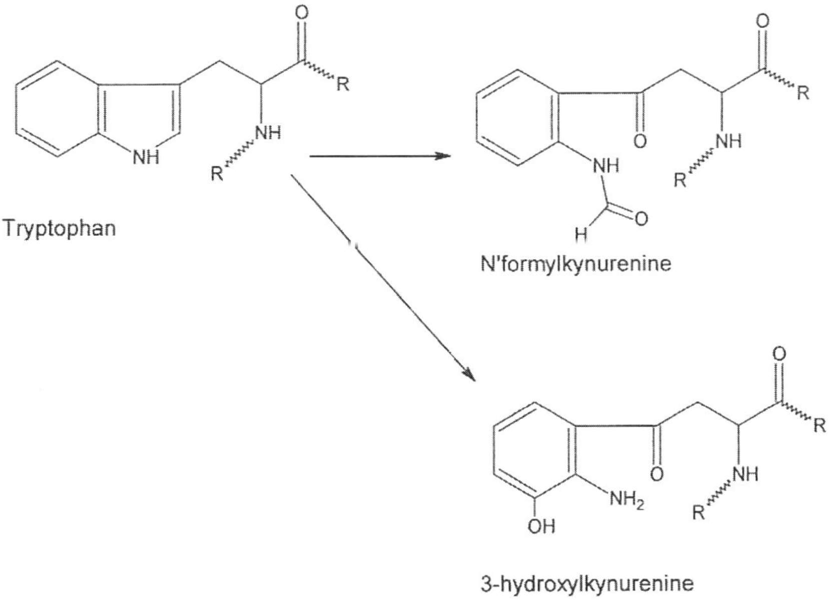

Figure 6 Tryptophan oxidation products.

it has been observed that Gly-Trp photooxidizes at a rate approximately 10-fold that of Trp-Gly. Similarly, Leu-Trp degrades at a rate approximately threefold greater than Trp-Leu. In tripeptides, Gly-Trp-Gly degrades more rapidly than Leu-Trp-Leu. The mechanistic basis of these observations is not clear (136), and additional work remains to be done (137). Under anaerobic conditions, the photolytic degradation rate of tryptophan in peptides is also observed, but it is slowed considerably from the rates observed in the presence of oxygen. In addition, Leu-Trp-Leu exhibits even greater stability over Gly-Trp-Gly. These data indicate that both in the presence and in the absence of oxygen, leucine (with its large side-chain) occupying the C-terminal position next to Trp tends to decrease the degradation rate. Although these data do not provide a sufficient base for a generalized rule, it can be speculated that steric effects have an influence on the rate of photodegradation (137).

Photooxidation of Trp in proteins is known to be directly dependent on the accessibility of the residue to oxygen and solvent water (139,140). Trp residues buried in the core of the protein are less rapidly oxidized than those located at the surface of the molecule (140). Micellar solubilization of hydrophobic peptides appears to protect only the Trp residues located in the core of the micelle (139).

Phenylalanine and Tyrosine

In the presence of copper ion, phenylalanine is oxidized to 2-, 3-, or 4- (tyrosine) hydroxyphenylalanine (141), as shown in Figure 7. Tyrosine may photo- or radio-oxidize to 3,4-dihydrophenylalanine (142), or cross-link with another tyrosine to form diotyrosine (143). The latter product may be protease resistant and stable to acid hydrolysis (143). Intermolecular cross-linking would result in increased molecular weight of the reaction product.

Proline

Hydroxyl radical oxidation by the hydroxyl radical of proline (144,145), as well as glutamic acid and aspartic acid (146), is characterized by site-specific cleavage of the polypeptide chain on the C-terminal end of the residue.

Formulation Factors and Oxidation

Overview of Excipient Effects

Shown in Table 2 is a list of first-order rate constants for the reaction of OH with selected formulation excipients that have proved useful in protein systems (101). As with the amino acids shown in Table 1, the rate constants of the excipients listed in Table 2 vary by up to three orders of magnitude. Naturally, ascorbate appears to be the most sensitive to the presence of OH, but so are the proteins (albumin and gelatin) and unsaturated long-chain carboxylate (linoleate). The sensitivity of the latter compound may give rise to concern about the reaction of OH with surfactants containing unsaturated carbon chains as well as lipid-based systems containing triglycerides. A relatively simple iodometric assay suitable for the determination of peroxide levels in surfactants has been published (147).

Iron-catalyzed oxidation has been shown to be rather sensitive to pH (148,149). Phosphate buffer appears to accelerate the reaction as compared to HEPES, but it is possible that trace-metal contamination of the excipients may

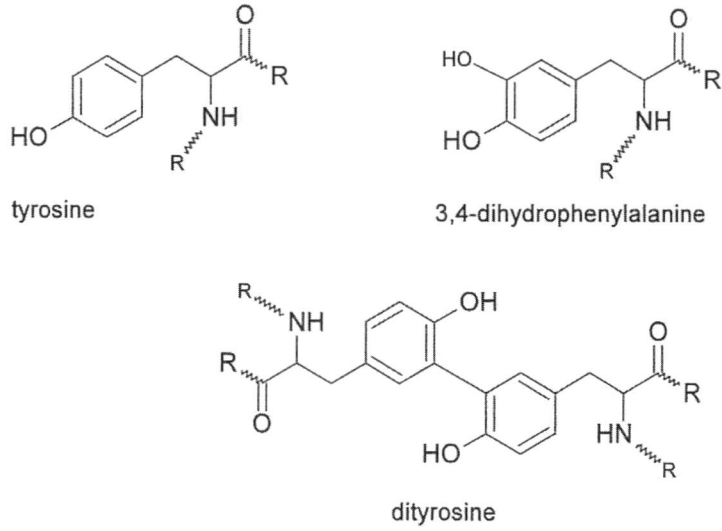

tyrosine

3,4-dihydrophenylalanine

dityrosine

Figure 7   Tyrosine oxidation products.

Table 2 Rate Constants Reported for the Reaction of OH$^\bullet$ with Selected Formulation Excipients

Excipient	pH	k (L/mole-s)
Ascorbate	7	1.3×10^{10}
Tartarate	7	6.8×10^8
EDTA	4	4×10^8
EDTA	9	2×10^9
Citric acid	1	5×10^7
Glucose	6.5	2.3×10^9
Glycerol	7	1.5×10^9
Linoleate	7	1.1×10^{10}
Sucrose	7	2.3×10^9
Albumin	7	7.8×10^{10}
Carboxymethylcellulose	NL	2×10^8
κ-Carrageenan	NL	2.2×10^8
Chondroitin-6-SO$_4$	NL	6.8×10^8
Dextran	7	1×10^8
Gelatin	NL	9.1×10^{10}
Polyethylene oxide	NL	3×10^8
Polyvinylpyrrolidone	7	5×10^7

Abbreviations: EDTA, ethylenediaminetetraacetic acid; NL, not listed.
Source: From Ref. 101.

be complicating the observations. In addition, pH effects on metal ion chelation, such as by EDTA, must also be taken into account.

Some of the methods proposed for addressing oxidative damage in a biological matrix may prove useful in the study of proteins and peptides in complex lipid-based drug delivery systems (150). Although lacking specificity, carbonyl assays are often employed as a convenient and rapid measure of the extent of protein damage by reactive oxygen species (151).

Polyethers

Many formulations take advantage of the ability of polyethylene glycol (PEG) and PEG-linked surfactants to stabilize proteins against aggregation and thermal denaturation. The potential ability of these adjuvants to promote oxidation becomes an important consideration. PEG and nonionic polyether surfactants are known to produce peroxides upon aging (152). These peroxides are responsible for drug degradation (153,153) in polyether-containing systems. The oxidation of one such polyether surfactant, polysorbate 80, has been shown to release formaldehyde, a potent protein cross-linking agent (154). Careful purification of PEG-containing adjuvants prior to formulation should minimize this potential degradation mechanism.

Sugars and Polyols

Sugars are often employed as lyoprotectants and as part of the vehicle in the formulation and administration of drugs. Literature reports indicate that moderately high concentrations of sugars and polyols seem to inhibit the oxidation of proteins, possibly by serving as hydroxyl radical scavengers (155–157). Although the rate constants for hydroxyl radical reactions with sugars and polyols are not as large as that of ascorbate (Table 2), the high concentrations of the carbohydrate and polymer excipients typically employed in protein/peptide formulations may be responsible for the protective effect.

It was reported in 1996 that various pharmaceutically acceptable sugars and polyols (glycerin, mannitol, glucose, and dextran) were successful in inhibiting ascorbate-Cu(II)–induced oxidation of the protein relaxin and of model peptides Gly-Met-Gly and Gly-His-Gly (158). Results of experiments with glycerin show that contrary to expectations, the protective effect is not the result of radical scavenging. Rather, these authors concluded that the protective effect of the sugars and polyols was due to complexation of transition metal ions. This is in accord with reports indicating weak, but stable, complexes of metal ions with sugars (159,160). Production of reactive oxygen species can be diminished or eliminated by means of competition with the peptide for binding of the metal ion (Eq. 7). Dextran also inhibited oxidation, but a detailed mechanism of the protective effect was not given. The safety and availability of these inexpensive additives make them very attractive as protective agents against oxidation.

The inclusion of glucose in a protein formulation is not without potential risks. Glucose has been shown to participate in oxidation reactions catalyzed by metal ions (161,162). Methionine oxidation products have been observed (163). Upon reaction of glucose with Fe(II), an enediol radical anion intermediate is formed that quickly reacts with molecular oxygen to form the ketoaldehyde (which itself can react with a free amino group on the protein, forming a keto aminomethylol) and the superoxide radical. As a reducing sugar, glucose has been shown to covalently modify relaxin by adding to the side-chains of lysine and arginine and by catalyzing the hydrolysis of the C-terminal serine amide bond. Neither nonreducing sugar (trehalose) nor polyhydric alcohol (mannitol) participates in these reactions (158).

Antioxidants

Antioxidants are commonly employed to protect both small-molecule and peptide/protein drugs from oxidation in pharmaceutical formulations. These are essentially sacrificial targets that have a great tendency to oxidize, consuming pro-oxidant species. The choice of an antioxidant is complicated in proteins and peptides because of the interaction chemistries possible between the antioxidant and the different amino acid side-chains (164). Even antioxidants that are themselves benign to proteins can become potent pro-oxidants in the presence of trace amounts of transition metal ions (e.g., ascorbic acid). In the absence of metal ions, cysteine, as a free amino acid, may act as an effective antioxidant (126,156). By virtue of its singlet oxygen scavenger activity, α-octocopherol has shown protective effects against photooxidation of proteins within lipid membranes (165). Whether this additive is effective in reducing oxidation of proteins or peptides in lipid-based delivery systems, such as liposomes or emulsions, remains unknown.

Processing and Packaging

Removal of oxygen from solution by degassing processes may be an effective means of inhibiting oxidation in protein and peptide solutions (164,166). Even very low concentrations of oxygen in the headspace will promote oxidation (145). To minimize foaming in protein solutions during degassing, Fransson et al. have suggested cyclic treatments of low temperature and low pressure, followed by exposure to atmospheric-pressure nitrogen gas (83). Packaging in a light-resistant container may be helpful in reducing light-catalyzed oxidation. It should be kept in mind that glass may release minute quantities of metal ions sufficient for metal ion–catalyzed oxidation (167).

Lyophilization

The influence of moisture content on oxidation and other protein-degradation reactions has been explored by a number of authors (4,168,169). Most often, residual moisture enhances the degradation of proteins (4). Hageman has listed both oxidation promotion and oxidation inhibition mechanisms of water (169). The pro-oxidant activities of moisture are believed to include mobilization of catalysts, exposing new reaction sites by swelling, and decreasing viscosity of the sorbed phase. The oxidation-promotion activities of water are thought to be initiated at or near monolayer coverage, where conformational flexibility of the protein is enhanced (168). The oxidation-inhibition activity of sorbed water is thought to arise from retardation of oxygen diffusion, promotion of radical recombination, decreased catalytic effectiveness of transition metals, and dilution of catalyst (169). Much higher water content is required for the modest oxidation-inhibitory effects to become manifest (169). The existence of both pro- and antioxidant effects of moisture would be consistent with the widely varying experimental results observed. For example, Fransson et al. (83) have observed no dependence on moisture of the second-order rate constants for methionine oxidation in insulin-like growth factor, while Pikal et al. (166) found a strong dependence on moisture for methionine oxidation in human growth factor. Luo and Anderson (170) have shown that the mechanism of oxidation of cysteine by hydrogen peroxide in a PVP matrix is significantly altered from that observed in solution. In particular, mobility effects on the cysteine, but not hydrogen peroxide, may be responsible for the observation that cystine is not the only reaction product in the solid state. The results of kinetic studies are consistent with a clustering of water in regions around the polymer and may be a determining factor for the reaction in the solid state.

Clear-cut mechanistic interpretation of these differing oxidation results is not yet possible because of the possible masking effects of formulation additives and moisture-dependent protein conformational states. Residual moisture values in lyophilized proteins of less than 1% tend to be associated with enhanced stability of the protein upon storage (168). In production lots, removal of water to attain such low levels of residual moisture is quite expensive.

Proteins and Peptides in Polymers

Care must be exercised when attempting to encapsulate pharmaceutical proteins or peptides within a polymer matrix. Cross-linking of methylated dextrans by addition of potassium peroxyodisulfate resulted in significant oxidation of methionine residues in interleukin-2 (171). Addition of tetramethylethylenediamine minimized, but did not eliminate, the oxidation of sensitive amino acids. Similar approaches may be necessary to minimize oxidation of proteins or peptides in other polymer-based systems (76).

ACYLATION IN THE SOLID STATE

Considerable efforts in biomedical research have been devoted to an understanding of posttranslational modification of proteins in biological systems. Other efforts have been geared to the chemical modification of food proteins as a means of controlling processing properties. In pharmaceutical dosage forms, our interest is in preventing modifications of the active ingredients by formulation excipients, such as polymers One of the most frequently employed polymers for this purpose is PLGA, a Food and Drug Administration–approved biocompatible material that releases the incorporated active ingredient by erosion. It has been established that drug substances incorporated into PLGA may be exposed to a highly acidic microenvironment that arises from the hydrolysis of ester bonds of the polymer. Microenvironment pH values as low as 2 have been reported (172). Acid-catalyzed denaturation and chain scission have been attributed to this low pH value (173). More recently, acylation of proteins by polymer components has been observed. Insulin (174), calcitonin (175), octreotide (176), and parathyroid hormone (175) all exhibited evidence of acylation to some extent when encapsulated in PLGA.

Houchin et al. (49) have proposed a mechanism of acylation of peptides by PLGA (Fig. 8). The backbone terminal amino group or the ε-amino of lysine is believed to carry out a nucleophilic attack on the polymer. Water potentiates the reaction, probably by enhancing molecular mobility as well as by promoting hydrolysis of polymer chains. It has been suggested that significant mass loss from the polymer is necessary before acylation can be observed (176). Other studies suggest acylation begins well before significant polymer mass loss (174). Results of model studies of peptide in lactic acid solution do suggest that, in addition to primary amines, other amino acids such as tyrosine and serine may also be sites of acylation (177). Attempts to prevent acylation by PEGylation either of the polymer (178) or of the peptide have been reported. In both cases, there appears to be some reduction in acylation compared to the peptide in the absence of PEGylation, but complete protection was not afforded. In addition, PEGylation of the peptide may result in loss of biological activity. Much additional work remains to be done before a clear strategy can be formulated to prevent acylation by PLGA.

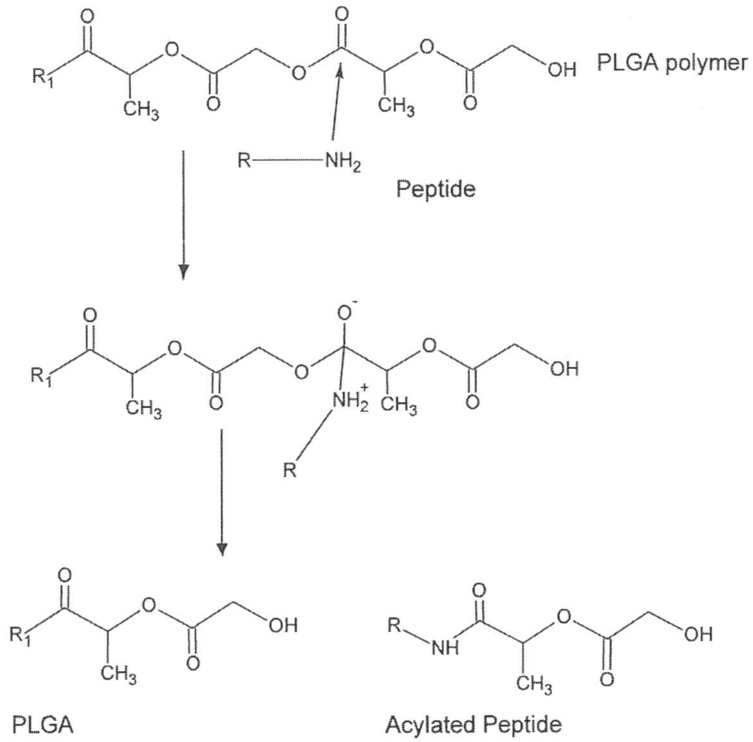

Figure 8 Acylation of peptide by PLGA polymer. *Abbreviation*: PLGA, polylactic-glycolic acid.

REFERENCES

1. Wakankar AA, Borchardt RT. Formulation considerations for proteins susceptible to asparagine deamidation and aspartate isomerization. J Pharm Sci 2006; 95:2321–2336.
2. Bischoff R, Kolbe HVJ. Deamidation of asparagine and glutamine residues in proteins and peptides: structural determination and analytical methodology. J Chromatogr B 1994; 662:261–278.
3. Wright HT. Nonenzymatic deamidation of asparaginyl and glutaminyl residues in proteins. Crit Rev Biochem Mol Biol 1991; 26:1–52.
4. Cleland JL, Powell MF, Shire SJ. The development of stable protein formulations: a close look at protein aggregation, deamidation, and oxidation. Crit Rev Ther Drug Carrier Syst 1993; 10:307–377.
5. Johnson BA, Asward DW. Deamidation and isoaspartate formation during in vitro aging of purified proteins. In: Asward DW, ed. Deamidation and Isoaspartate Formation in Peptides and Proteins. Boca Raton, FL: CRC Press, 1995:91.
6. Teshima G, Hancoch WS, Canova-Davis E. Effect of deamidation and isoaspartate formation on the activity of proteins. In: Asward DW, ed. Deamidation and Isoaspartate Formation in Peptides and Proteins. Boca Raton, FL: CRC Press, 1995:167.

7. Meyers JD, Ho B, Manning MC. Effects of conformation on the chemical stability of pharmaceutically relevant polypeptides. In: Carpenter JF, Manning MC, ed. Rational Design of Stable Protein Formulations. New York, NY: Kluwer Academic/ Plenum Press, 2002:85.

8. Gracy RW, Yuksel KU, Gomez-Puyou A. Deamidation of triosephosphate isomerase in vitro and in vivo. In: Asward DW, ed. Deamidation and Isoaspartate Formation in Peptides and Proteins. Boca Raton, FL: CRC Press, 1995:133.

9. Wright TH. Amino acid abundance and sequence data: clues to the biological significance of non-enzymatic asparagine and glutamine deamidation in proteins. In: Asward DW, ed. Deamidation and Isoaspartate Formation in Peptides and Proteins. Boca Raton, FL: CRC Press, 1995:229.

10. Geiger T, Clarke S. Deamidation, isomerization, and racemization at asparaginyl and aspartyl residues in peptides. J Biol Chem 1987; 262:785–794.

11. Stephenson RC, Clarke S. Succinimide formation from aspartyl and asparaginyl peptides as a model for the spontaneous degradation of proteins. J Biol Chem 1989; 264:6164–6170.

12. Tyler-Cross R, Schrich V. Effects of amino acid sequence, buffers, and ionic strength on the rate and mechanism of deamidation of asparagine residues in small peptides. J Biol Chem 1991; 266:22549–22556.

13. Tomizawa H, Yamada H, Wada K, Imoto T. Stabilization of lysozyme against irreversible inactivation by suppression of chemical reactions. J Biochem 1995; 117:635–639.

14. Zhao M, Bada JL, Ahern TJ. Racemization rates of asparagine-aspartic acid residues in lysozyme at 100°C as a function of pH. Bioorg Chem 1989; 17:36–40.

15. De Boni S, Oberthür C, Hamburger M, Scriba GKE. Analysis of aspartyl peptide degradation products by high-performance liquid chromatography and high-performance liquid chromatography-mass spectrometry. J Chromatogr A 2004; 1022:95–105.

16. Patel K, Borchardt RT. Chemical pathways of peptide degradation. II Kinetics of deamidation of an asparaginyl residue in a model hexapeptide. Pharm Res 1990; 7:703–711.

17. Brennan TV, Clarke S. Deamidation and isoaspartate formation in model synthetic peptides. The effects of sequence and solution environment. In: Asward DW, ed. Deamidation and Isoaspartate Formation in Peptides and Proteins. Boca Raton, FL: CRC Press, 1995:65.

18. Voorten CEM, deHaard-Hoekman WA, van den Oeelaar PJM, Bloemendal H, deJong WW. Spontaneous peptide bond cleavage in aging α-crystalline through a succinimide intermediate. J Biol Chem 1988; 263:19020–19023.

19. Bhatt NP, Patel K, Borchardt RT. Chemical pathways of peptide degradation. I Deamidation of adrenocorticotropic hormone. Pharm Res 1990; 7:593–599.

20. Darrington RT, Anderson BD. Evidence for a common intermediate in insulin deamidation and covalent dimer formation: effects of pH and aniline trapping in dilute acidic solutions. J Pharm Sci 1995; 84:275–282.

21. Peters B, Trout BL. Asparagine Deamidation: pH-dependent mechanism from density functional theory. Biochemistry 2006; 45:5384–5392.

22. DiDonato A, D'Alessio G. Heterogeneity of bovine seminal ribonuclease. Biochemistry 1981; 20:7232–7237.

23. Senderoff RI, Wootton SC, Boctor AM, et al. Aqueous stability of human epidermal growth factor 1–48. Pharm Res 1994; 11:1712–1720.

24. Robinson AB, Rudd CJ. Deamidation of glutaminyl and asparginyl residues in peptides and proteins. Curr Top Cell Regul 1974; 8:247–295.

25. Scotchler JW, Robinson AB. Deamidation of glutaninyl residues: dependence on pH, temperature, and ionic strength. Anal Biochem 1974; 59:319–322.

26. DiBiase MD, Rhodes CT. The design of analytical methods for use in topical epidermal growth factor product development. J Pharm Pharmacol 1991; 43:553–558.

27. Georg-Nascimento C, Lowenson J, Borissenko M, et al. Replacement of a labile aspartyl residue increases the stability of human epidermal growth factor. Biochemistry 1990; 29:9584–9591.

28. DeDomato A, Ciardiello MA, deNigris M, Piccoli R, Mazzarella L, D'Alessio G. Selective deamidation of ribonuclease A. J Biol Chem 1993; 268:4745–4751.

29. Aliyai C, Patel JP, Carr L, Borchardt RT. Solid state chemical instability of an asparaginyl residue in a model hexapeptide. PDAJ Pharm Sci Technol 1994; 48:167–173.

30. McKerrow JH, Robinson AB. Deamidation of asparaginyl residues as a hazard in experimental protein and peptide procedures. Anal Biochem 1971; 42:565–568.

31. Capasso S, Mazzarella I, Zagari A. Deamidation via cyclic imide of asparaginyl peptides: dependence on salts, buffers and organic solvents. Peptide Res 1991; 4:234–238.

32. Johnson BA, Shirokawa JM, Aswad DW. Deamidation of calmodulin at neutral and alkaline pH: quantitative relationships between ammonia loss and the susceptibility of calmodulin to modification by protein carboxyl methyltransferase. Arch Biochem Biophys 1989; 268:276–286.

33. Tomizawa H, Yamada H, Tanigawa K, Imoto T. Effect of additives on irreversible inactivation of lysozyme at neutral pH and 100°C. J Biochem 1995; 117:364–369.

34. Zheng JY, Janis LJ. Influence of pH, buffer species, and storage temperature on physicochemical stability of a humanized monoclonal antibody LA298. Int J Pharm 2006; 308:46–51.

35. Capasso S, Mazzarella L, Sica F, Zagari A, Salvadori S. Kinetics and mechanism of succinimide ring formation in the deamidation process of asparagines residues. J Chem Soc Perkin Trans 1993; 2:679–682.

36. Capasso S, Kirby AJ, Salvadori S, Sica F, Zagari A. Kinetics and mechanism of the reversible isomerization of aspartic acid residues in tetrapeptides. J Chem Soc Perkins Trans 1995; 2:437–442.

37. Lura R, Schrich V. Role of peptide conformation in the rate and mechanism of deamidation of asparaginyl residues. Biochemistry 1988; 27:7671–7677.

38. Song Y, Schowen RL, Borchardt RT, Topp EM. Formaldehyde production by Tris buffer in peptide formulations at elevated temperature. J Pharm Sci 2001; 90:1198–1203.

39. Brennan TV, Clarke S. Spontaneous degradation of polypeptides at aspartyl and asparaginyl residues: effects of the solvent dielectric. Protein Sci 1993; 2:331–338.

40. Nozaki Y, Tanford C. The solubility of amino acids and two glycine peptides in aqueous ethanol and dioxane solutions. Establishment of a hydrophobicity scale. J Biol Chem 1971; 246:2211–2217.

41. Catak S, Monard G, Aviyente V, Ruiz-López MF. Reaction mechanism of deamidation of asparaginyl residues in peptides: effect of solvent molecules. J Phys Chem A 2006; 110:8354–8365.

42. Panyam J, Labhasetwar V. Biodegradable nanoparticles for drug and gene delivery to cells and tissue. Adv Drug Deliv Rev 2003; 55:329–347.

43. Cleland JL, Lam X, Kendrick B, et al. A specific molar ratio of stabilizer to protein is required for storage stability of a lyophilized monoclonal antibody. J Pharm Sci 2001; 90:310–321.

44. Lai MC, Hageman MJ, Schowen RL, Borchardt RT, Topp EM. Chemical stability of peptides in polymers. 1. Effect of water on peptide deamidation in poly(vinyl alcohol) and poly(vinyl pyrrolidone) matrixes. J Pharm Sci 1999; 88:1073–1080.

45. Schwendeman S. Recent advances in the stabilization of proteins encapsulated in injectable PLGA delivery systems. Crit Rev Ther Drug Del Syst 2002; 19:73–102.

46. Tamber A, Johansen P, Merkle HP, Gander B. Formulation aspects of biodegradable polymeric microspheres for antigen delivery. Adv Drug Deliv Rev 2005; 57:357–376.

47. Li B, Schowen RL, Topp EM, Borchardt RT. Effect of N-1 and N-2 residues on peptide deamidation rate in solution and solid state. AAPSJ 2006; 8:E166–E173.

48. D'Souza AJM, Schowen RL, Borchardt RT, et al. Reaction of a peptide with poly-vinylpyrrolidone in the solid state. J Pharm Sci 2003; 92:585–593.

49. Houchin ML, Heppert K, Topp EM. Deamidation, acylation and proteolysis of a model peptide in PLGA films. J Control Release 2006; 112:111–119.

50. Fu K, Pack DW, Klibanov AM, Langer R. Visual Evidence of acidic environment within degrading poly(lactic-co-glycolic acid) (PLGA) microspheres. Pharm Res 2000; 17:100–106.

51. Xiang T-X, Anderson BD. A molecular dynamics simulation of reactant mobility in an amorphous formulation of a peptide in poly(vinylpyrrolidone). J Pharm Sci 2004; 93:855–876.

52. Li B, O'Meara MH, Lubach JW, et al. Effects of sucrose and mannitol on asparagine deamidation rates of model peptides in solution and in the solid state. J Pharm Sci 2005; 94:1723–1735.

53. Lai MC, Topp EM. Solid-state chemical stability of proteins and peptides. J Pharm Sci 1999; 88:489–500.

54. Leach SJ, Lindy H. The kinetics of hydrolysis of the amide group in proteins and peptides. Trans Faraday Soc 1953; 49:915–920.

55. Patel K. Stability of adrenocorticotropic hormone (ACTH) and pathways of deami-dation of asparginyl residue in hexapeptide segments. In: Wang YJ, Pearlman R, eds. Characterization of Protein and Peptide Drugs. New York: Plenum Press, 1993:201.

56. Radkiewicz JL, Zipse H, Clarke S, Houk KN. Neighboring side-chain effects on aspa-raginyl and aspartyl degradation: an ab initio study of the relationship between peptide conformation and backbone NH acidity. J Am Chem Soc 2001; 123:3499–3501.

57. Brennan TV, Clarke S. Effects of adjacent histidine and cysteine residues on the spontaneous degradation of asparaginyl- and aspartyl-containing peptides. Int J Protein Pept Res 1995; 45:547–553.

58. Capasso S. Estimation of the deamidation rate of asparagine side-chains. J Pept Res 2002; 55:224–229.

59. Robinson NE, Robinson AB. Prediction of protein deamidation rates from primary and three-dimensional structure. Proc Natl Acad Sci USA 2001; 98:4367–4372.

60. Robinson NE, Robinson ZW, Robinson BR, et al. Structure-dependent nonenzy-matic deamidation of glutaminyl and asparaginyl pentapeptides. J Pept Res 2004; 63:426–436.

61. Robinson NE, Robinson AB. Prediction of primary structure deamidation rates of asparaginyl and glutaminyl peptides through steric and catalytic effects. J Pept Res 2004; 63:437–448.

62. Chazin WJ, Kossiakoff AK. The role of secondary and tertiary structure in intramolecular deamidation of proteins. In: Aswad DW, ed. Deamidation and Isoaspartate Formation in Peptides and Proteins. Boca Raton, FL: CRC Press, 1995:193.

63. Riha WE, Izzo HV, Zhang J, Ho C-T. Nonenzymatic deamidation of food proteins. Crit Rev Food Sci Nutr 1996; 36:225–255.

64. Creighton TE. Proteins: Structures and Molecular Properties. New York: WH Freeman, 1984.

65. Habibi AE, Khajeh K, Naderi-Manesh H, Ranjbar B, Nemat-Gorgani M. Thermostabilization of *Bacillus amyloliquefaciens* α-amylase by chemical cross-linking. J Biotechnol 2006; 123:434–442.

66. Kossiakoff AK. Tertiary structure is a principal determinant to protein deamidation. Science 1988; 240:191–194.

67. Toniolo C, Bonora GM, Mutter M, Rajasekharan-Pillai VN. The effect of the insertion of a proline residue on the solution conformation of host peptides. Makromol Chem 1981; 182:2007–2014.

68. Darrington RT, Anderson BD. The role of intramolecular nucleophilic catalysis and the effects of self-association on the deamidation of human insulin at low pH. Pharm Res 1994; 11:784–793.

69. Darrington RT, Anderson BD. Effects of insulin concentration and self-association on the partitioning of its A-21 cyclic anhydride intermediate to desamido insulin and covalent dimer. Pharm Res 1995; 12:1077–1084.

70. Sun AQ, Yuksel KU, Gracy RW. Terminal marking of triosephosphate isomerase. Consequences of deamidation. Arch Biochem Biophys 1995; 322:361–368.

71. Norde W. The behavior of proteins at interfaces, with special attention to the role of the structure stability of the protein molecule. Clin Mater 1992; 11:85–91.

72. Zhang Y, Yuksel KU, Gracy RW. Terminal marking of avian triosephosphate isomerase by deamidation and oxidation. Arch Biochem Biophys 1995; 317:112–120.

73. Tomizawa H, Yamada H, Imoto T. The mechanism of irreversible inactivation of lysozyme at pH 4 and 100°C. Biochemistry 1994; 33:13032–13037.

74. Volkim DB, Verticelli AM, Bruner MW, et al. Deamidation of polyanion-stabilized acidic fibroblast growth factor. J Pharm Sci 1995; 84:7–11.

75. Strickley RG, Anderson BD. Solid state stability of human insulin II Effect of water on reaction intermediate partitioning in lyophiles from pH 2–5 solutions: stabilization against covalent dimer formation. J Pharm Sci 1997; 86:645–653.

76. Li S, Schoneich C, Borchardt RT. Chemical instability of protein pharmaceuticals. Mechanism of oxidation and strategies for stabilization. Biotechnol Bioeng 1995; 48:490–500.

77. Nguyen TH. Oxidation in degradation of protein pharmaceuticals. In: Cleland JL, Langer R, eds. Formulation and Delivery of Proteins and Peptides, ACS Symposium 567. Washington, DC: American Chemical Society, 1994:59.

78. Johnson DM, Gu LC. Autoxidation and antioxidants. In: Swarbrick J, Boyland JC, eds. Encyclopedia of Pharmaceutical Technology. Vol. 1. New York: Marcel Dekker, 1994:415–449.

79. Yang YC, Hanson MA. Parenteral formulation of proteins and peptides: stability and stabilizers. J Parenter Sci Technol 1988; 42:s3–s26.

80. Cross CM, Reznick AZ, Packers L, Davis PA, Suzuki YJ, Halliwell B. Oxidative damage to human plasma proteins by ozone. Free Radical Res Commun 1992; 15:347–352.

81. Oliver CN, Ahn B, Wittenberger ME, Levine RL, Stadtman ER. Age-related alterations of enzymes may involve mixed-function oxidation reactions. In: Adelman RC, Dekker EE, eds. Modifications of Proteins During Ageing. New York: Alan R Liss, 1985:39.

82. Babior BM. Oxygen-dependent microbial killing by phagocytes. New Engl J Med 1978; 298:659–668.

83. Fransson J, Florin-Robertsson E, Axelsson K, Nyhlen C. Oxidation of human insulin-like growth factor I in formulation studies: kinetics of methionine oxidation in aqueous solution and in solid state. Pharm Res 1996; 13:1252–1257.

84. Cipolla DC, Shire SJ. Analysis of oxidized human relaxin by reversed phase HPLC, mass spectroscopy and bioassay. In: Villafranca JJ, ed. Techniques in Protein Chemistry. Vol. 2. New York: Academic Press, 1990:543–555.

85. Stadtman ER. Oxidation of free amino acids and amino acid residues in proteins by radiolysis and by metal-catalyzed reactions. Annu Rev Biochem 1993; 62:797–821.

86. Vogt W. Oxidation of methionyl residues in proteins: tools, targets and reversal. Free Radical Biol Med 1995; 18:93–105.

87. Riesz P, Kondo T. Free radical formation induced by ultrasound and its biological implications. Free Radical Biol Med 1992; 13:247–270.

88. Nguyen TH, Burnier J, Meng W. The kinetics of relaxin oxidation by hydrogen peroxide. Pharm Res 1993; 10:1563–1571.

89. Li S, Nguyen TH, Schoneich C, Borchardt RT. Aggregation and precipitation of human relaxin induced by metal-catalyzed oxidation. Biochemistry 1995; 34:5762–5772.

90. Maier KL, Matejkova E, Hinze H, Leuschel L, Weber H, Beck-Speier I. Different selectivities of oxidants during oxidation of methionine residues in the α_1-proteinase inhibitor. FEBS Lett 1989; 250:221–256.

91. Schoneich C, Zhao F, Wilson GS, Borchardt RT. Iron-thiolate induced oxidation of methionine to methionine sulfoxide in small model peptides. Intra molecular catalysis by histidine. Biochem Biophys Acta 1993; 1168:307–322.

92. Li S, Schoneich C, Borchardt RT. Chemical pathways of peptide degradation. VIII Oxidation of methionine in small model peptides by prooxidant/transition metal ion systems: influence of selective scavengers for reactive oxygen intermediates. Pharm Res 1995; 12:348–355.

93. Halliwell B. Superoxide-dependent formation of hydroxyl radicals in the presence of iron chelates. FEBS Lett 1978; 92:321–326.

94. Halliwell B, Gutteridge JMC. Formation of a thiobarbituric-acid-reactive substance from deoxyribose in the presence of iron salts. FEBS Lett 1981; 128:347–352.

95. Gutteridge JMC, Richmond R, Halliwell B. Oxygen free-radicals in lipid peroxidation: inhibition by the protein caeruloplasmin. FEBS Lett 1980; 112:269–272.

96. Stadtman ER. Metal ion-catalyzed oxidation of proteins: biochemical mechanisms and biological consequences. Free Radical Biol Med 1990; 9:315–325.

97. The LC, Murphy LJ, Huq NL, et al. Methionine oxidation in human growth hormone and human chorionic somatomammotropin. J Biol Chem 1987; 262:6472–6477.

98. Luo D, Smith SW, Anderson BD. Kinetics and mechanism of the reaction of cysteine and hydrogen peroxide in aqueous solution. J Pharm Sci 2005; 94:304–316.

99. Keck RG. The use of t-butyl hydroperoxide as a probe for methionine oxidation in proteins. Anal Biochem 1996; 236:56–62.

100. Neumann NP. Oxidation with hydrogen peroxide. Methods Enzymol 1972; 25:393–400.

101. Buxton GV, Greenstock CL, Helman WP, Ross AB, Tsang W. Critical Review of rate constants for reactions of hydrated electrons. Chemical kinetic data

base for combustion chemistry. Part 3: Propane. J Phys Chem Ref Data 1988; 17:513–886.

102. Davies MJ. The oxidative environment and protein damage. Biochim Biophys Acta 2005; 1703:93–109.

103. Hawkins CL, Davies MJ. Generation and propagation of radical reactions on proteins Biochim Biophys Acta 2001; 1504:196–219.

104. Paik WK, Kim S. Protein methylation: chemical, enzymological, and biological significance. Adv Enzymol 1975; 42:227–286.

105. Sysak PK, Foote CS, Ching T-Y. Chemistry of singlet oxygen. XXV Photooxygenation of methionine. Photochem Photobiol 1977; 26:19–27.

106. Schoneich C, Aced A, Asmus K-D. Halogenated peroxyl radicals as two-electron-transfer agents. Oxidation of organic sulfides to sulfoxides. J Am Chem Soc 1991; 113:375–376.

107. Sawyer DT, Kang C, Llobet A, Redman C. Fenton Reagents (1:1 Fe(II)L$_x$mOOK) React via [L$_x$Fe(II)OOH(BH$^+$)] (1) as hydroxylases (RH => ROH), not as generators of free hydroxyl radicals (HOC). J Am Chem Soc 1993; 115:5817–5818.

108. Yamazaki J, Piette LH. EPR spin-trapping study on the oxidizing species formed in the reaction of the ferrous ion with hydrogen peroxide. J Am Chem Soc 1991; 113:7588–7593.

109. Li S, Schoneich C, Wilson GS, Borchardt RT. Chemical pathways of peptide degradation. V Ascorbic acid promotes rather than inhibits the oxidation of methionine to methionine sulfoxide in small model peptides. Pharm Res 1993; 10:1572–1579.

110. Good NE, Winget GD, Winter W, Connolly TN, Izawa S, Singh RMM. Hydrogen ion buffers for biological research. Biochemistry 1966; 5:467–477.

111. Hicks M, Gebicki JM. Rate constants for reaction of hydroxy radicals with tris, tricine and hepes buffer. FEBS Lett 1986; 199:92–94.

112. Miskoski S, Garcia NA. Influence of the peptide bond on the singlet molecular oxygen-mediated (O$_2$[$^1\Delta_g$]) photooxidation of histidine and methionine dipeptides. A kinetic study. Photochem Photobiol 1993; 57:447–452.

113. Linetsky M, Ortwerth BJ. Quantitation of the reactive oxygen species generated by the UVA irradiation of ascorbic acid-glycated lens proteins. Photochem Photobiol 1996; 63:649–655.

114. Funakoshi T, Abe M, Sakata M, Shoji S, Kubota Y. The functional side of placental anticoagulant protein: essential histidine residue of placental anticoagulant protein. Biochem Biophys Res Commun 1990; 168:125–134.

115. Stuart J, Pessah IN, Favero TG, Abramson JJ, Photooxidation of skeletal muscle sarcoplasmic reticulum induces rapid calcium release. Arch Biochem Biophys 1992; 292:512–521.

116. Tomita M, Irie M, Ukita T. Sensitized photooxidation of histidine and its derivatives. Products and mechanism of the reaction. Biochemistry 1969; 8:5149–5160.

117. Batra PP, Skinner G. A kinetic study of the photochemcial inactivation of adenylate kinases of *Mycobacterium marinum* and bovine heart mitochondria. Biochim Biophys Acta 1990; 1038:52–59.

118. Giulive C, Sarcansky M, Rosenfeld E, Boveris A. The photodynamic effect of rose bengal on proteins of the mitochondrial inner membrane. Photochem Photobiol 1990; 52:745–751.

119. Ortwerth BJ, Linetsky M, Olesen PR. Ascorbic acid glycation of lens proteins produces UVA sensitizers similar to those in human lens. Photochem Photobiol 1995; 62:454–462.

120. Amici A, Levine RL, Tsai L, Stadtman ER. Conversion of amino acid residues in proteins and amino acids homopolymers to carbonyl derivatives by metal-catalyzed oxidation reactions. J Biol Chem 1989; 264:3341–3346.

121. Cheng R-Z, Kawakishi S. Site-specific oxidation of histidine residues in glycated insulin mediated by Cu^{+2}. Eur J Biochem 1994; 223:759–764.

122. Ueda J-I, Ozawa T, Miyazaki M, Fujiwara Y. Activation of hydrogen peroxide by copper(II) complexes with some histidine-containing peptides and their SOD-like activities. J Inorganic Biochem 1994; 55:123–130.

123. Uchida K, Kawakishi S. Identification of oxidized histidine generated at the active site of Cu, Zn-superoxide dismutase exposed to H_2O_2. J Biol Chem 1994; 269:2405–2410.

124. Lewisch SA, Levine RL. Determination of 2-oxohistidine by amino acid analysis. Anal Biochem 1995; 231:440–446.

125. Engelka KA, Maciaj T. Inactivation of human fibroblast growth factor- 1 (FGF-1) activity by interaction with copper ions involves FGF-1 dimer formation induced by copper-catalyzed oxidation. J Biol Chem 1992; 267:11307–11315.

126. Munday R. Toxicity of thiols and disulphides: involvement of free-radical species. Free Radical Biol Med 1989; 7:659–673.

127. Tsai PK, Volkin DB, Dabora JM, et al. Formulation design of acidic fibroblast growth factor. Pharm Res 1993; 10:649–659.

128. Muslin EH, Li D, Stevens FJ, Donnelly M, Schiffer M, Anderson LE. Engineering a domain-locking disulfide into a bacterial malate dehydrogenase produces a redox-sensitive enzymes. Biophys J 1995; 68:2218–2223.

129. Little C, O'Brien PJ. Mechanism of peroxide-inactivation of the sulphydryl enzyme glyceraldehyde-3-phosphate dehydrogenase. Eur J Biochem 1969; 10:533–538.

130. Itakura K, Uchida K, Kawakishi S. A novel tryptophan dioxygenation by superoxide. Tetrahedron Lett 1992; 33:2567–2570.

131. Nakagawa M, Watanabe H, Kodato S, et al. A valid model for the mechanism of oxidation of tryptophan to formylkynurenine—25 years later. Proc Natl Acad Sci USA 1977; 74:4730–4733.

132. Maskos Z, Rush JD, Koppenol WH. The hydroxylation of tryptophan. Arch Biochem Biophys 1992; 296:514–520.

133. Kell G, Steinhart H. Oxidation of tryptophan by H_2O_2 in model systems. J Food Sci 1990; 55:1120–1123.

134. Itakura K, Uchida K, Kawakishi S. Selective formation of oxindole- and formylkynurenine-type products from tryptophan and its peptides treated with a superoxide-generating system in the presence of iron(III)-EDTA: a possible involvement with iron-oxygen complex. Chem Res Toxicol 1994; 7:185–190.

135. Savige WE. Isolation and identification of some photo-oxidation products of tryptophan. Aust J Chem 1975; 24:1285–1293.

136. Holt LA, Milligan B, Rivett DE, Stewart FHC. The photodecomposition of tryptophan peptides. Biochim Biophys Acta 1977; 499:131–138.

137. Tassin JD, Borkman RF. The photolysis rates of some di- and tripeptides of tryptophan. Photochem Photobiol 1980; 32:577–585.

138. Templer H, Thistlewaite PJ. Flash photolysis of aqueous tryptophan, alanyl tryptophan and tryptophyl alanine. Photochem Photobiol 1976; 23:79–85.

139. McKim S, Hinton JF. Direct observation of differential UV photolytic degradation among the tryptophan residues of gramicidin A in sodium dodecyl sulfate micelles. Biochim Biophys Acta 1993; 1153:315–321.

140. Pigault C, Gerard D. Influence of the location of tryptophanyl residues in proteins on their photosensitivity. Photochem Photobiol 1984; 40:291–296.
141. Huggins TG, Wells-Knecht MC, Detorie NA, Baynes JW, Thorpe SR. Formation of *o*-tyrosine and dityrosine in proteins during radiolytic and metal-catalyzed oxidation. J Biol Chem 1993; 268:12341–12347.
142. Stadtman ER. Role of oxidized amino acids in protein breakdown and stability. Methods Enzymol 1995; 258:379–393.
143. Heinecke JW, Li W, Daehnke HL III, Goldsten JA. Dityrosine, a specific marker of oxidation, is synthesized by the myeloperoxidase-hydrogen peroxide system of human neutrophils and macrophages. J Biol Chem 1993; 268:4069–4077.
144. Uchida K, Kato Y, Kawakishi S. A novel mechanism for oxidative cleavage of prolyl peptides induced by the hydroxyl radical. Biochem Biophys Res Commun 1990; 169:265–271.
145. Kato Y, Uchida K, Kawakishi S. Oxidative fragmentation of collagen and prolyl peptide by $Cu(II)/H_2O_2$. J Biol Chem 1992; 267:23646–23651.
146. Garrison W. Reaction mechanisms in the radiolysis of peptides, polypeptides, and proteins. Chem Rev 1987; 87:381–398.
147. Magill A, Becker AR. Spectrophotometric method for quantitation of peroxides in sorbitan monooleate and monostearate. J Pharm Sci 1984; 73:1663–1664.
148. Welch KD, Davis TZ, Aust SD. Iron autoxidation and free radical generation: effects of buffers, ligands, and chelators. Arch Biochem Biophys 2002; 397:360–366.
149. Harris DC, Aisen P. Facilitation of Fe(II) oxidation by Fe(III) complexing agents. Biochim Biophys Acta 1973; 329:156–158.
150. Halliwell B, Whiteman M. Measuring reactive species and oxidative damage in vivo and in cell culture: how should you do it and what do the results mean? Br J Pharmacol 2004; 142:231–255.
151. Levine RL, Williams JA, Stadtman EP, Shacter E. Carbonyl assays for determination of oxidatively modified proteins. Methods Enzymol 1994; 233:346–357.
152. Nishinaga A, Shimizu T, Matsuura T. Base-catalyzed oxygenation of *tert*-butylated phenols. J Org Chem 1979; 44:2983–2988.
153. McGinity JW, Hill JA, LaVia AL. Influence of peroxide impurities in polyethylene glycols on drug stability. J Pharm Sci 1975; 64:356–357.
154. Chafetz L, Hong W-H, Tsilifonis DC, Taylor AK, Philip J. Decrease in the rate of capsule dissolution due to formaldehyde from polysorbate 80 autoxidation. J Pharm Sci 1984; 73:1186–1187.
155. Labrude P, Chaillo B, Bonneaux F, Vigneron C. Freeze-drying of haemoglobin in the presence of carbohydrates. J Pharm Pharmacol 1980; 32:588–589.
156. Rowley DA, Halliwell B. Superoxide-dependent and ascorbate-dependent formation of hydroxyl radicals in the presence of copper salts: a physiologically significant reaction? Arch Biochem Biophys 1983; 225:279–284.
157. Braughler JM, Duncan LA, Chase RL. The involvement of iron in lipid peroxidation. J Biol Chem 1986; 261:10282–10289.
158. Li S, Patapoff TW, Nguyen TH, Borchardt RT. Inhibitory effect of sugars and polyols on the metal-catalyzed oxidation of human relaxin. J Pharm Sci 1996; 85:868–875.
159. Angyal SJ. Complexing of carbohydrates with copper ions: a reappraisal. Carbohydr Res 1990; 200:181–188.
160. Franzini E, Sellak H, Hakim J, Pasquier C. Comparative sugar degradation by (OH)· produced by the iron-driven Fenton reaction and gamma radiolysis. Arch Biochem Biophys 1994; 309:261–265.

161. Wolff SP, Dean RT. Glucose autoxidation and protein modification. Biochem J 1987; 245:243–250.

162. Hunt JV, Dean RT, Wolff SP. Hydroxyl radical production and autoxidative glycosylation. Biochem J 1988; 256:205–212.

163. Hunt PK, Roberts RC. Methionine oxidation and inactivation of α_1-Proteinase inhibitor by Cu^{+2} and glucose. Biochim Biophys Acta 1988; 1121:325–330.

164. Fujimoto S, Nakagawa T, Ishimitsu S, Ohara A. On the mechanism of inactivation of papain by bisulfite. Chem Pharm Bull 1983; 31:992–1000.

165. Fryer MJ. Evidence for the photoprotective effects of vitamin E Photochem Photobiol 1993; 58:304–312.

166. Pikal MJ, Dellman K, Roy ML. Formulation and stability of freeze-dried proteins: effects of moisture and oxygen on the stability of freeze-dried formulations of human growth hormone. Dev Biol Stand 1992; 74:21–38.

167. Enever RP, Li Wan Po A, Shotton E. Factors influencing decomposition rate of amitriptyline hydrochloride in aqueous solution. J Pharm Sci 1977; 66:1087–1089.

168. Town C. Moisture content in proteins: its effects and measurement. J Chromatogr A 1995; 705:115–127.

169. Hageman MJ. The role of moisture in protein stability. Drug Dev Ind Pharm 1988; 14:2047–2070.

170. Luo D, Anderson BD. Kinetics and mechanism for the reaction of cysteine with hydrogen peroxide in amorphous polyvinylpyrrolidone lyophiles. Pharm Res 2006; 23:2239–2253.

171. Cadée JA, van Steenbergen MJ, Versluis C, et al. Oxidation of recombinant human interleukin-2 by potassium peroxodisulfate. Pharm Res 2001; 18:1461.

172. van de Weert M, Hennink WE, Jiskoot W. Protein instability in poly(Lactic-co-Glycolic Acid) microparticles. Pharm Res 2000; 17:1159–1167.

173. Ibrahim MA, Ismail A, Fetouh MI, Göpferich A. Stability of insulin during the erosion of poly(lactic acid) and poly(lactic-co-glycolic acid) microspheres. J Control Release 2005; 106:241–252.

174. Na DH, Youn YS, Lee SD, et al. Monitoring of peptide acylation inside degrading PLGA microspheres by capillary electrophoresis and MALDI-TOF mass spectrometry. J Contol Release 2003; 92:291–299.

175. Murty SB, Na DH, Thanoo BC, DeLuca PP. Impurity formation studies with peptide-loaded polymeric microspheres: Part II in vitro evaluation. Int J Pharam 2005; 297:62–72.

176. Lucke A, Kiermaier J, Göpferich A. Peptide acylation by poly(α-hydroxy esters). Pharm Res 2002; 19:175–181.

177. Lucke A, Fustella E, Teßmar J, Gazzaniga A, Göpferich A. The effect of poly(ethylene glycol)–poly(D-L-lactic acid) diblock copolymers on peptide acylation. J Control Release 2002; 80:157–168.

178. Na DH, DeLuca PP. PEGylation of octreotide: I separation of positional isomers and stability against acylation by poly(D,L-lactide-co-glycolide). Pharm Res 2005; 22:736–742.

3

Physical Considerations in Protein and Peptide Stability

Sandy Koppenol

Nastech Pharmaceutical Company, Inc., Bothell, Washington, U.S.A.

PHYSICAL CONSIDERATIONS

This section considers the physical stability of the peptide or protein including conformational changes, aggregation, and adsorption. The physical stability of the protein, and thus protein function, can be affected by a number of environmental parameters. Relatively small changes in temperature, pressure, pH, or concentration of denaturing agents (e.g., guanidine hydrochloride, surfactants), or exposure of the macromolecule to mechanical disruption can lead to an irreversible loss of protein function. The primary mechanism for protein inactivation by these agents or processes involves the denaturation or unfolding of the protein macromolecule. Protein unfolding refers to the loss of tertiary structure and the formation of a disordered protein in which the proper intramolecular contacts within the protein no longer exist. The intermolecular recognition events necessary for proper protein folding are usually cooperative and reversible upon removal of the denaturing agent. However, unfolding can be followed by secondary irreversible inactivating processes such as the chemical changes described in Chapter 2, or by other physical processes such as aggregation of the protein to form higher-order oligomers or "aggregates" and adsorption to surfaces. Therefore, the preservation of protein tertiary structure becomes paramount for preventing losses in protein function. Although scientists have made significant progress in understanding and predicting protein folding, the numerous molecular determinants that drive intraprotein recognition events make protein stabilization an interesting challenge for the protein formulator.

The tertiary structure of proteins is driven by two classes of noncovalent interactions, electrostatic and hydrophobic. In this review, electrostatic interactions include ion pairs (1), H bonds, weakly polar interactions, and van der Waals forces. "Hydrophobic interactions" refer to actions and hydration effects of nonpolar groups. Protein folding is opposed by conformation entropy. The interplay between enthalpy and entropy makes individual changes in a physical parameter complex, and the exact causes are still controversial. In this section, we attempt to sort out experimentally accessible parameters that may be useful to the protein formulator for conducting preformulation studies, formulation of the protein for a desired extent of stability, and ways to test for stability.

Protein Folding Thermodynamics and Aggregation Kinetics

Thermodynamics of Protein Folding

Proteins exist in unique conformations under physiological conditions. Conformation is driven by intramolecular energy and solvation free energy (2). The simplest model used to describe reversible protein folding is a two-state model in which an equilibrium exists between the native (N) and denatured states (D) as shown in Equation 1.

$$N \rightleftharpoons D \qquad (1)$$

This model provides a useful tool for studying proteins and can provide essential initial insight into stability in vitro. The equilibrium constant for unfolding is thus given in Equation 2.

$$K = \frac{D}{N} \qquad (2)$$

The change in Gibbs free energy of folding, regardless of any intermediate complexes, can then be deduced from the equilibrium constant for the folding reaction, as given in Equation 3.

$$\Delta G = (G_D - G_N) = -RT \ln K \qquad (3)$$

where R is the gas constant and T is the absolute temperature. The free energy for the equilibrium under physiological conditions is typically 5 to 20 kcal/mol, reflecting a remarkably low stability of the native protein (3,4). The temperature coefficient of the equilibrium constant is given by Equation 4.

$$\frac{\partial \ln K}{\partial T} = \frac{\Delta H}{RT^2} \qquad (4)$$

which is commonly referred to as the van't Hoff equation. The change in Gibbs energy as a function of temperature is given as in Equation 5.

$$\Delta G = \Delta H - T\Delta S \tag{5}$$

where ΔH and ΔS are the changes in enthalpy and entropy, respectively, at a given temperature. Equation 5 reveals that the free energy is a balance between the enthalpy and entropy changes, which are typically large numbers (e.g., 50–200 kcal). Since the small changes in free energies associated with the transition from the native to the denatured state are the difference of large numbers, this makes the difference in free energy very sensitive to small perturbations in the attractive and repulsive intramolecular interactions. This makes protein stability easy to manipulate but makes exact predictions of protein stability difficult. The dependency of enthalpy and entropy on temperature can be described in terms of the heat capacity (ΔC_p) at a constant pressure (Eqs. 6 and 7)

$$\Delta H = \Delta H^0 + \Delta C_p(T - T_0) \tag{6}$$

$$\Delta S = \Delta S^0 + \Delta C_p \ln(T / T_0) \tag{7}$$

where ΔH^0 and ΔS^0 are the enthalpy and entropy at a given reference temperature T^0. The change in heat capacity, ΔC_p, is the difference in heat capacity between the N and D states and, in most cases involving proteins, is large and relatively constant within experimental error (5). The large value for ΔC_p is driven by the exposure of nonpolar groups to water upon protein unfolding (6,7). Robertson and Murphy have summarized the thermodynamic variables for globular proteins of known structure (8). Combining Equations 4, 5, and 6, yields Equation 8, which is the modified Gibbs–Helmholtz equation (for an excellent review, see Ref. 8).

$$\Delta G = \Delta H^0 - T\Delta S^0 + \Delta C_p(T - T_0 - T \ln (T / T_0)) \tag{8}$$

It is important to remember that Equation 8 holds true only for the two-state model, in which no stable folding intermediates exist. This assumption has been shown to be valid for many small globular proteins, although exceptions do exist (9). In some of these cases, the existence of intermediates that are not significantly populated or are short-lived cannot be ruled out.

The value of ΔG is thus the fundamental measure of the stability of the protein. In cases that allow the assumption of a two-state model, the ΔG can be determined from measurements of K as a function of temperature by means of spectroscopic methods (bioassays, chromatography, electrophoresis, etc.) to determine relative populations of the native and denatured proteins (10). At the melting temperature T_m, ΔG is zero and by Equation 5, ΔS is equal to $\Delta H/T_m$. Therefore, Equation 8 can be used to fit the experimental values of data on ln K versus T, to extract the values for ΔH_m, T_m, and ΔC_p at T_m. Once these parameters have been determined, ΔH and ΔS can be calculated over a wide range of temperatures. Typically, a maximum in stability is observed between 0°C and 25°C (11) (to be discussed further). However, the range of temperatures in which the protein is stable will change depending on the presence and concentration of other agents.

Reversible versus Irreversible Protein Denaturation

Reversible denaturation or unfolding can be followed by processes that create irreversibly inactivated forms of the protein (12). While K is the equilibrium constant for reversible denaturation, which correlates to a protein's thermodynamic stability as described above, k is the rate constant for the irreversible inactivation of the protein, which is a measure of a protein's long-term stability (13). Inactivation is typically observed after unfolding, however, inactivation can proceed directly from the native state. Unfolded proteins are more prone to proteolysis than tightly packed, globular proteins (14), often tend to aggregate into insoluble masses, and are able to become kinetically trapped in improperly folded conformations. Chemical reactivity, i.e., hydrolysis and reduction of disulfide bonds, can also lead to unfolding. Formulation scientists need to understand, when possible, the relationship between chemical and physical instabilities. Stabilization of the native state should be paramount and is discussed in the subsections that follow.

Kinetics of Protein Aggregation

Irreversible protein aggregation occurs when nonnative forms of the protein associate to form higher-order multimers. These higher-order multimers can form insoluble precipitates (15). This can happen during many protein-processing operations such as purification, freezing and thawing, and lyophilization. A kinetic model for protein aggregation has been described by Lumry and Eyring (16). The model (Eq. 9) assumes a reversible change in the native structure (N) followed by an irreversible aggregation of aggregate competent species (A).

$$N \rightleftharpoons A$$
$$A + A_m \rightleftharpoons A_{m+1}$$
$$(9)$$

If the association of A to form A_{m+1} is rate limiting, then the kinetics are second order (or higher) with respect to protein concentration. Thus, lowering the protein concentration should significantly reduce aggregation kinetics (Chapter 7). However, it is generally well accepted that aggregation, in many cases, can be controlled by maximizing the thermodynamic stability as mentioned in the previous section.

Physical Protein Stability in Solution

Pharmaceutical formulations are complex systems, and it is often difficult to separate the effects of any single variable that can account for a shift in the equilibrium to favor the denatured protein state. Losses in protein physical stability most likely are due to a mixture of many different stabilizing [e.g., hydrogen bonding and the hydrophobic effect (17)] and destabilizing [e.g., configurational entropy (18)] effects. A large range in each physical parameter exists to which proteins have adapted: for temperature, $-5°C$ to $110°C$; water activity, 1 to 0.6 [corresponding to ≤ 6 M salt (19)]; pH, 4 to 12. Although alteration of these parameters often leads to a reversible denatured state, it is important to remember that unfolding is usually the first step in irreversible losses in protein structure, and thus keeping the protein folded is paramount in stabilizing protein structure.

Temperature

The stability of proteins to thermal stress is an important variable to the protein formulator. Changes in temperature may accompany processing (e.g., lyophilization or spray-drying) (20) and serves as a convenient thermodynamic variable to probe the stability and shelf life for the protein pharmaceutical through accelerated protein stability testing. Furthermore, calorimetry can be used to obtain accurate measurements of the thermodynamic properties of unfolding as a function of temperature.

Thermally induced unfolding is highly cooperative and often reversible (21,22). Protein stability curves (23) refer to plots of ΔG versus temperature and can be described by the Gibbs–Helmholtz equation (Eq. 8), which can be written as follows (Eq. 10):

$$\Delta G(T) = \Delta H_m \left(1 - \frac{T}{T_m}\right) - \Delta C_p \left[T_m - T + T \ln\left(\frac{T}{T_m}\right)\right] \tag{10}$$

where $\Delta G(T)$ is the ΔG at temperature T, T_m is the midpoint of the thermal unfolding curve, and ΔH_m is the enthalpy change for unfolding measured at T_m. At T_m, $\Delta G = 0$ and ΔS has just been replaced by $\Delta H/T_m$. The temperature for maximum stability (T_s) occurs at the temperature when $\Delta S = 0$, and is given by Equation 11:

$$T_s = T_m \exp\left(\frac{\Delta H_m}{T_m \Delta C_p}\right) \tag{11}$$

Typically T_s is between −10°C and 35°C for most proteins. Figure 1 is a plot of ΔG versus temperature for the protein RNase T1 at pH 7 (24,25). Equation 11 was used to calculate the solid line, with the parameters listed in the figure legend. The plot illustrates that minimal changes in temperature, by heating or cooling, can have profound effects on protein stability. Temperature-induced destabilization is driven by the gain in conformational entropy of the protein chain and usually begins with partial unfolding of the protein (26,27). Although the unfolding is typically reversible, many proteins will undergo chemical reactions or other irreversible denaturation processes such as aggregation, leading to loss of protein function. This is especially true at elevated temperatures where reaction kinetics and collision rates are increased (28).

Much work in the area of protein stability with respect to temperature has been derived from an interest in understanding how psychrophilic (29) and thermophilic organisms survive (for reviews spanning the 1990s, see Refs. 30–33). There is a strong effort in the development of thermally stable proteins and understanding how subtle changes in the amino acid structure can significantly alter

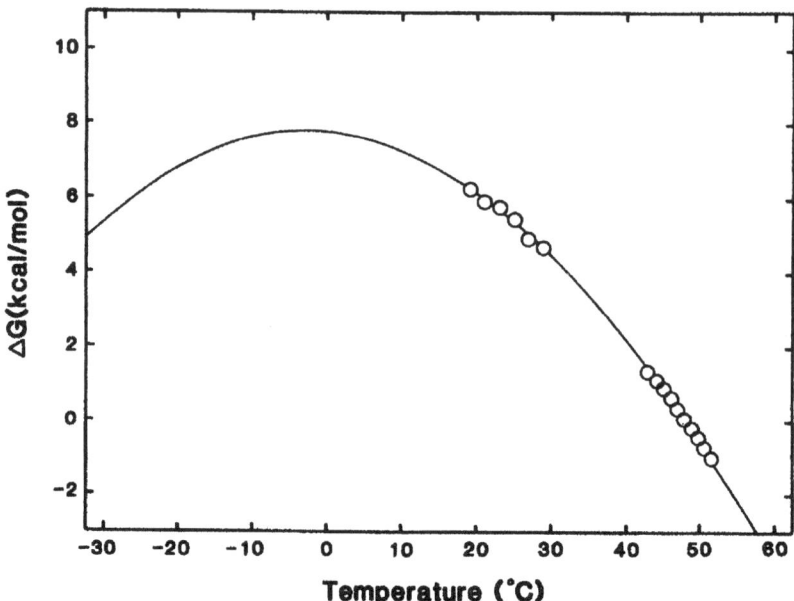

Figure 1 ΔG as a function of temperature for the unfolding of RNase T1 and pH 7.0 (30 mM MOPS). The points in the 19°C to 30°C temperature range are from urea denaturation experiments, while those above 40°C are from thermal unfolding curves and Equations 11, 12, and 13. The solid curve was calculated from Equation 19 using $T_m = 48.25°C$, $\Delta H_m = 95.2$ kcal/mol, and $\Delta C_p = 1.72$ kcal/mol deg. *Source*: From Ref. 10.

protein stability. Primary sequence analyses of proteins belonging to thermophylic organisms reveal several consistencies.

First, the positively charged residue of choice appears to be arginine over lysine. Reasons for such a substitution in heat-stable proteins include arginine's higher pK_a, which would prevent this amino acid from losing its charge at higher temperatures (pK_a decreases with increasing temperature) (34), the increased surface area of arginine, and its shorter hydrocarbon attachment to the polypeptide backbone (13). There is a decrease in asparagine content, which is especially susceptible to deamidation at high temperatures and neutral pH. Finally, an increase in proline content appears to correlate with increased thermal stability, likely because of the increase in rigidity in protein structure (35). However, enhancement of protein rigidity to increased stability may come at a cost, since protein flexibility is required for protein function. The ease with which protein engineers can manipulate primary protein structure makes enhancing protein thermal stability with small sequence changes a feasible mechanism.

Many of the proteins belonging to thermophilic organisms have tertiary structure (36,37), similar to that of their nonthermophilic counterparts. However, it has been observed that thermophilic proteins have additional salt bridges, hydrogen bonds (38), and hydrophobic interactions (36). Of these interactions, the major driving force that serves to enhance protein thermal stability appears to come about by an increase in the hydrophobic interactions of the hydrophilic core (27,39). Increases in the overall hydrophobic character of the thermophilic protein are brought about by loss of surface loops, increase in helix-forming amino acids, and restriction of *N*-terminal residues (19,33). This enhancement of stability is believed to come from a large increase in the enthalpy of unfolding, which increases with increasing hydrophobicity (40,41). The enthalpic increase is attributed to the melting of water cages that surround the exposed nonpolar side chains (26).

In addition to optimization of the protein structure, nature protects proteins against thermal stress, in particular via cryoprotection, by introducing free amino acids (42), organic salts (e.g., trehalose, sucrose), and polymers such as polyethylene glycol (19,43–45). The low-molecular-weight compounds are hypothesized to be excluded from the protein and thus to increase water activity around the protein. However, these additives may also operate through a mechanism whereby the chemical and physical processes are kinetically hindered by the high viscosity of the additives. The reader is referred to Chapter 9, which specifically addresses stabilization of proteins with osmolytes. Addition of stabilizers should be considered carefully, since both reduced and elevated temperatures can cause the crystallization and inactivation of added excipients that might hinder their stabilizing properties (46,47). Therefore, independent from the thermal stability of the protein, the formulator needs to consider the stability of the entire formulation to changes in temperature.

Lastly, the long-term storage temperatures for proteins have been related to the glass transition temperature (T_g) or the dynamical transition temperature (T_d). T_g is defined as the temperature where an equilibrium exists between protein in a glassy and a rubbery state. Unfortunately, for native globular proteins, this transition is weak and is therefore not easily measured. The T_d defines a temperature where there is a substantial change in protein dynamics, i.e., large-scale segmental motions that lead to unfolding. The T_d is typically near 200 K for native protein. Additives that raise T_g or T_d have been incorporated with some success into formulations to decrease molecular mobility in lyophilized protein products (48–53). Spectroscopic methods that measure protein motions have been used to identify additives that interact with the protein, i.e., hydrogen bond to limit protein internal dynamics. Examples of these additives include sucrose, trehalose, and polyvinylpyrrolidone. The impact of the addition of these additives and their subsequent effect on protein hydration must also be considered when developing stable solid formulations (54).

pH

Proteins are typically most stable (and often least soluble) at their isoelectric points (pI), where opposing charges serve to stabilize the protein structure (1,55,56). Figure 2 is a plot of the free energy [denoted as $\Delta G(H_2O)$ as a function of pH] for RNase T1 (24). The maximum in stability occurs at a pH where the net charge

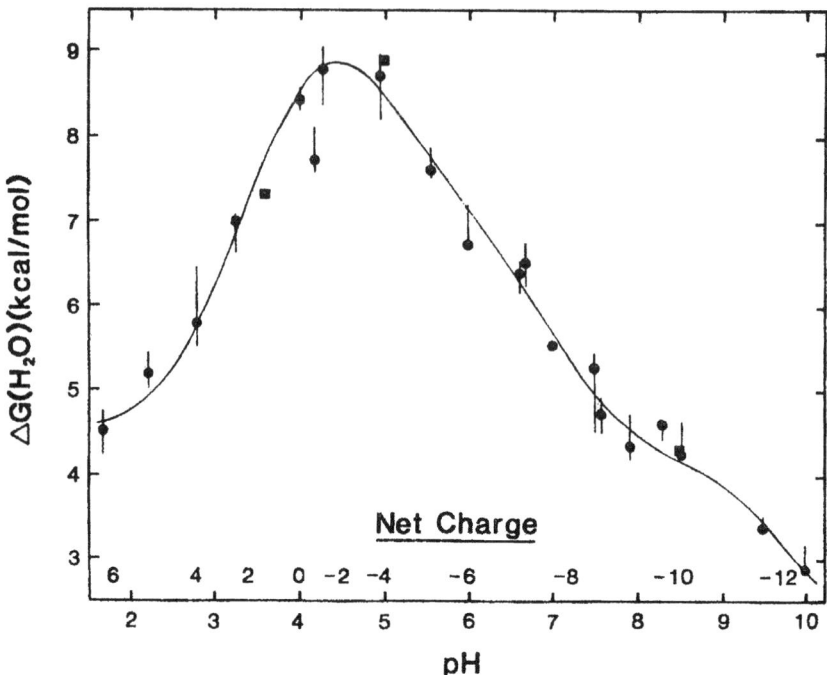

Figure 2   The free energy for unfolding $\Delta G(H_2O)$ for Rnase Tl plotted as a function of pH. An estimate of the net charge is given at each pH based on titration studies. The solid line has no theoretical significance. *Source*: From Ref. 57.

on the protein is zero. Acid or base environments can influence protein stability primarily through changes in the electrostatic free energy of the protein as the ionization states of acidic and basic groups of the protein are changed. Through the use of site-directed mutagenesis, it was found that acid and base denaturation is primarily caused by buried charges rather than by changes in the overall surface charge–charge repulsion (58–61). For example, acid denaturation of a protein below its pI is governed by a net increase in positive charges that, as pH is lowered, generates repulsive forces that exceed the stabilizing forces and lead to the loss of the native state (62). The pH stability of protein mutants also suggested that the electrostatic free energy difference between native and denatured states can be brought about through changes in the pK_a values of a relatively small number of amino acids (25,57,63,64).

The issue of pH stability is complicated by the fact that the pK_a values of the ionizable groups in the native and denatured state change (55). For example, nuclear magnetic resonance (NMR) studies of RNase T1 have shown that certain ionizable groups have higher pK_a values in the folded state, so that decreasing pH would allow the functional groups in the native state to bind protons more tightly (26), shifting the equilibrium to the native state. However, decreasing pH further raises the pK_a of aspartic acid residues in an unfolded conformation so that the unfolded becomes the most thermodynamically stable state (65).

The pK_as of the protein side chains in the 3D structure can be determined and can serve as a tool for formulators to optimize solution pH. Various empirical methods may be employed to determine the pK_a of the ionizable groups in proteins. Ultraviolet (UV), infrared (IR), NMR, and fluorescence spectroscopies and capillary electrophoresis (66) have all been used to obtain these values. Mathematical models have also been developed to calculate pK_a. These models rely on electrostatic calculations to determine the pK_a of each ionizable group on the protein (56,67–69) and require an accurate representation of the protein structure in native and denatured states. Knowledge of the pK_as allows for the calculation of the titration curve for the unfolding free energies as a function of pH. These methods have been found to be accurate enough to provide a useful tool in the interpretation of experimental results (56,67).

The pH of the protein solution can also affect aggregation between protein molecules. At the pI, the charge on the protein is neutralized and the association of neutral protein molecules can occur. Changes in other formulation variables, i.e., salt concentration, can lead to subtle protein conformation changes such that the protein association becomes irreversible. Solution pH conditions may also lead to favorable intermolecular interactions. There are two major driving forces for interactions between proteins, electrostatic and van der Waals interactions. A more detailed discussion of colloidal stability is given in Ref. (70).

Individual protein folding stability and intermolecular interactions should be carefully considered when selecting an optimal pH.

Solvation

Ions can interact with proteins in a variety of ways and can affect protein stability dramatically. Many empirical studies exist for solvation-induced perturbation of protein structure (for a review, see Ref. 71). In general, protein structural stability in the solution is associated with agents that are preferentially excluded from the native protein; i.e., in the presence of these agents, proteins in the folded conformation are preferentially hydrated. Denaturants, in contrast, are preferentially bound to the denatured protein. The stabilizing, destabilizing, precipitating, and solubilizing properties of these agents are manifested through affecting the balance between the affinities of the proteins for water and the agent (72).

Salting-in/salting-out: Increasing the solubility of a protein with low salt concentrations, salting-in, is due to Debye–Hückel screening. At these concentrations, the ion interactions with the protein are nonspecific and electrostatic. At higher ionic strengths (typically >1.0), protein salting-out dominates, which is linearly related with increasing ionic strength (72) and is relatively ion specific (73). In general, salting-out is hypothesized to be due to a preferential exclusion of the solute from the protein domain, which results in a change in the hydrogen-binding properties of the water (74). However, the exact molecular mechanisms are not understood (75). As a result, predictions of how molecular interactions affect protein stability tend to be protein specific and dependent on the system conditions such as temperature and counterion identity (i.e., size and charge).

Stabilization/denaturation: Organisms in nature face environmental stress (viz., freezing or osmotic stress) by accumulating "compatible" organic solutes that stabilize protein structure. These cryoprotectants or osmolytes are low-molecular-weight molecules that are produced by the living organisms in response to stress (76). Studies of sugars (sucrose, glucose, lactose, trehalose), polyhydric alcohols (mannitol, ethylene glycol, sorbitol, xylitol, inositol), and glycerol, which are associated with the stressed organisms, showed that such solutes stabilize the native structure of proteins by means of their preferential exclusion from the surface of the proteins (12,71,77–80). This preferential exclusion could be due to steric, solvophobic effects and/or increases in surface tension (81–83). Baskakov and Bolen have shown the extraordinary stabilization ability of some osmolytes that have been illustrated to be able to force thermodynamically unstable proteins to fold (84).

The exclusion of the solutes from immediately outside the domain of the protein causes a preferential enrichment of water around the protein (Fig. 3) (85). The result is that these additives will increase the chemical potential of the protein. This increase in the free energy of the system is proportional to surface area and is thus more unfavorable for the unfolded state (Fig. 3), which has a greater surface area exposed to the solvent. This shifts the equilibrium back to the folded, native state (71). Changing protein hydration due to additives is one of the simplest approaches to stabilizing proteins (86). Sucrose is the most studied example of this type of additive (80,87). However, not all additives that cause preferential hydration work to stabilize proteins under all conditions (different temperatures, pH, etc.); a thorough study of the dependence of the additive being used for stabilization on solution conditions must be made (88). A classification has been established that attempts to accurately describe which additives can be classified as stabilizers

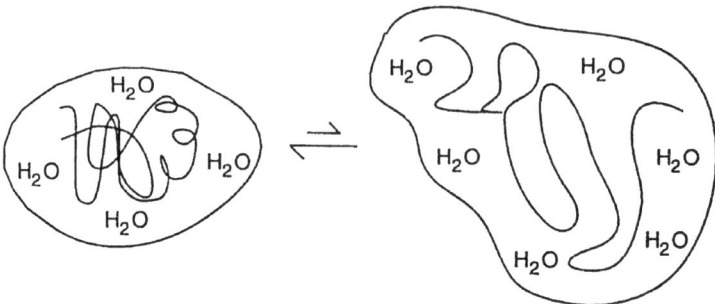

Figure 3 Schematic representation of preferential hydration of proteins. The greater exposed surface area in the denatured state shifts the equilibrium to the left, favoring the native state. *Source*: From Ref. 72 .

(77,78,88). The first class includes solutes for which the preferential hydration of the protein is independent of the solution conditions (i.e., pH, solute concentration). These are solutes where the predominant interaction is that of cosolvent exclusion, and the identity of the protein does not play a role. The second class includes solutes that vary with conditions because they bind with the protein and are thus influenced by the chemical nature of the protein surface. Solutes belonging to the first class stabilize proteins, while the others do not always do so. Therefore, thorough measurements (89,90) of the preferential hydration of the native state of a protein should be performed to verify a molecule's stabilizing action.

Conformational stability can also be increased by preferential binding of ions to a folded molecule (12,27). In the case of RNase T1 (91), there exists a cation and anion binding site on the surface of the protein. As might be expected, this can be used to advantage by protein engineers, who can engineer into the native protein binding sites to favor the folded protein. Furthermore, advances in performing theoretical calculations are allowing accurate predictions of changes in protein stability after mutation (92).

Denaturation of proteins by chaotropic agents such as guanidine hydrochloride and urea are brought about by direct specific interaction of the small molecule with the protein. Experimental evidence from thermodynamic equilibrium and calorimetric titration techniques has shown that unfolding is favored because a favorable free energy gain is produced upon interaction of the denaturants with residues in the protein that become exposed on denaturation (11,23,72,93,94). In some cases, the denaturing solute can be repelled by charges (95) of the protein in the folded state, but, as the protein unfolds, the charges are dispersed and the solute interacts with the newly exposed nonpolar residues (96). As mentioned above, salting-out sometimes can be followed by denaturation. In these cases, increases in ionic strength most likely induce denaturation by decreasing attractive charge–charge interactions (56).

Surfactants

The interaction of sodium dodecyl sulfate (SDS) with proteins is well known to cause the unfolding of proteins. Furthermore, SDS can cause unfolding at relatively low concentrations (<0.1%). This disruption of structural order occurs after the binding of SDS to the protein (97,98). Experiments with soybean trypsin inhibitor (99) have shown that only a few SDS molecules are needed to perturb the protein's structure and that the loss of structure gradually increases as SDS is added. After binding 7 to 10 molecules of SDS, the protein displays only a limited amount of α-helical order. SDS's long aliphatic tail is believed to penetrate hydrophobic pores, while shorter alkyl sulfates cannot. Charge–charge interactions between the negatively charged SDS and positively charged sites are also presumed to play a role in this interaction, as no destabilizing effects on proteins with large negative charge densities have been observed with SDS (100).

In contrast to the SDS–protein interactions, surfactants are commonly being exploited for their solubilizing and stabilizing properties (e.g., to increase refolding yields in protein purification) (101). Protein formulations commonly contain non-ionic (polysorbates and polyethers) or anionic surfactants as stabilizing agents. The bulk of protein–surfactant interactions are not well understood. Some mechanisms that may prevent losses in protein stability include prevention of surface-induced deactivation of proteins (102,103) and inhibition of aggregation and precipitation (101,104–106). Thus, preferential binding of surfactants to the native, intermediate or denatured states can influence conformational stability (107). In all of these cases, it is believed that the surfactant binds the protein and reduces the protein's available hydrophobic surface area, thus reducing the protein's self-association and any deleterious interactions with nonspecific hydrophobic surfaces.

Empirical methods can be used to determine the extent of surfactant stabilization and binding. Prevention of denaturation at interfaces by surfactants is typically studied in shaking studies. Test formulations are mechanically agitated and examined for aggregation/precipitation. Protein stabilization can be studied by monitoring phase transition temperatures with differential scanning calorimetry. Techniques used to determine surfactant–protein stoichiometry include surface tension (108), viscosity (109), dye solubilization (98), dialysis (110), ion-selective electrodes (111–113), electron paramagnetic resonance spectroscopy (114), and analytical ultracentrifugation (115). These measurements have shown that protein–surfactant aggregates form well below the critical micelle concentration for the surfactant and that the aggregates are typically smaller than micelles, although some evidence suggests that micelles do form in some systems (116,117).

Processing

Shear: Protein formulation and manufacturing often employ processes in which the molecule is subject to shear forces. These processes include mixing, flow both in solution and in powder form, filtration, and passage through pumps (118,119). Empirical modeling of data obtained on shearing proteins shows that for many proteins, there is a logarithmic relationship between loss in activity and shearing time (120–122). Conformational changes and accelerated aggregation (118) have been attributed to high shear rates. Burgess et al. have developed a method to determine protein stability to shear by measuring the interfacial shear rheology of adsorbed protein layers and the effects of additives (123).

Dehydration and lyophilization: Destabilization of protein structure by lyophilization may be brought about by several mechanisms. There may be effects due to cold denaturation, concentration/crystallization of salts, changes in pH, and the creation of solid–liquid interfaces if ice forms in the solution. All these topics have been or will be addressed in this section. Protein dehydration through lyophilization (43,124–126) or spray-drying (127,128) can also lead to irreversible denaturation and aggregation upon rehydration. This has been measured using IR, Raman, and NMR spectroscopies (129,130). Whether the changes are solely due to dehydration remains somewhat controversial, although overwhelming data suggest that dehydration alone is responsible. All proteins appear to have water associated with the native form. Removal of all water disrupts intramolecular hydrogen bonding and hydrophobic interactions responsible for the native structure. This may promote intermolecular interactions. Loss of structure and aggregation due to drying is often irreversible, making avoidance of this process or discovering ways of stabilizing proteins to dehydration matters of serious concern in the manufacture of protein pharmaceuticals (125). When considering a formulation that is going to be lyophilized, the formulator should take precautions for preserving protein structure due to dehydration (i.e., sucrose or mannitol) as well as cryoprotection (43). The problem of dehydration may be thought to be solved by lyophilizing with residual moisture. However, reports of moisture-induced aggregation document several case studies in which residual moisture had sufficient mobility for noncovalent aggregation as well as for chemical reactions (12,54,131–135). A more thorough evaluation of the effects of lyophilization and the use of excipients is given in later chapters.

Surface-Associated Mechanisms

Liquid–Air Interface

Many proteins have been found to denature at the air–water interface (136–139). This property can have an impact during shipment of protein solutions, reconstitution of lyophilized protein products, administration through intravenous tubing, or any processing that could result in foaming of the protein solution (102,138). Once at the interface, the protein can unfold and the chains can become entangled as the interface is further perturbed. It has been suggested that primary aggregation can occur at an interface, which can seed the bulk formation of larger insoluble aggregates (140,141). Thus, surface-induced denaturation will often lead to precipitation (136,138).

The driving force responsible for surface accumulation of proteins is the reduction in the interfacial surface energy, which reduces the overall free energy of the entire system. It has been observed that the less stable proteins tend to be more surface active, presumably owing to exposure of nonpolar groups upon denaturation. Changes in the surface tension of a protein solution with time are a direct measure of the affinity of the molecule for the interface and have been shown to be a useful indicator for predicting the stability of proteins in aqueous solutions and in designing formulations that may prevent surface denaturation (138). A systematic study by Wang and McGuire (139) on T4 phage lysozyme showed that among mutants with decreased free energies of unfolding, less-stable variants decreased surface tension to a greater extent and occupied more surface area. Levine et al. (138) were also able to correlate the surface tension decrease with the ease of denaturation and precipitation. Surfactants can often aid in stabilization against such denaturation, presumably via preferential adsorption at the denaturing interfaces. Surfactants that have been used include polysorbate 80 and SDS (142).

Solid–Liquid Interface

Nonspecific protein adsorption or fouling of surfaces can lead to denaturation of the protein and inactivation of the protein due to irreversible binding (143) and/or subsequent aggregation events (105,119,144). Surfaces for potential inactivation of proteins that are important for consideration in protein formulation include delivery pumps (e.g., infusion sets) (119,145–147), silicone rubber tubing (105,148), glass and plastic containers (e.g., intravenous bags) (102,104,105,118,149–151), and polymeric surfaces in protein drug delivery devices (148). Undesirable adsorption of protein can also lead to the failure of devices.

The interaction of a protein with a solid surface depends on the electrostatic and hydrophobic character of both the surface and the protein (143,148). Matsuno et al. (152), in a study done with γ-crystalline water-soluble proteins, found that hydrophobic surfaces tend to interact with the internal apolar regions of this protein, thus favoring denaturation. The degree of unfolding was observed to increase with surface hydrophobicity: see also Ref. (144). Protein adsorption to solid surfaces may also be facilitated by specific functional groups on the protein. Andrade and Hlady (153) show that the carbohydrate moiety of plasma proteins is responsible for the surface-induced structural instabilities that dominate the adsorption process for these proteins.

The best way to prevent surface-induced denaturation requires a knowledge of the protein adsorption process and a characterization of the solid surface, including hydrophobicity, flexibility, and porosity (151,153). Polyethylene oxide (PEO) has been shown to resist nonspecific protein adsorption. Possible mechanisms for PEO's inertness are its low interfacial free energy, its water solubility due to a unique hydrogen-bonding network, and its steric stabilizing effects (154). Coatings containing PEO have been employed as a treatment in rendering hydrophobic surfaces protein resistant. Surfactants containing PEO functional groups have been used to treat a variety of commonly used commercial polyethylenes. The treated surfaces were found to be resistant to protein adsorption as studied by X-ray photoelectron spectroscopy and through the use of radiolabeled proteins (154). Copolymers of alkyl methacrylates with methoxy (PEO) also have been studied for use as coatings or as cleaners for the removal of proteins already adsorbed (155). Protein adsorption onto plastic bags can be reduced by depositing a thin layer of triethylene glycol monoallyl ether. Beyer et al. (156) deposited this layer by means of a plasma-aided manufacturing technique that involves plasma polymerization of the monomer during the deposition process onto solid substrates. Radio-labeled bovine serum albumin adsorption studies with these stable, modified surfaces showed that there was significantly less protein adsorbed to the modified substrate. Saturation of adsorption is often observed, so preadsorption of a protein layer may also prevent adsorption of the protein therapeutic of interest (143,147). Finally, the use of long-chain surfactants has also been shown to interfere with insulin inactivation due to surface-induced instabilities (105).

Another solid–liquid interface frequently encountered by the protein formulator is the ice–water interface that is generated during the freezing of protein solutions. Ice can be formed during routine storage conditions for the protein or through lyophilization protocols. This type of surface-induced denaturation has been reported to correlate with the amount of ice surface area present during the freezing process (157). In this report, the authors report that denaturation induced by the ice–liquid interface can be reduced by the use of surfactants.

Immobilization onto a Solid Surface

Immobilization can be used as a protein-stabilizing strategy. Much of the research in this area comes from the study of enzyme immobilization (158,159). Immobilization on poly(lactic-co-glycolic acid) microspheres has also been used for controlled protein delivery (160). A number of options can be used to immobilize proteins including covalent chemical linkage (161), lipid-mediated immobilization via ligand binding, metal chelation, or electrostatic interaction (162), and physical immobilization. In addition, proteins that do not readily adsorb noncovalently to surfaces can be fused with proteins that do (158), without interfering with either protein's activity.

The exact mechanisms for enhancement of protein stability upon immobilization are not well understood. Stabilization may be due to altered protein conformation. A protein that is rigidly fixed on a support may be more difficult to unfold (163). Ribonuclease A covalently immobilized on silica beads was found to have a greater stability to thermal denaturation, which was hypothesized to occur through a decoupling of structural domains that, before being immobilized, underwent cooperative unfolding (159). Alterations in protein microenvironment may also aid in stabilizing the protein (163).

STABILITY TESTING

Measuring and Predicting Thermodynamic Protein Stability

The change in free energy ΔG between the native and denatured states is the fundamental measure of protein stability. Using calorimetric techniques such as differential scanning calorimetry (164–166), protein folding stability can be measured directly and is therefore model independent. Typically in these experiments, the protein is heated until it undergoes denaturation (although cold denaturation can also be used). The observable parameters are the enthalpy of the unfolding transition, ΔH_m, which is obtained from the area under the curve of the measured excess heat capacity as a function of temperature, and the change in heat capacity between the states, ΔC_p, which is obtained from a shift in the baseline from the native to the denatured states (Fig. 4). Heat capacity is slightly temperature dependent but is usually assumed to be constant, a practice that does not introduce significant amounts of error (17). An advantage

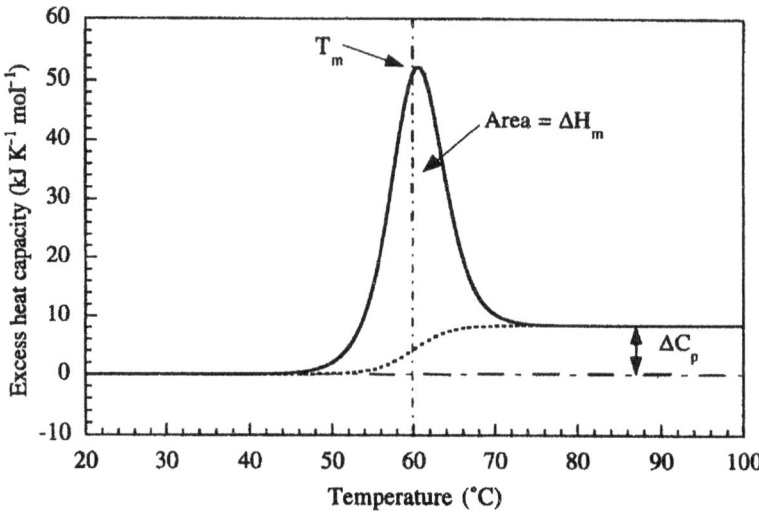

Figure 4 Simulated differential scanning calorimetry experiment for the two-state unfolding of a globular protein. The simulation assumed the following values: $T_m = 60°C$, $\Delta H_m = 418$ kJ/mol, and $\Delta C_p = 8.4$ kJ/K/mol. *Source*: From Ref. 8.

in using calorimetric methods is that the measured ΔH_m can be compared with that determined from a fit to the van't Hoff equation (Eq. 4), which assumes a two-state model; discrepancies indicate the presence of intermediates. Finally, using the calorimetry data, the Gibbs–Helmholtz equation (Eq. 8), can be used to calculate ΔG. The melt temperature T_m, the temperature at which the native and denatured states are in equilibrium, is also used as a measure of protein stability (Fig. 4). This can be especially useful in comparisons to clarify the factors that may affect protein stability in the presence of additives (47) but are not always predictive (167).

Potential problems with this analysis arise in determining the ΔC_p. It is difficult to get accurate measurements of ΔC_p from shifts in the baseline. Typically, careful measurements of the unfolding need to be done and should be performed at several values of T_m, which requires a perturbation of T_m. This is commonly done using pH to shift T_m and then measuring the corresponding ΔH (168). A plot of ΔH as a function of T_m will yield a slope that is the ΔC_p. This assumes that there is no change in ΔH as a function of pH: that is, all the effects are due to changes in ΔS_m. Care also needs to be taken to be sure that the temperature-induced denaturation is reversible, which is necessary to apply this thermodynamic analysis. Reversibility can be checked by multiple scans of the same sample and by checks to be sure that the change in the measured excess heat is independent of scan rate (8). Other limitations of this technique include the high concentrations (i.e., >1 mg/mL) and large sample volumes (>1–2 mL) required for each experiment. With some proteins, these concentrations enter into the range where aggregation is a problem. Additional sources of uncertainty come from the determination of the protein concentration. Typically, the reproducibility is on the order of 5%, given a 2% variability in the determination of the extinction coefficient for the protein (169). Experimentally determined standard deviation in ΔC_p and ΔH_m are 4% to 10% (10,170) and 2% to 10% (171,172), respectively.

The equilibrium constant and thus free energies for unfolding can also be measured by means of a variety of indirect methods as long as the unfolding for the protein follows a two-state model (8,27,173) (some complex cases may also be considered, although analysis is more difficult). Two-state denaturation does not always exist, as has been demonstrated with staphylococcal nuclease mutants (171). However, it has been suggested that multiple unfolded states may be considered as a single "macrostate" (174,175).

Indirect methods of measuring protein unfolding stability involve using optical techniques of various types [e.g., fluorescence and UV spectroscopies, circular dichroism (CD), light scattering, bioassay, immunoassay, enzyme assay, chromatography, sedimentation, electrophoresis] to monitor the effects of a perturbant on protein structure. All of these methods are sensitive to three-dimensional protein structure, and since it is common for the unfolding transition to be cooperative, the transition measured by these techniques is sharp, occurring over a narrow temperature or denaturant-concentration range. The ΔG for unfolding can be obtained from thermal or urea-guanidine hydrochloride denaturation curves. As can be seen in Figure 5, the curves have linear portions at the extremes (i.e., the lowest and highest temperature or denaturant concentrations) and a sharp transition. The potential for error in experiments of these types is brought about by collection of too few points at the extremes used to extrapolate into the transition region. The extrapolated points are used to calculate the

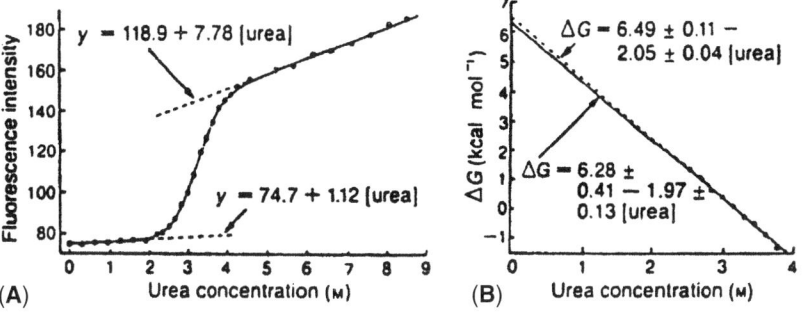

Figure 5 (**A**) Urea denaturation curve for Rnase A in 30 mM formate buffer, pH 3.55, 25°C based on fluorescence measurements at 305 nm. The lines and equations are least-square fits to the pre- and post-transition baseline. (**B**) ΔG calculated using Equations 3, 12, and 13 as a function of urea concentration. *Source*: From Ref. 27.

fraction of denatured (F_d) and native (F_n) protein at each point. This is given by Equation 12:

$$F_d = \frac{y_n - y_{observed}}{y_n - y_d} \tag{12}$$

where $y_{observed}$ is the point from the experimental curve, and y_n and y_d are the values for the native and denatured states, respectively, which have been extrapolated into the transition region (Fig. 5). The equilibrium constant can be determined from Equation 13

$$K = \frac{F_n}{1 - F_n} \tag{13}$$

and ΔG from Equation 3. The problem with this determination is that the ΔG obtained is in the presence of perturbant, while the desirable value is that in the absence of perturbant. The ΔG values can be extrapolated to zero perturbant concentration, but this often involves a long extrapolation and assumes that ΔG is linear with perturbant concentration. Mathematical methods for extrapolating data rely on sorting out individual amino acid contributions to the energy to transfer an amino acid from an aqueous solvent to a denaturant solvent (176).

The advantage of these indirect techniques for determining the value of ΔG is that often much less protein is needed to perform the experiment, especially in measurements that involve the use of spectroscopy. Sample concentrations are typically 0.01 mg/mL, and a range of concentrations can be used to detect self-association. Also, solvent denaturation experiments are fast and simple. Commercially available instruments carry out automated chemical and thermal denaturation using fluorescence, CD, and ultraviolet–visible spectroscopy as a probe of protein structure (177). The disadvantage of indirect methods is that a two-state model must be followed. These techniques do not allow for the direct measure of an intermediate state. Finally, as mentioned above, a lack of data points or inaccuracies of data points in determining the baselines can significantly affect the thermodynamic parameters that are extracted.

A model-independent approach to analyze denaturant-induced unfolding has been developed. Native protein is titrated with a perturbant, and denaturation is monitored with an optical technique. However, rather than assuming that ΔG has a linear dependence with perturbant concentration, a nonlinear extrapolation is used. The method involves measurement of denaturant profiles at several temperatures. All parameters can be obtained from the experimental solvent denaturation data (41). This technique appears to show promise, but its validity needs to be further established before it can necessarily replace the linear extrapolation method described earlier.

Statistical mechanical theory has also been used to predict a protein's T_m as a function of pH (178). From physical properties of the constitutive amino acids, values for apomyoglobin were generated that were in agreement with experimental data from the literature. The calculations are based on the number of nonpolar groups, the protein chain length, the temperature-dependent free energy of transfer of each amino acid, and the values of pK_a as a function of temperature.

Predicting Long-Term Protein Stability

Accelerated stability testing is a common method for the prediction of the long-term storage conditions for pharmaceutical products (173,179). This test typically involves storing the product at elevated temperatures, above normal storage conditions, and monitoring the product's structural and/or functional integrity with time (13,180). Often, the denaturation is a first-order process, that is,

$$A_t = A_0 e^{-kt} \tag{14}$$

where A_t and A_0 are the activities or concentrations at time t and time zero. From these data, the rate of degradation can be calculated at each of the different temperatures. The rates may then be fit to the Arrhenius equation:

$$\ln k - \ln \alpha - \frac{\Delta E_a}{RT} \tag{15}$$

which can be used to predict rates at any temperature, most importantly, normal storage conditions (179). Arrhenius kinetics have been observed for a number of reactions involving peptides and proteins (181–187), including protein folding stability (19). For example, Arrhenius kinetics have been applied to protein hydrolysis and deamidation pathways. When Arrhenius kinetics apply, accelerated testing provides an accurate method for the determination of product shelf life. More recently, evaluation of accelerated aggregation kinetics (high temperature) using the Lumry–Eyring model (Eq. 10) have also been used to predict real-time protein stability (187–189).

However, protein degradation may involve inactivation by several processes (4), and their complexity cannot be described by Arrhenius kinetics (181,190). Accelerated testing cannot be used to accurately predict shelf lives for protein formulations in which the loss of protein activity involves two or more consecutive simultaneous reactions that have different activation energies. In this reaction scheme, the rate of disappearance of the parent compound will not be identical to the rate of accumulation of the final product, making the plots of activity versus time complex (173). In this case, measurements of the parent compound, intermediate species, and final product must be made. Arrhenius kinetics also cannot be applied to protein formulations that degrade by different mechanisms based on temperature. In this case, formulations optimized for elevated temperatures may not be optimized for storage temperature.

Potential errors can also arise in products that have inhomogeneous compositions, i.e., lyophilized formulations created by poor process design (173). For example, a nonuniform distribution of phases, i.e., crystalline amorphous phases, can be set up in the freeze-dried cake, leading to a biphasic stability profile. This can lead to a product appearing to have a fast rate of degradation followed by a slower one at longer times (or vice versa).

Finally, accelerated stability testing can be misleading when the protein's glass transition temperature T_g falls within the range over which the testing is being performed (173). During lyophilization, many components of the formulation are rendered amorphous. When these products are subject to accelerated stability testing, some of the samples may be stored at temperatures above their T_g while other temperatures are below. The kinetics of reactions would not be expected to be the same in the glassy and the rubbery (i.e., viscoelastic) states. Rather, the kinetics at temperatures above T_g may follow empirical relationships such as Vogel–Tamman–Fulcher (VTF) equation or Williams–Landel–Ferry (WLF) kinetics (191). WLF kinetics are described using Equation 16:

$$\log(k) = -\frac{C_1\left(T - T_g\right)}{C_2 + \left(T - T_g\right)} \tag{16}$$

where C_1 and C_2 are constants related to the free volume (which would naturally play a bigger role above T_g). Thus the reaction rate correlates with $T - T_g$ rather than T (192,193). Likewise, any other phase transition (i.e., crystallization of an amorphous component) within a protein formulation that occurs in the temperature range of the stability tests would require care in applying appropriate rate equations for the prediction of product shelf life. Stability testing should be performed over a wide range in temperature to be sure that the kinetic behavior is well understood. Linear fits to Equation 15 can be obtained even if non-Arrhenius kinetics are followed in cases of an experimental temperature range that is too small.

Accelerated stability testing can be a useful technique; however, there are many pitfalls. One should acquire a thorough understanding and characterization of the degradative mechanism(s) before attempting to interpret degradation profiles and the prediction of activity under different conditions. The Food and Drug Administration reminds the biologics community that accelerated testing can be used as supportive data but cannot be substituted for real-time data for product approval and labeling (180).

REFERENCES

1. Anderson DE, Becktel WJ, Dahlquist FW. pH-induced denaturation of proteins: a single salt bridge contributes 3–5 kcal/mol to the free energy of folding of T4 lysozyme. Biochemistry 1990; 29:2403–2408.
2. Lazaridis T, Karplus M. Thermodynamics of protein folding: a microscopic view. Biophys Chem 2003; 100:367–395.
3. Pace CN. The stability of globular proteins. CRC Crit Rev Biochem 1975; 3:1–43.

4. Aune K, Tanford C. Thermodynamics of the denaturation of lysozyme by guanidine hydrochloride. II. Dependence on denaturant concentration at 25 degrees. Biochemistry 1969; 8:4586–4590.

5. Privalov P, Khechinashvilli N. A thermodynamic approach to the problem of stabilization of globular protein structure: a carlorimetric study. J Mol Biol 1974; 86:665–684.

6. Madan B, Sharp KJ. Heat capacity changes accompanying hydrophobic and ionic solvation: a Monte Carlo and random network model study. J Phys Chem 1996; 100:7713–7721.

7. Edsall JT. Apparent molal heat capacities of amino acids and other organic compounds. J Am Chem Soc 1935; 57:1506.

8. Robertson AD, Murphy KP. Protein structure and the energetics of protein stability. Chem Rev 1997; 97:1251–1267.

9. Lumry R, Biltonen R, Brandts J. Validity of the "two-state" hypothesis for conformational transitions of proteins. Biopolymers 1966; 4:917–944.

10. Pace CN, Laurents DV. A new method for determining the heat capacity change for protein folding. Biochemistry 1989; 28:2520–2525.

11. Schellman JA. The thermodynamic stability of proteins. Annu Rev Biophys Biophys Chem 1987; 16:115–137.

12. Chen B-L, Arakawa T. Stabilization of recombinant human keratinocyte growth factor by osmolytes and salts. J Pharm Sci 1996; 85:419–422.

13. Fágáin C. Understanding and increasing protein stability. Biochim Biophys Acta 1995; 1252:1–14.

14. Parsell DA, RT Sauer. The structural stability of a protein is an important determinant of its proteolytic susceptibility in *Escherichia coli*. J Biol Chem 1989; 264:7590–7595.

15. Glatz CE. Modelling of aggregation-precipitation phenomena. In Ahern TJ, Manning MC, eds. Stability of Protein Pharmaceuticals, Part A: Chemical and Physical Pathways of Protein Degradation. New York: Plenum Press, 1992:135–166.

16. Lumry R, Eyring H. Conformation changes of proteins. J Phys Chem 1954; 58:110–125.

17. Privalov PL, Gill SJ. Stability of protein structure and hydrophobic interaction. Adv Protein Sci 1988; 39:191–234.

18. Dill KA, Alonsa DOV. In: Winnacker EL, Huber R, eds. Protein Structure and Protein Engineering. Berlin, Springer-Verlag, 1988:51.

19. Scalley ML, Baker D. Protein folding kinetics exhibit an Arrhenius temperature dependence when corrected for the temperature dependence of protein stability. Proc Natl. Acad. Sci. USA 1997; 94:10636–10640.

20. Dong A, Prestrelski SJ, Allison SD, Carpenter JF. Infrared spectroscopic studies of lyophilization- and temperature-induced protein aggregation. J Pharm Sci 1995; 84:415–424.

21. Tanford C, Pain RM, Otchin NS. Unfolding of hen egg white lysozyme by guanidine hydrochloride. J Mol Biol 1966; 15:489.

22. Privalov PL, Khechinashvili NN, Atanasov BP. Thermodynamic analysis of thermal transitions in globular proteins. I. Calorimetric study of ribotrypsinogen, ribonuclease and myoglobin. Biopolymers 1971; 10:1865.

23. Schellman JA. Selective binding and solvent denaturation. Biopolymers 1987; 26:549–559.

24. Pace CN, Laurents DV. RNases A and T1: pH dependence and folding. Biochemistry 1990; 29:2520–2525.

25. Hu CQ, Sturtevant JM, Thomson JA, Erickson RE, Pace CN. Thermodynamics of ribonuclease Tl denaturation. Biochemistry 1992; 31:4876–4882.

26. Pace CN. Conformational stability of globular proteins. Trends Biol Sci 1990; 15:14–17.

27. Pace CN. Measuring and increasing protein stability. Trends Biotechnol 1990; 8:93–98.

28. Chi EY, Krishnan S, Randolf TW, Carpenter JF. Physical stability of proteins in aqueous solution: mechanism and driving forces in nonnative protein aggregation. Pharm Res 2003; 20:1325–1336.

29. Feller G, Gerday C. Psychrophilic enzymes: molecular basis of cold adaptation. Cell Mol Life Sci 1997; 53:830–841.

30. Jaenicke R. Protein stability and protein folding. CIBA Found Symp 1991; 161:206–216.

31. Jaenicke R. Protein stability and molecular adaptation to exteme conditions. Eur J Biochem 1991; 202:715–728.

32. Somero GN. Temperature and proteins: little things can mean a lot. News Physiol Sci 1996; 11:72–77.

33. Cowan DA. Protein stability at high temperatures. Essays Biochem 1995; 29:193–207.

34. Volkin DB, Middaugh CR. The effect of temperature on protein structure. In: Ahern TJ, Manning MC, eds. Stability of Protein Pharmaceuticals, Part A: Chemical and Physical Pathways of Protein Degradation. New York: Plenum Press, 1992:215–247.

35. Suzuki Y. Proc Jpn Acad Sci Ser B. A general principle of increasing protein thermostability. 1989:146–148.

36. Perutz MF, Raidt H. Stereochemical basis of heat stability in bacterial ferredoxins and in haemoglobin A2. Nature 1975; 255:256–259.

37. Shirley BA. Protein conformational stability estimated from urea, guanidine hydrochloride, and thermal denaturation curves. In: Ahern TJ, Manning MC, eds. Stability of Protein Pharmaceuticals, Part A: Chemical and Physical Pathways of Protein Degradation. New York: Plenum Press, 1992:167–194.

38. Pfeil W, Gesierich U, Kleemann GR, Sterner R. Ferredoxin from the hyperthermophile *Thermotoga maritima* is stable beyond the boiling point of water. J Mol Biol 1997; 272:591–596.

39. Baldwin RL. Temperature dependence of the hydrophobic interaction in protein folding. Proc Natl Acad Sci USA 1986; 83:8069–8072.

40. Privalov PL. Stability of proteins: small globular proteins. Adv Protein Chem 1979; 33:167–241.

41. Ibarra-Molero B, Sanchez-Ruiz JM. A model independent, nonlinear ex trapolation procedure for the characterization of protein folding energetics from solvent-denaturation data. Biochemistry 1996; 35:14689–14702.

42. Rishi V, Anjum F, Ahmad F, Pfeil W. Role of non-compatible osmolytes in the stabilization of proteins during heat stress. Biochem J 1998; 329:137–143.

43. Prestrelski J, Arakawa T, Carpenter JF. Separation of freezing- and drying-induced denaturation of lyophilized proteins using stress-specific stabilization. Arch Biochem Biophys 1993; 303:465–473.

44. Somero GN. Proteins and temperature. Annu Rev Physiol 1995; 57:43–68.

45. Hottiger T, Boiler T, Wiemken A. Rapid change of heat and desiccation tolerance correlated with changes of trehalsoe content in *Saccharomyces cerevisiae* cells subjected to temperature shifts. FEBS Lett 1987; 22:113–115.

46. Akers MJ, Milton N, Byrn SR, Nail SL. Glycine crystallization during freezing: the effects of salt form, pH, and ionic strength. Pharm Res 1995; 12:1457–1461.

47. Izutsu K, Yoshioka S, Takeda Y. Protein denaturation in dosage forms measured by differential scanning calorimetry. Chem Pharm Bull 1990; 38:800–803.

48. Yoshioka S, Aso Y, Nakai Y, Kojima S. Effect of high molecular mobility of poly(vinyl alchohol) on protein stability of lyophilized γ-globulin formulations. J Pharm Sci 1998; 87:147–151.

49. Yoshioka S, Aso Y, Kojima S. Dependence of the molecular mobility and protein stability of freeze-dried γ-globulin formulations on the molecular weight of dextran. Pharm Res 1997; 14:736–741.

50. Prestrelski SJ, Pikal KA, Arakawa T. Optimizations of lyophilization conditions for recombinant human interleukin-2 by dried-state conformational analysis using Fourier-transform infrared spectroscopy. Pharm Res 1995; 12:1250–1259.

51. Bell LN, Hageman MJ, Muraoka LM. Thermally induced denaturation of lyophilized bovine somatotropin and lysozyme as impacted by moisture and excipients. J Pharm Sci 1995; 84:707–712.

52. Oksanen CA, Zografi G. Molecular mobility in mixtures of absorbed water and solid poly(vinylpyrrolidone). Pharm Res 1993; 10:791–799.

53. Hancock BC, Shamblin SL, Zografi G. Molecular mobility of amorphous pharmaceutical solids below their glass transition temperatures. Pharm Res 1995; 12:799–806.

54. Hill JJ, Shalaev EY, Zografi G. Thermodynamic and dynamic factors involved in the stability of native protein structure in amorphous solids in relation to levels of hydration. J Pharm Sci 2005; 94(8):1636–1667.

55. Tan Y-J, Oliveberg M, Davis B, Fersht AR. Perturbed pK_a-values in the denatured state of proteins. J Mol Biol 1995; 254:980–992.

56. Yang A-S, Honig B. Structural origins of pH and ionic strength effects on protein stability. J Mol Biol 1994; 237:602–614.

57. Pace CN, Laurents DV, Thomsom JA. pH dependence of the urea and guanidine hydrochloride denaturation of ribonuclease A and ribonuclease Tl. Biochemistry 1990; 29:2564–2572.

58. Stites WE, Gittis AG, Lattman E, Shortle D. In a staphylococcal nuclease mutant the side-chain of a lysine replacing valine 66 is fully buried in the hydrophobic core. J Mol Biol 1991; 221:7–14.

59. Rashin A, Honig B. On the environment of ionizable groups in globular proteins. J Mol Biol 1984; 173:515–521.

60. Perutz M. Mechansim of denaturation of hemoglobin by alkali. Nature 1974; 247:341–344.

61. Langsetmo K, Fuchs JA, Woodward C, Sharp KA. Linkage of thioredoxin stability to titration of ionizable groups with perturbed pK_a. Biochemistry 1991; 30:7609–7614.

62. Creighton TE. Proteins: Structures and Molecular Properties. New York: Freeman WH, 1993:1–48.

63. Yang AS, Honig B. Electrostatic effects on protein stability. Curr Opin Struct Biol 1992; 2:40–45.

64. Pace CN, Laurents DV, Erickson RE. Urea denaturation of barnase: pH dependence and characterization of the unfolded state. Biochemistry 1992; 31:2728–2734.

65. Inagaki F, Kawano T, Shimada I, Takahashi K, Miyazawa T. Nuclear magnetic resonance study on the microenvironments of histidine residues of ribonuclease Tl and carboxymethylated ribonuclease Tl. J Biochem 1981; Tokyo 89:1185–1195.

66. Gao J, Mrksich M, Gomex FA, Whitesides GM. Using capillary electrophoresis to follow the acetylation of the amino groups of insulin and to estimate their basicities. Anal Chem 1995; 67:3093–3100.

67. Yang A-S, Honig B. On the pH dependence of protein stability. J Mol Biol 1993; 231:459–474.

68. Schaefer M, Sommer M, Karplus M. pH-dependence of protein stability: absolute electrostatic free energy differences between conformations. J Phys Chem 1997; 101:1663–1683.

69. Antosiewicz J, McCammon JA, Gilson MK. Prediction of pH-dependent properties of proteins. J Mol Biol 1994; 238:415–436.

70. Israelachvili J. Intermolecular and Surface Forces. San Diego, CA: Academic Press, 1992.

71. Arakawa T, Kita Y, Carpenter JF. Protein-solvent interactions in pharmaceutical formulations. Pharm Res 1991; 8:285–291.

72. Timasheff SN. The control of protein stability and association by weak interactions with water: how do solvents affect these processes? Annu Rev Biophys Biomol Struct 1993; 22:67–97.

73. Franks F. Protein hydration. In: Franks F, ed. Protein Biotechnology. Cambridge: Parfa Ltd., 1993:395–436.

74. Collins KD, Washabaugh MW. The Hofmeister effect and the behavior of water at interfaces. Q Rev Biophys 1985; 18:323–422.

75. Baldwin RL. How Hofmeister ion interactions affect protein stability. Biophys J 1996; 71:2056–2063.

76. Yancey PH, Clark ME, Hand SC, Bowlus RD, Somero GN. Living with water stresses: evolution of osmolyte systems. Science 1982; 217:1214.

77. Xie G, Timasheff SN. Mechanism of the stabilization of ribonuclease A by sorbitol: preferential hydration is greater for the denatured than for the native protein. Protein Sci 1997; 6:211–221.

78. Xie G, Timasheff SN. The thermodynamic mechanism of protein stabilization by trehalose. Biophys Chem 1997; 64:25–43.

79. Gekko K, Timasheff SN. Mechanism of protein stabilization by glycerol: preferential hydration in glycerol–water mixtures. Biochemistry 1981; 20:4667–4676.

80. Lee JC, Timasheff SN. The stabilization of proteins by sucrose. J Biol Chem 1981; 256:7193–7201.

81. Timasheff SN, Arakawa T. Stabilization of protein structure by solvents. In: Creighton TE, ed. Protein Structure: A Practical Approach. Oxford: IRL Press, 1989:331.

82. Lin TY, Timasheff SN. On the role of surface tension in the stabilization of globular proteins. Protein Sci 1996; 5:372–381.

83. Cioci FR, Lavecchia R, Marrelli L. Perturbation of surface tension of water by polyhydric additives: effect on glucose oxidase stability. Biocatalysis 1994; 10:137.

84. Baskakov I, Bolen DW. Forcing thermodynamically unfolded proteins to fold. J Biol Chem 1998; 273:4831–4834.

85. Timasheff SN, Arakawa T. In: Creighton TE, ed. Protein Function: A Practical Approach. Oxford: IRL Press, 1988.
86. Schellman A. Selective binding and solvent denaturation. Biopolymers 1987; 26:549.
87. Kendrick BS, Chang BS, Arakawa T, et al. Preferential exclusion of sucrose from recombinant interleukin-1 receptor antagonist: role in restricted conformational mobility and compaction of native state. Proc Natl Acad Sci USA 1997; 94:11917–11922.
88. Arakawa T, Bhat R, Timasheff SN. Why preferential hydration does not always stabilize the native structure of globular proteins. Biochemistry 1990; 29:1924–1931.
89. Lee C, Timasheff SN. The calculation of partial specific volumes of proteins in guanidine hydrochloride. Arch Biochem Biophys 1974; 165:268–273.
90. Arakawa T, Timasheff SN. Stabilization of protein structure by sugars. 1982; 21:6536–6544.
91. Pace CN, Grimsley GR. Ribonuclease T1 is stabilized by cation and anion binding. Biochemistry 1998; 27:3242–3246.
92. Gillis D, Rooman M. Predicting protein stability changes upon mutation using database-derived potentials: Solvent accessibility determines the importance of local versus non-local interactions along the sequence. J Mol Biol 1997; 272:276–290.
93. Tanford C. Isothermal unfolding of globular proteins in aqueous urea solutions. J Am Chem Soc 1964; 86:2050–2059.
94. Tanford C. Protein denaturation. C. Theoretical models for the mechanism of denaturation. Adv Protein Chem 1970; 24:1–95.
95. Pittz EP, Timasheff SN. Interaction of ribonuclease A with aqueous 2-methyl-2,4-pentanediol at pH 5.8. Biochemistry 1978; 17:615–623.
96. Pittz EP, Bello J. Studies on bovine pancreatic ribonuclease A and model compounds in aqueous 2-methyl-2,4-pentanediol. I. Amino acid solubility, thermal reversibility of ribonuclease A, and preferential hydration of ribonuclease A crystals. Arch Biochem Biophys 1971; 146:513–524.
97. Jirgensons B. Factors determining the reconstructive denaturation of proteins in sodium dodecyl sulfate solutions. Further circular dichroism studies on structural reorganization of seven proteins. J Protein Chem 1982; 1:71–84.
98. Steinhardt J, Scott JR, Birdi KS. Differences in the solubilizing effectivness of the sodium dodecyl sulfate complexes of various proteins. Biochemistry 1977; 16:718–725.
99. Mori E, Jirgensons B. Effect of long-chain alkyl sulfate binding on circular dichroism and conformation of soybean trypsin inhibitor. Biochemistry 1981; 20:1630–1634.
100. Jirgensons B. Circular dichroism study on structural reorganization of lectins by sodium dodecyl suflate. Biochim Biophys Acta 1980; 623:69–76.
101. Tandon S, Horowitz PM. Detergent-assisted refolding of guanidine chloride–denatured rhodanese. J Biol Chem 1987; 262:4486–4491.
102. Twardowski ZJ, Nolph KD, McGray TJ, Moore HL. Nature of insulin binding to plastic bags. Am J Hosp Pharm 1983; 40:579–581.
103. Bohnert JL, Horbett TA. Changes in adsorbed fibrinogen and albumin interactions with polymers indicated by decreases in detergent elutability. J Colloid Interface Sci 1986; 111:363–377.
104. Chawala AS, Hinberg I, Blais P, Johnson D. Aggregation of insulin, containing surfactants, in contact with different materials. Diabetes 1985; 34:420–424.
105. Loughheed WD, Albisser AM, Martindale HM, Chow JC, Clement JR. Physical stability of insulin formulations. Diabetes 1983; 32:424–432.

106. Piatigorsky J, Horowitz J, Simpson T. Partial dissociation and renaturation of embryonic Chick d-crystalline. Characterization by ultracentrifugation and circular dichroism. Biochim Biophys Acta 1977; 490:279–289.

107. Balasubramanian SV, Bruenn J, Straubinger RM. Liposomes as formulation excipients for protein pharmaceuticals: a model protein study. Pharm Res 2000; 17(3):344–350.

108. Schwuger MJ, Bartnik FG. In: Gloxhuber C, ed. Anionic Surfactants. Vol. 10. New York: Marcel Dekker, 1980:1–52.

109. Greener J, Constable BA, Bale MM. Interaction of anionic surfactants with gelatin: viscosity effects. Macromolecules 1987; 20:2490–2498.

110. Makino S. Interaction of proteins with amphiphilic substances. In: Kotani M, ed. Advances in Biophysics. Vol. 12. Baltimore: University Park Press, 1979:131–184.

111. Rendall HM. Use of a surfactant selective electrode in the measurement of the binding of anionic surfactants to bovine serum albumin. J Chem Soc Faraday Trans 1976; 1:481–484.

112. Kreschek GC, Constandinidis I. Ion-selective electrodes for octyl and decyl sulfate surfactants. Anal Chem 1984; 56:152–156.

113. Hayakawa K, Ayub AL, Kwak JCT. The application of surfactant selective electrodes to the study of surfactant adsorption in colloidal suspensions. Colloids Surf 1982; 4:389–397.

114. Bam NB, Randolf TW, Cleland JL. Stability of protein formulations: investigation of surfactant effects by a novel EPR spectroscopic technique. Pharm Res 1995; 12:2–11.

115. Lustig A, Engel A, Zulauf M. Density determination by analytical ultracentrifugation in a rapid dynamical gradient: application to lipid and detergent aggregates containing proteins. Biochem Biophys Acta 1991; 1115:89–95.

116. Smith ML, Muller N. Fluorine chemical shifts in complexes of sodium trifluoroalkylsulfates with reduced proteins. Biochem Biophys Res Commun 1975; 62:723–728.

117. Tsuji K, Takagi T. Proton magnetic resonance. J Biochem 1975; 77:511–519.

118. Sato S, Ebert CD, Kim SW. Prevention of insulin self-association and surface adsorption. J Pharm Sci 1983; 72:228–232.

119. Brennan JR, Gebhart WG. Pump-induced insulin aggregation. A problem with the biostatror. Diabetes 1985; 34:353–359.

120. Charm SE, Wong BL. Enzyme inactivation with shearing. Biotechnol Bioeng 1970; 12:1103–1109.

121. Charm SE, Wong BL. Shear inactivation of fibrinogen in the circulation. Science 1975; 170:466–468.

122. Charm SE, Wong BL. Shear inactivation of heparin. Biorheology 1975; 12:275–278.

123. Burgess DJ, Yoon JK, Sahin NO. A novel method of determination of protein stability. J Parenter Sci Technol 1992; 46:150–155.

124. Prestrelski SJ, Tedeschi N, Arakawa T, Carpenter JF. Dehydration-induced conformational transitions in proteins and their inhibition by stabilizers. Biophys J 1993; 65:661–671.

125. Allison SD, Dong A, Carpenter JF. Counteracting effects of thiocyanate and sucrose on chymotrypsinogen secondary structure and aggregation during freezing, drying, and rehydration. Biophys J 1996; 71:2022–2032.

126. Yu H-T, Jo BH. Comparison of protein structure in crystals and in solution by laser raman scattering. I. Lysozyme. Arch Biochem 1973; 156:469–474.

127. Maa YF, Nguyen PA, Andya J, et al. Effect of spray drying and subsequent processing conditions on residual moisture content and physical/biochemical stability of protein inhalation powders. Pharm Res 1998; 15(5):768–775.

128. Maa YF, Sellers SP. Solid-state protein formulation: methodologies, stability and excipient effects. Meth Mol Biol 2005; 308:265–285.

129. Poole PL, Finney JL. Sequential hydration of a dry globular protein. Biopolymers 1983; 22:255–260.

130. Poole PL, Finney JL. Sequential hydration of dry proteins: a direct difference IR investigation of sequence homologs lysozyme and α-lactoal-bumin. Biopolymers 1984; 23:1647–1666.

131. Schwendeman SP et al. Stabilization of tetanus and diphtheria toxoids aganist moisture-induced aggregation. Proc Natl Acad Sci USA 1995; 92:11234–11238.

132. Costantino HR. Moisture-induced aggregation of lyophilized insulin. Pharm Res 1994; 11:21–29.

133. Costantino HR, Griebenow K, Mishra P, et al. Fourier-transform infrared spectroscopic investigation of protein stability in the lyophilized form. Biochim Biophys Acta 1995; 1253:69–74.

134. Costantino HR, Shieh L, Klibanov AM, Langer R. Heterogeneity of serum albumin samples with respect to solid-state aggregation via thiol–disulfide interchange—implications for sustained release from polymers. J Control Release 1997; 44:255–261.

135. Chang BS, Beauvais RM, Dong A, Carpenter JF. Physical factors affecting the storage stability of freeze-dried interleukin-1 receptor antagonist: glass transition and protein conformation. Arch Biochem Biophys 1996; 331:249–258.

136. MacRitchie F. Chemistry at Interfaces. San Diego, CA: Academic Press, 1990.

137. Donaldson TL, Boonstra EF, Hammond JM. Kinetics of protein denaturation at gas–liquid interfaces. J Colloid Interface Sci 1980; 74:443.

138. Levine HL, Ransohoff TC, Kawahata RT, McGregor WC. The use of surface tension measurement in the design of antibody-based product formulations. J Parenter Sci Technol 1991; 45:160–165.

139. Wang J, McGuire J, Surface tension kinetics of the wild type and four synthetic stability mutants of T4 phage lysozyme at the air-water interface. J Colloid Interface Sci 1997; 185:317–323.

140. Henson AF, Mitchell JR, Musselwhite PR. The surface coagulation of proteins during shaking. J. Colloid Interface Sci 1970; 32:162.

141. Cumper CWN. The surface chemistry of proteins. Trans Faraday Soc 1950; 46:235.

142. Wang WJ, Hanson MA. Parenteral formulations of proteins and peptides. J Parenter Sci Technol 1988; 42(suppl):2S.

143. Feng L, Andrade JD. Protein adsorption on low temperature isotropic carbon. V. How is it related to its blood compatibility? J Biomater Sci Polym 1995; 7:439–452.

144. Sefton MV, Antonacci GM. Adsorption isotherms of insulin onto various materials. Diabetes 1984; 33:674–680.

145. Lougheed WD, Woulfe-Flannagen H, Clement JR, Albisser AM. Insulin aggregation in artificial delivery systems. Diabetologia 1980; 19:1–9.

146. James DE, Jenkins AB, Kraegen EW, Chisholm DJ. Insulin precipitation in artificial infusion devices. Diabetologia 1981; 21:554–557.

147. Peterson L, Caldwell J, Hoffman J. Insulin adsorbance to polyvinylchloride surfaces with implications for constant-infusion therapy. Diabetes 1976; 25:72–74.

148. Tzannis ST, Hrushesky WJM, Wood PA, Przybycien TM. Adsorption of a formulated protein on a drug delivery device. J Colloid Interface Sci 1997; 189:216–228.
149. Twardowski ZJ, Nolph KD, McGray TJ, Moore HL. Influence of temperature and time on insulin adsorption to plastic bags. Am J Hosp Pharm 1983; 40:583–586.
150. Twardowski ZJ et al. Insulin binding to plastic bags: a methodologic study. Am J Hosp Pharm 1983; 40:575–579.
151. Dong DE, Andrade JD, Coleman DL. Adsorption of low density lipoproteins onto selected biomedical polymers. J Biomed Mater Res 1987; 21:683–700.
152. Matsuno K, Lewis RV, Middaugh CR. The interaction of γ-crystallins with model surfaces. Arch Biochem Biophys 1991; 291:349–355.
153. Andrade JD, Hlady V. Plasma protein adsorption: the big twelve. Ann NY Acad Sci 1987; 516:158–172.
154. Lee JH, Kopecek J, Andrade JD. Protein-resistant surfaces prepared by PEO-containing block copolymer surfactants. J Biomed Mater Res 1989; 23:351–368.
155. Lee JH, Kopeckova J, Andrade JD. Surface properties of copolymers of alkyl methacrylates with methoxy(polyethylene oxide) methacrylates and their applications as protein-resistant coatings. Biomaterials 1990; 11:455–464.
156. Beyer D et al. Reduced protein adsorption on plastics via direct plasma deposition of triethylene glycol monallyl ether. J Biomed Mater Res 1997; 36:181–189.
157. Chang BS, Kendrick BS, Carpenter JF. Surface-induced denaturation of proteins during freezing and its inhibition by surfactants. J Pharm Sci 1996; 85:1325–1330.
158. Stempfer G, Holl-Neughbauer B, Kipetzki E, Rudolf R. A fusion protein designed for noncovalent immobilization: stability, enzymatic activity, and use in an enzyme reactor. Nat Biotechnol 1996; 14:481–484.
159. Rialdi G, Battistel E. Decoupling of melting domains in immobilized ribo-nuclease A. Proteins Struct Funct Gene 1994; 19:120–131.
160. Jiang G, Woo BH, Kang F, et al. Assessment of protein release kinetics, stability and protein polymer interaction of lysozyme encapsulated poly (D, L-lactice-co-glycolide) microspheres. J Control Release 2002; 79:137–145.
161. Hermanson GT, Mallia AK, Smith PK. Immobilization affinity ligand techniques. San Diego, CA: Academic Press, 1992.
162. Koppenol S, Stayton PS. Engineering two-dimensional protein order at surfaces. J Pharm Sci 1997; 86:1204–1209.
163. Gupta MN. Thermostabilization of proteins. Biotechnol Appl Biochem 1991; 14:1–11.
164. Privalov PL, Potekhin SA. Scanning microcalorimetry in studying temperature-induced changes in proteins. Methods Enzymol 1986; 131:4.
165. Freire E. Differential scanning colorimetry. In: Shirley BAP, ed. Protein Stability and Folding. Vol. 40. Totowa, NJ: Humana Press, 1995.
166. Shnyrov VL, Sanchz-Ruiz JM, Boiko BN, Zhadan GG, Permyakov EA. Applications of scanning microcalorimetry in biophysics and biochemistry. Thermochim Acta 1997; 302:165–180.
167. Cromwell MEM, Patapoff T, Shire SJ. Correlation of physical and chemical stability of rhDNase with melting temperature. Abstracts of Papers, 225th ACS National Meeting, New Orleans, LA, March 23–27, 2003, Washington DC: American Chemical Society, 2003.
168. Swing L, Robertson AD. Thermodynamics of unfolding for turkey ovomucoid third domain: thermal and chemical denaturation. Protein Sci 1993; 2:2037–2049.

169. Pace CN, Vajdos F, Fee L, Grimsley G, Gray T. How to measure and predict the molar adsorption coefficient of a protein. Protein Sci 1995; 4:2411–2423.
170. Becktel WJ, Schellman JA. Protein stability curves. Biopolymers 1987; 26:1859–1877.
171. Carra JH, Privalov PL. Thermodynamics of denaturation of staphylococcalnuclease mutants: an intermediate state in protein folding. FASEB J 1996; 10:67–74.
172. DeKoster GT, Robertson AD. Thermodynamics of unfolding for Kazal-type serine protease inhibitors: entropic stabilization of ovomucoid first domain by glycosylation. Biochemistry 1997; 36:2323–2331.
173. Franks F. Accelerated stability testing of bioproducts: attractions and pitfalls. Trends Biotechnol 1994; 12:114–117.
174. Eftnik MR, Ionescu R. Thermodynamics of protein unfolding: questions pertinent to testing and validity of the two-state model. Biophys Chem 1997; 64:175–197.
175. Shirley BA. Protein conformational stability estimated from urea, guanidine hydrochloride, and thermal denaturation curves. In: Ahern TJ, Manning MC, eds. Stability of Protein Pharmaceuticals. Part A: Chemical and Physical Pathways of Protein Degradation. New York: Plenum Press, 1992:167–194.
176. Nguyen TH, Ward C. In: Wang YJ, Pearlman R, eds. Stability and Characterization of Protein and Peptide Drugs. New York: Plenum Press, 1993.
177. Stites WE, Byrne MP, Aviv J, Kaplan M, Curtis PM, Instrumentation of automated determination of protein stability. Anal Biochem 1995; 277:112–122.
178. Alonso DO, Dill KA, Stigter D. The three states of globular proteins: acid denaturation. Biopolymers 1991; 31:1631–1649.
179. Kirkwood TBL. Design and anlysis of accelerated degradation tests for the stability of biological standards. III. Principles of design. J Biol Stand 1984; 12:215–224.
180. U.S. Food and Drug Administration. Points to consider in the manufacture and testing of monoclonal antibody products for human use. J Immunother 1997; 20:214–243.
181. Andreotti G et al. Stability of a thermophilic TIM-barrel enzyme: indole-3-glycerol phosphate synthase from the thermophilic Archaeon sulfolobus solfa-taricus. Biochem J 1997; 323:259–264.
182. Tokumitsu H et al. Degradation of a novel tripeptide, *tert*-butoxycarbonyl-Tyr-Leu-Val-CH$_2$C1, with inhibitory effect on human leukocyte elastase in aqueous solution and in biological fluids. Chem Pharm Bull Tokyo 1997; 45:1845–1850.
183. Patel K, Borchardt RT. Chemical pathways of peptide degradation. III. Effect of primary sequence on the pathways of deamidation of asparaginyl residues in hexapeptides. Pharm Res 1990; 7:787–793.
184. Lee KC, Lee YJ, Song HM, Chun CJ, DeLuca PP. Degradation of synthetic salmon calcitonin in aqueous solution. Pharm Res 1992; 9:1253–1256.
185. Kearney AS, Mehta SC, Radebaugh GW. Aqueous stability and solubility of CI-988, a novel "dipeptoid" cholescystokinin-B receptor antagonist. Pharm Res 1992; 9:1092–1095.
186. Motto MG, Hamburg PF, Graden DA, Shaw CJ, Cotter ML. Characterization of the degradation products of luteinizing hormone releasing hormone. J Pharm Sci 1991; 80:419–423.
187. Yoshioka S, Aso Y, Izutsu K. Is stability prediction possible for protein drugs? Denaturation kinetics of β-galactosidase in solution. Pharm Res 1994; 11:1721–1725.

188. Roberts CJ. Kinetics of irreversible protein aggregation: analysis of extended lumry-eyring models and implications for predicting protein shelf life. J Phys Chem B 2003; 107:1194–1207.

189. Roberts CJ, Darrington RT, Whitley MB. Irreversible aggregation of recombinant bovine granulocyte-colony stimulating factor (bG-CSF) and implications for predicting protein shelf life. J Pharm Sci 2003; 92(5):1095–1111.

190. Gu LC, Erdös EA, Chiang HS, et al. Stability of interleukin 1β (IL-1β) in aqueous solution: analytical methods, kinetics, products, and solution formulation implications. Pharm Res 1991; 8(4):485–490.

191. Williams ML, Landel RF, Ferry JD. The temperature dependence of relaxation mechanisms in amorphous polymers and other glass-forming liquids. J Am Chem Soc 1995; 77:3701–3707.

192. Peleg M. On modeling changes in food and biosolids at and around their glass transition temperatures. Crit Rev Food Sci Nutr 1996; 36:49–67.

193. Roy ML, Pikal MJ, Rickard EC, Maloney AM. The effects of formulation and moisture on the stability of a freeze-dried monoclonal anti body—*Vinca* conjugate: a test of the WLF glass transition theory. Dev Biol Stand 1992; 74:323–340.

4

Analytical Methods and Stability Testing of Biopharmaceuticals

Helmut Hoffmann and Sandra Pisch-Heberle
*Boehringer Ingelheim Pharma GmbH & Co. KG,
Biberach an der Riss, Germany*

INTRODUCTION

Proteins and monoclonal antibodies (mABs) derived from recombinant DNA are of increasing economic importance for the pharmaceutical industry. More than 165 biopharmaceuticals are presently on the market, with an estimated market size of US$33 billion in 2004 (1). The projection for the end of the decade is ranging around US$70 billion (2). Presently about 2500 biotech drugs are in discovery, 900 in preclinical trials, and 1600 are under clinical investigation (1). Biopharmaceuticals are developed and applied for a broad spectrum of applications, such as tumor therapy and diagnostics, AIDS and other immunological disorders, infectious diseases, neurology, cardiovascular diseases, hematology, wound healing, ophthalmics, skin disorders, diabetes, and respiratory diseases. Besides therapeutic applications, protein-based biopharmaceuticals are applied as vaccines, as drug carriers, and as diagnostic tools (in vivo and in vitro).

A large variety of protein classes are currently being investigated for use as drugs or devices including (recombinant humanized) mABs, hormones, growth factors, interleukins (ILs), immune modulators, blood factors, enzymes, and soluble receptors. With respect to product categories, an interesting trend for biotech drug approvals during the last three years was the fact that no interferon, IL, thrombolytic, or anticoagulant-based biopharmaceutical was approved, and proportionally larger numbers of growth factors, mAB-based products and enzymes came on the market, with a parallel shift in the therapeutic area and with cancer being a prominent target (1).

The potential for biogenerics has been widely discussed in recent years, and the first product Omnitrope (Sandoz, Holzkirchen, Germany) is now approved both in the United states and in Europe.

The appropriate production system for such protein drugs depends on the nature of the desired protein. Bacteria as host cells for recombinant DNA are suitable for nonglycosylated proteins with low molecular weight and a small number of disulfide bridges. Especially, mammalian cell culture systems are more appropriate and will continue to constitute the major production cell line for the production of complex glycoproteins with high molecular weight and numerous disulfide bridges (3), and significant advances in cell culture technology have led to protein levels approaching 5 g/L, which is a 10-fold increase compared to some years ago (4). Alternatively, yeast, insect, and plant production cell lines are capable of producing glycosylated proteins, with the major drawback being that the glycosylation reactions of these systems yield a glycosylation pattern which significantly differs from the human glycosylation profile (5–7). However, considerable advancement has been made in humanizing the glycosylation profile as reported for glycoengineered yeast (8).

Proteins as drug substances differ from conventional chemical drugs with respect to many properties such as size, shape, and conformation, multiplicity of functional groups, amphotericity, physical form of bulk drug, and heterogeneity of structure. Therefore, proteins represent complexity that is an order of magnitude greater than that of traditional drugs. This complexity of protein drugs has an impact on the number of analytical methods that must be applied for protein characterization and for the development of stability-indicating assays. In spite of these differences, the general principles of chemical drug stability testing are also applicable to protein drugs. The requirements and the goals are the same: the drugs must be designed to withstand long periods of transport and storage. The full effectiveness and safety of industrially manufactured drugs must be guaranteed until the end of the declared shelf life. This can be achieved and ensured only by development of a stable formulation, supported by extensive product characterization and an appropriate stability testing program.

This chapter describes the methods used to characterize protein drugs and to support formulation development and stability testing. This presentation of methods also focuses on the criteria an analytical method must meet to be stability indicating and to be useful in screening formulations and in formal stability programs. Because the process used to produce a biopharmaceutical often has a strong influence on the resulting characteristics and stability of the protein, we begin with a brief discussion of the manufacturing process. The discussion focuses on the importance of producing a consistent product, which is achieved by strict control of the cell culture and purification processes.

THE IMPORTANCE OF A WELL-CONTROLLED MANUFACTURING PROCESS

The basis for protein stability testing is a comprehensive product characterization, with emphasis on homogeneity of preparation and demonstration of lot-to-lot consistency. All possible aspects of protein heterogeneity (protein variants) and possible influence of the manufacturing process on the physicochemical and biological properties in relation to the biological activity should be characterized. Characterization of the variants (e.g., glycoforms, deamidation) comprises identification of the modification including the respective site(s) and the mechanism of accumulation (9).

During development changes in the manufacturing process (active substance and/or finished product) are likely to occur in most cases scale-up of the process may be required. This necessitates an assessment of comparability to ensure that these manufacturing changes do not affect the identity, purity, safety, or efficacy, including immunogenicity of the product.

An underlying principle of comparability is that under certain conditions, protein products may be considered comparable on the basis of analytical testing results alone. The confirmation of comparability does not essentially require the prechange and postchange biological product to be identical. To conclude, comparability, high similarity, and an accumulated expertise that is adequately prognostic to demonstrate that any differences in the quality profile have no adverse effect upon safety and/or efficacy is imperative (10,11).

Lot-to-lot consistency of biopharmaceuticals can be ensured by validated and well-controlled biotechnical production processes (12). Adequate design of a process and knowledge of its capabilities are part of the strategy used to develop a manufacturing process that is controlled and reproducible, yielding a drug substance/drug product that meets specifications. The manufacturing of the active ingredient comprises a large number of different steps during cell culture, scale-up, fermentation, and protein recovery (13–15). Because there are so many different process steps that can influence the product quality and because protein products are so complex, the quality of a biopharmaceutical is always linked to the process by which it is produced. Therefore, the manufacturing process must be tightly controlled by defining standard procedures for all operations and setting limits for all critical process parameters.

The ultimate stability of a protein can often be a function of the physical and chemical conditions to which it was exposed during processing (Chapters 2 and 3). A number of critical components of the cell culture process may influence product quality and stability: media conditions, the producing organism, the type of fermentation technology employed, and the techniques used for cell separation and harvesting to isolate the protein prior to the start of purification. To minimize the effects of the cell culture process on product quality and stability, the cells used for production are grown under defined conditions in a fixed and validated fermentation process. After separation from the producing organisms, the protein is subjected to a series of filtration and chromatography steps to remove other protein contaminants. The aim of the first steps in purification is to transfer the desired protein to a high-purity, stable form within a short period of time, to avoid potential degradation due to proteolysis. The stability of product intermediates needs to be demonstrated throughout downstream processing to rule out potential degradation that can be caused by such steps as the ultrafiltration, microfiltration, and concentration steps, and by elution conditions during chromatography. Later stages of the purification process serve to remove trace protein impurities and DNA in addition to removing/ inactivating potential viral contaminants. For most biologics, final bulk formulation is the last step in the protein recovery scheme, and at this point, the purified protein is exchanged into the desired buffer with all appropriate excipients that have been chosen based on the formulation development program. Drug product manufacturing usually encompasses the filling of the formulated bulk protein into final product containers, with subsequent lyophilization if required (e.g., for reasons of stability, drug delivery).

During the development phase of the production process, analytical methods are developed, optimized, and applied on a routine basis to detect deviations of a given protein structure. In many cases, the results of these methods reflect the heterogeneity of the protein, for example, in the profile of a chromatogram or in the banding pattern of an electrophoretic gel. A reference standard is established, which is representative of the product that has been tested in preclinical and clinical trials. This reference material is used in many of the analytical techniques as a basis for comparison, to demonstrate product consistency during process development and scale-up.

After scale-up to the commercial scale, three to five consistency runs are performed, including process validation, to demonstrate the reliability and reproducibility of the process. The product from these consistency runs is used in clinical Phase III studies, and the reference standard is established with material from these lots. An extensive analytical characterization of reference material from these consistency runs is performed, and a consistent product quality must be demonstrated for material from these runs. The final product from the consistency runs is put on real-time stability, and the results are used to define the shelf life of the product. All the data obtained from the consistency runs form the basis for approval of the process and product by regulatory authorities.

The concept of lot-to-lot consistency is based on the goal of obtaining material from each production cycle that meet the specified heterogeneity profile, to ensure that the commercial lots possess the same safety, efficacy, and stability profile as the lots that were used in the clinical trials. This goal can be achieved by the establishment of a controlled and validated manufacturing process and by an extensive characterization of the protein drug during quality control and lot-release testing. Thus, the existing regulatory concept of a "specified biological" is considered, requiring that the product is produced on a consistent basis and that the natural molecular heterogeneity, impurity profile, identity, and potency is assessed by suitable state-of-the-art bioanalytical methodology with a high degree of confidence.

In addition, extensive stability testing is performed to establish the expiration date and guarantee full effectiveness and safety throughout the declared shelf life of the biopharmaceutical product.

ANALYTICAL METHODS FOR PROTEIN CHARACTERIZATION OF THE DRUG SUBSTANCE

The characterization of a protein drug is a complex undertaking, requiring the use of a wide range of methods to establish such properties of the drug substance as structural integrity, consistency, activity, purity, and safety (Fig. 1). The complexity of protein molecules means that there are many potential degradation pathways, each with its individual dependencies on such parameters as pH, ionic strength, and temperature. Each protein may represent a unique combination of such pathways and dependencies. It is therefore critical that a broad spectrum of methods be used to evaluate the effects of processing and storage to assure optimal maintenance of safety and efficacy of the drug (16–18). The characterization of a protein drug substance, formulation experiments, and stability testing of a given biopharmaceutical starts when sufficient quantities of highly purified protein are available. The analytical methods applied should have the capability to detect and quantitate different forms of the active ingredient from each other and from their degradation products. Frequently, the protein preparation can be purified to the point of showing only one component by many analytical techniques. However, large and complex proteins often exhibit "micro heterogeneity," which in the case of glycoprotein due to a number of very similar molecules that possess the same amino acid sequence

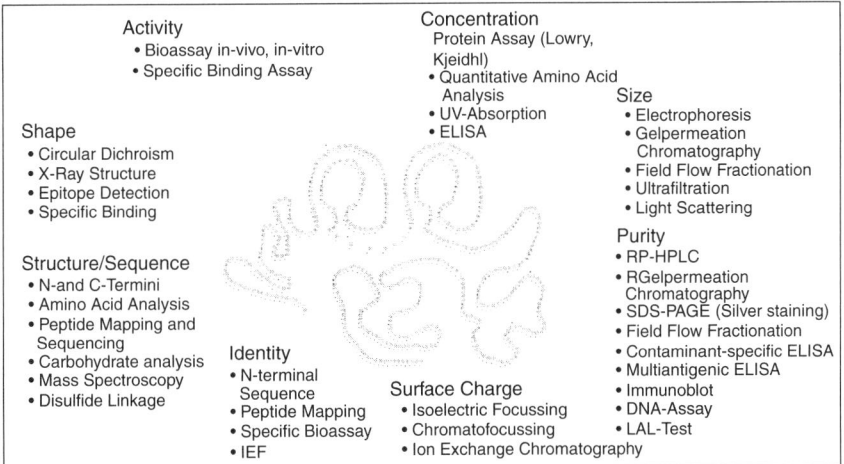

Figure 1 Analytical methods for characterization of proteins.

but differ in the details of their glycan structures. The larger and more complex a protein molecule is, the more difficult it is to separate the major product from minor degradation forms by large-scale purification. It is therefore crucial to optimize the analytical methods for maximum resolution; this requires a thorough understanding of the principles of each method and the influence of the operating parameters on method performance (19). In case of product heterogeneity, the different product forms must be characterized, and batch-to-batch consistency of this heterogeneity needs to be demonstrated for production batches (20).

We now turn to a detailed description of the wide range of analytical methods typically used to characterize protein drug substances. The application of analytical methods to formulation development and stability testing is discussed later in the chapter.

Primary Structure and Sequence

The amino acid sequence of a given recombinant protein can be derived from the nucleic acid sequence of the gene in the expression vector. The verification of the amino acid sequence on the protein level is most often achieved by peptide mapping and characterization (sequencing) of the isolated peptides by mass spectroscopic analysis. Peptide mapping is a very powerful and widely applicable tool for protein characterization. The high specificity of certain proteolytic enzymes in cleaving polypeptide chains only at certain residues results in a very characteristic set of peptides (21). Trypsin, for example, cleaves proteins only at lysine and arginine residues, resulting in peptides that end either with lysine or with arginine (except the C-terminal peptide), and the same pattern will be obtained for each digest of a given protein. These peptides are then resolved by reverse-phase high-performance liquid chromatography (RP-HPLC). The HPLC pattern, therefore, results in a "fingerprint" that is characteristic for a given protein.

Enzymatic cleavage, e.g., by trypsin, as described above in combination with mass spectrometry (MS) is performed for a detailed characterization of the primary structure. Molecular mass differences of 1 Da as introduced by deamidation (22,23) can be resolved.

For modifications that give rise to higher differences in molecular mass compared to the native sequence, or for proteins with a less-complex microheterogeneity profile application of, e.g., electrospray ionization, time-of-flight MS (ESI-TOF-MS) techniques without enzymatic digestion are also feasible (24).

This method has the capability to detect changes in the primary structure at a 5% to 10% level for large proteins such as tissue plasminogen activator (t-PA) (25), and below 5% for smaller proteins such as bovine growth hormone (bGH) or peptides (26). Compared to earlier methods for peptide identification and sequencing, such as amino acid analysis and sequencing (see below), peptide identification by MS techniques is much faster and more generally applicable.

Peptide sequencing is a complementary tool to identify the site of a chemical degradation or enzymatic cleavage in degradation products generated during processing or storage of the biopharmaceutical product. Direct N-terminal sequencing has long been used as a chemical procedure known as Edman degradation, which derivatizes the amino-terminal amino acid in such a way as to release and identify the amino acid and expose the amino terminal to the next cycle of degradation (27). This method has been applied to the intact protein for a limited number of residues, primarily to confirm the integrity of the N-terminus of the protein. In a similar way, carboxy-terminal sequencing has been applied for confirmation of the identity/integrity of the carboxy terminus for final products. The identity/integrity of the N-terminal and C-terminal ends of proteins is often confirmed by identifying and tracking the corresponding peptides in the peptide maps, which are routinely applied for release testing of the products (28).

MS is now playing the major role in peptide sequencing; however, Edman degradation is still used as an alternative to address specific questions (29,30).

Secondary and Tertiary Structure

The secondary and tertiary structure of proteins must be considered to be equally important in the maintenance of the overall native status of the protein, and changes in the conformation of proteins may lead to aggregation, decrease in biological activity, and, in some cases, immunogenicity. Therefore, the design of a formulation for optimal stability of proteins must consider all aspects of the structure of a biopharmaceutical. Methods for evaluating secondary structure are primarily physical, spectroscopic methods, and their application to bGH has been published (31).

The secondary structure can be evaluated by a technique known as circular dichroism (CD) (32). Amino acids (except glycine) are asymmetric owing to the presence of the chiral carbon. The optical properties of polypeptides are due to the asymmetric centers of their constituent amino acids. Polypeptides, thus, interact differently with right- and left-circularly polarized light. CD is a technique that measures the unequal absorption of left- and right-circularly polarized light. In the far-ultraviolet (UV) region (<250 nm), CD spectroscopy can be used for the prediction of secondary structures in a protein that are expressed as percentages of a-helix, β-pleated sheet, and random structures (33,34). Because of the sensitivity of signals to detect changes in the environment of the aromatic amino acids (Trp, Tyr, Phe), changes in the tertiary structure of proteins are observed in the near-UV CD region (240–320 nm) (35). Disulfide bonds are also chromophores, which can give rise to CD bands in this near-UV region, while free thiol (SH–) groups do not (32,36). The near-UV CD spectrum is commonly used as a "fingerprint" of the tertiary structure to provide confirmation that the correct structure is present (37) or that refolding efforts were successful when the product was isolated from an insoluble starting material (38).

Fourier transform infrared spectroscopy (FT-IR) has also provided an estimate of secondary structure composition (39–41). This method uses special deconvolution methods to separate and integrate overlapping amide I infrared absorption bands associated with a-helix, β-pleated sheet, and random structures. In this method, the spectrum is related to the subtle effects of the regular secondary structure on the energetics (i.e., vibrational frequency) of amide groups in the peptide linkage. FT-IR also has the advantage of being able to evaluate the aspects of protein structure in the solid state (42). Both this method (43) and calorimetry (44) have been used to study protein conformation at various stages of the freeze-drying process. This information has been used to optimize formulations and processes by maintaining the native structure of the protein.

Many proteins exhibit fluorescence in the 300 to 400 nm range when excited at 250 to 300 nm, as a result of the presence of aromatic amino acids (Trp, Tyr, Phe). Fluorescence spectroscopy can yield information regarding the microenvironments of these aromatic amino acids (45). Thus a buried tryptophan is usually in a hydrophobic environment and will fluoresce with a wavelength maximum in the 325 to 330 nm range, while an exposed residue (or free amino acid) fluoresces at around 350 to 355 nm (46). Unfolding of a given protein may result in a shift of the fluorescence spectrum due to conformational changes that induce modified emission patterns of the aromatic amino acids in their modified environment.

Protein Glycosylation

Glycosylation of proteins is one of the most common and important posttranslational modifications found in eukaryotic secretory proteins. The type and extent of N-glycosylation contributes to the physicochemical and recognition properties of glycoproteins (47,48). The biological activity of glycoprotein hormones frequently depends on the attached N-linked oligosaccharides (49,50). In recent years, the role of N-glycosylation in human proteins expressed in Chinese Hamster Ovary cells (the most commonly used eukaryotic cells for expression of human recombinant proteins) has been studied extensively (51,52). In many glycoproteins, the oligosaccharides contribute to solubility (53) and influence the in vivo circulatory lifetime of the product (54,55). Other possible functions of the oligosaccharides, such as facilitating secretion, affecting biological activity, or increasing stability, must be investigated for each individual protein. For most mABs, the efficacy arises from their target-binding activity and their effector functionality. A variety of parameters influence effector functionality, including glycosylation (56–58). Although the type of host cell and the primary structure of the expressed glycoprotein mainly govern N-glycosylation, environmental factors also influence glycosylation (59–63). Alterations in oligosaccharide structures occur either by affecting intracellular synthesis or by changing glycosidase activity after secretion (64,65). Because changes in the glycosylation pattern may have important consequences in glycoprotein pharmaceuticals, carbohydrate analysis is of great importance in product characterization and batch analysis to ensure consistency in product quality.

The microheterogeneity of glycoproteins frequently complicates the interpretation of results from analytical methods for product characterization. Complex carbohydrate structures often contain a varying number of sialic acid residues, which result in a heterogeneity of charge distribution that is manifested by several bands on an isoelectric focusing (IEF) gel or multiple peaks in ion exchange chromatography (IEC). The charge heterogeneity of glycopeptides that have been isolated by RP-HPLC after enzymatic cleavage of the protein (peptide mapping) can be separated into individual peaks by capillary electrophoresis (CE).

Substantial progress in the analysis of the oligosaccharide structures of recombinant proteins was made with the introduction by Townsend and Hardy (66,67) of high pH anion exchange chromatography with pulsed amperometric detection. This technology was superior to refractive index detection or radioactive labeling of carbohydrates and was for a long time a valuable tool in many laboratories for the characterization of carbohydrate structures of biopharmaceuticals.

Currently, glycosylation heterogeneity is assessed by a combination of several chromatographical and mass spectrometrical techniques (68,69). The chromatographical techniques include enzymatic or chemical release of the N-linked oligosaccharides followed by fluorescence derivatization (e.g., 2-aminobenzamide labeling) and a subsequent chromatographic or capillary electrophoretic separation. The released oligosaccharides can be characterized by comparison to known standards. Their structural composition can be confirmed by exoglycosidase digestion studies, and molecular masses are easily determined by matrix-assisted laser desorption ionization (MALDI)-TOF-MS. To identify the site occupancy of the oligosaccharides and determine their composition, ESI-TOF-MS is performed on the enzymatically digested protein. Several proteases are available; typically used enzymes are trypsin, lys-C, AspN, Papain (to generate Fab and Fc fragments), or a combination thereof.

The glycan structures on a protein are generally quite stable and usually do not change under conditions used for formulation and storage of protein pharmaceuticals. Therefore, glycosylation analysis usually does not belong to a standard set of stability-indicating methods. To analyze for underlying degradation processes, however, it may occasionally be beneficial or necessary to remove the carbohydrates and the associated analytical heterogeneity.

Protein Concentration

One of the most fundamental measurements made on a protein is the determination of concentration. Peptides and proteins that do not absorb visible light can react with reagents to form colored compounds. The widely used reagent ninhydrin reacts with amino groups of amino acids and peptides to produce an intensely colored product that has a maximum absorbance at 570 nm. A relatively more specific, but less sensitive, method of quantitatively determining proteins is the biuret reaction of copper in a basic solution in the Lowry assay. The resulting blue product is quantitated in the visible region at 540 to 560 nm, and the assay is linear at microgram protein levels (70). The Bradford colorimetric assay depends on the binding to the product in an acid environment of the dye Coomassie brilliant blue (71). The reaction can take place in just two minutes, with good color stability for one hour, to allow measurement at 595 nm. These colorimetric methods can provide relative protein concentrations, when compared with defined standards such as bovine serum albumin, that can be compared between different labs. The binding of the reagents may be different for individual proteins, however, and it also can be influenced by buffer conditions.

UV absorption spectroscopy is considered to be the most convenient and accurate measure of the protein concentration (72,73). The absorption spectrum of a protein in the UV wavelength range is the net result of absorption of light by the carbonyl group of the peptide bond (190–210 nm), the aromatic amino acids (250–320 nm), and the disulfide bonds (250–300 nm). Protein concentration can be determined from the absorption spectrum for a purified protein on the basis of its specific absorption coefficient (74). This absorption coefficient is derived from a known protein concentration, usually measured on a 1 mg/mL solution at the wavelength maximum near 280 nm. An accurate independent measurement of the protein concentration (for the sample whose spectrum is recorded) is a critical part of the determination of the extinction coefficient and is often performed by quantitative amino acid analysis, nitrogen assay (Kjeldahl), or dry weight measurements. Since, however, the amino acid sequence is known for most protein pharmaceuticals, the theoretical extinction coefficient may also be calculated from the content of the aromatic acid components (75). The effects of the protein structure on the UV spectrum are not considered by this theoretical calculation. The magnitude of these effects can be measured by recording the spectrum when the protein is digested with enzymes until the spectrum stabilizes to that of free aromatic amino acids in the resulting peptides, and their concentration can be determined from their known extinction coefficients (76). This spectroscopic method avoids the errors associated with dry weight analysis and amino acid analysis and is very simple with computer-assisted spectrophotometers.

Surface Charge

The acidic and basic groups of proteins make them polyelectrolytes, and they can therefore be separated in an applied electric field. The use of this phenomenon, known as electrophoresis, is one of the most common methods of separating mixtures of proteins on an analytical scale. Many different forms of electrophoresis are applied, and details on the theoretical principles behind these electromigration techniques have been published (77,78).

In native electrophoresis, performed in the absence of denaturing agents, the major factors controlling the electrophoretic mobility of a macromolecule are its net charge (at the pH of separation) and its Stokes radius, a hydrodynamic parameter determined primarily by the size and, to a lesser extent, the shape of the molecule. The basis for separation is therefore determined mainly by differences in mass-to-charge ratio. The use of an anticonvective matrix, such as a gel, can enhance the resolving power of this method. If the pores of the gel are comparable to the dimensions of the protein, they will present resistance to the movement of the molecules in a size-dependent fashion. Since native gel electrophoresis provides information on both size and charge, independent data are required to resolve these two factors.

The charge of a native protein is dependent on the pH of the solution. At the isoelectric point (pI), the protein's net charge will be zero, and the mobility in the electric field will be zero. Thus, if the electric field is also a pH gradient, the protein will migrate to the point at which the pH is the same as the pI and the migration will stop. This phenomenon is known as "isoelectric focusing" (79). Stable pH gradients are established by means of carrier ampholytes with appropriate pI and buffering capacity, which are included in large-pored agarose or acrylamide gels (80,81). Such IEF systems can separate protein species that differ in pI by as little as 0.02 pH units (82). This method has been applied to separate different glycoforms of proteins that differ in the degree of sialylation of the complex carbohydrates (83,84). The method is also widely used as a stability-indicating method to assess deamidation during stability studies by quantitative densitometry (85).

IEF was successfully established in the capillary format with the improvements made in CE. Two separation modes are available for the resolution of charge variants, capillary zone electrophoresis (CZE) and capillary IEF (cIEF) (86,87). Both modes are now widely used techniques in protein characterization and quality control. In most CE applications, UV or UV–visible absorbance is employed as the primary mode of detection with on-tube detection. In cIEF, a stable pH gradient is formed in the capillary using carrier ampholytes, and proteins become focused in the gradient at their pIs. Diffusion cIEF gives a resolution comparable to the slab gel IEF but may result in a more complex pattern because smaller peptides that diffuse out of the gel or peptides that do not stain well are also detected. In CZE, the inlet and outlet reservoir are filled with the same buffer and components injected at the inlet migrate towards the detection point according to their mass-to-charge ratio.

IEC is a powerful separation technique at the preparative scale and it is also an analytical tool for assessment of charge heterogeneities in protein preparations (88). A protein with a net positive charge tends to bind to a matrix with a net negative charge by ionic interactions, and vice versa, provided the ionic strength of the buffer is sufficiently low. The passage of a salt gradient over a column to which a mixture of proteins has bound will cause the elution of each protein at its own critical salt concentration, where the protein binds less tightly than the salts. Thus, the proteins, or differently charged variants of a protein, are separated from each other, and the components of the mixture can be individually quantitated. The charge distribution on the surface of a given protein is dependent on the solution pH, which dictates the binding and elution conditions of the IEC method. If two forms of the same protein, which differ by one charged amino acid, are analyzed at a pH where that group is uncharged, this difference may not be detected by IEC. Therefore, a single homogeneous peak on ion exchange is not a guarantee that the sample is homogeneous.

Chromatofocusing, as the name implies, combines aspects of IEF and IEC. It is performed under low salt conditions and can use the same ampholytes used in IEF (89,90). The sample is loaded onto an ion exchange column at a pH where it binds. The column is then eluted with an ampholyte or buffer mixture selected to generate a pH gradient that gradually flows down the column. When the pl of a bound protein is reached, that protein is released from the resin. The method has a high resolving power, and the proteins are eluted in highly focused peaks from the column. This method has one significant disadvantage: because most proteins exhibit minimum solubility at low ionic strength around their pl, this means that neutral surfactants or urea may be required to keep the protein in solution.

Protein Size

The most common form of protein electrophoresis is polyacrylamide gel electrophoresis with the denaturing agent sodium dodecyl sulfate (SDS-PAGE). Reduced proteins tend to bind a relatively constant amount of SDS on weight basis: approximately 1.4 of SDS per gram of protein (91). The SDS molecule carries a negative charge, and complexes of proteins with SDS have very similar mass-to-charge ratios and therefore free electrophoretic mobility (92). When a mixture of SDS-saturated proteins is electrophoresed in a gel matrix with the correct pore size, the major factor determining their migration rate is their effective size. SDS-PAGE is commonly used for assessing purity and as a tool for determining the apparent molecular weight of a protein (93,94). The most reliable estimate of molecular weight is obtained from analyses in which the disulfide bonds have been reduced and the polypeptide chains are truly random (95). A widely used system is the discontinuous buffer system with SDS, published by Laemmli (95). This system may be used for proteins ranging in molecular weight from 10,000 to 300,000 Da by varying the concentrations of the acrylamide gel and the cross-linker, bisacrylamide, to vary the pore size of the gel (95,96).

The resolving power of SDS-PAGE can be further increased by using gradients of acrylamide to vary the pore sizes in the gel. Determination of the apparent molecular weight of a protein can be influenced by sample treatment (e.g., heating of samples and use of reducing agents, such as mercaptoethanol or dithiothreitol or carboxymethylation with iodoacetic acid), which then makes the results difficult to interpret. It is also common to find that proteins with significant carbohydrate content running as diffuse bands, probably because of a combination of heterogeneity of molecular weight, charge (as a result of variable extents of sialylation), and SDS binding (97). Protein quantitation by SDS-PAGE can be complicated by the variable dye-binding properties of individual proteins (98,99) and the denaturing effects of SDS.

The apparent molecular weight of a protein can also be determined by gel filtration or size exclusion chromatography (SEC) on a column that has been calibrated with molecular weight standards (100). This method is of relatively low resolution primarily because it is based on the hydrodynamic properties of the protein and consequently gives accurate estimates only for spherical proteins. Since separation by gel filtration is based on apparent molecular weight, this method is frequently used to assess the aggregation state of the protein and to quantitate dimers, oligomers, and aggregates in a protein preparation (101,102). While yielding lower resolution than SDS-PAGE, this method does allow the quantitation of concentration.

The incorporation of specialized ionization methods has extended the applications of MS to protein characterization and accurate determination of molecular weight: ESI-MS and MALDI-MS (103). ESI-MS has extremely high precision because it generates multiple-charged ions through the abstraction or association of protons or via association of anions or cations from solution. Because the ionization process generates multiple charged ions, even very large proteins can be analyzed, since quadrupole and magnetic mass detectors measure the mass-to-charge ratio. Even for large proteins, the molecular weight can be determined with high precision (104). This precision arises by transforming the mass/charge scale of the obtained ESI spectra to the real mass scale and by using a simple algorithm averaging all the signals of one series of multiply charged ions.

This approach was used to characterize recombinant γ-interferon and its C-terminal degradation products simultaneously, yielding an estimate of the molecular weight of 16908.4 ± 1.2 while the theoretical mass is 16907.3 (105). MALDI-MS was pioneered by Hillenkamp and Karas (106,107), who showed that if a high concentration of a chromophore is added to the sample, a high-intensity laser pulse will be absorbed by the matrix and the energy absorbed will volatize a portion of the matrix, carrying the protein sample with it into the vapor phase essentially intact. The resulting ions are then analyzed in a TOF-MS. The "gentle" nature of the ionization may be responsible for the ability of the method to provide information on quaternary structure. A major extension of the TOF-MS method was developed by Beavis and Chait (106), who showed that the method is relatively insensitive to large amounts of buffer salts and inorganic contaminants. This type of methodology may have a wide utility for several reasons: it requires only picomole amounts of sample; it is very fast (<15 minutes from start to finish) and does not fragment the molecules; the sample can be a crude mixture of proteins; and the result is in principle as easy to interpret as (and indeed resembles) a densitometric scan of an SDS-PAGE gel, with a mass range well above 100 kDa.

Bioassay, Potency Assays

All the methods described thus far are universally applicable to different proteins, although the selection and focus of the analytical methods applied will be influenced by the specific properties of the protein. Potency assays, however, are required to mimic the specific biological activity of the biopharmaceutical and hence are usually protein specific. For many proteins, in vivo bioassays in animals have been developed to measure the "true" biological activity including the pharmacodynamics of the product. For example, the bioassay for human growth hormone (hGH) measures the daily weight gain in hypophysectomized (hypoxed) rats given daily injections of hGH (109). The rats respond to exogenous growth hormone (even from different species). Usually 10 rats are used per group, and two doses are compared for the sample, with inclusion of a reference standard and a blank as controls. The dosing and the recording phase of the study take 10 days. As can be seen from the design of this bioassay, the analysis of a single sample requires a tremendous effort. Therefore, such a bioassay is not suitable for analyzing large numbers of samples derived from formulation screening or from stability studies. In addition, such in vivo bioassays have the drawback of animal use and additionally suffer from poor reproducibility. In many cases, cell-culture–based in vitro bioassays that mimic the biological activity of the protein have replaced in vivo bioassays. The murine thymocyte proliferation assay that is widely used for routine analysis of human IL-1 (hIL-1) activity is an example of such a surrogate assay (110). The proliferation by hIL-1 is mediated via IL-2 release of hIL-1–stimulated T-cells, since antibodies specific for hIL-2 can block this response (111). The assay, which is performed in microtiter plates, is sensitive to 10 to 50 pg/mL IL-1. Proliferation is assessed by [^{3}H] thymidine incorporation after 72 hours of culture. A major problem with this assay is its lack of specific-ity: it can be stimulated by hIL-2 and compounds used to induce hIL-1 produc-tion (e.g., lipopolysaccharide and phorbol myristate acetate). Similar proliferation assays were developed for hIL-1 on the basis of different cell lines. Three differ-ent amino-terminal variants of recombinant hIL-1β were demonstrated to differ in their activity by 3- to 10-fold in these cell-based bioassays (112).

The biological activity of interferons is measured by their dose-dependent inhibition of the cytopathic effect due to virus infection of cell cultures in microti-ter plates (113). The activity is measured in comparison to international standards and is expressed in international units (114). Although cell-based in vitro assays have a much higher sample throughput and reproducibility than in vivo assays in animals, the assay variability of the former is still comparatively high, with coef-ficients of variation in the range of 20% to 50%.

For some proteins, especially for enzymes, biomimetic in vitro test systems with good reproducibility have been developed. An example is recombinant t-PA (rt-PA), a serine protease that cleaves plasminogen to plasmin and thereby initi-ates the lysis of fibrin clots and blood coagulates. This biopharmaceutical is used for several indications, such as myocardial infarction, stroke, lung embolism, and deep venous thrombosis. The enzyme activity of rt-PA can be measured in a chro-mogenic assay using a synthetic substrate, a tripeptide linked to *p*-nitroanilide (S-2288). The rate of cleavage is monitored following the formation of *p*-nitroaniline spectrophotometrically at 405 nm. The one- and two-chain forms of rt-PA have different affinities for the substrate and therefore differ in their specific enzymatic activity in this assay (80). The concentrations of one- and two-chain rt-PA can be determined based on the difference in amidolytic activity between the two forms toward S-2288 substrate. For this purpose, the assay is performed with the mixture of one- and two-chain rt-PA and also after all the rt-PA in the mixture has been transformed to the two-chain form (with higher specific enzymatic activity) by adding a trace amount of plasmin to the mixture.

A more specific and relevant enzyme assay is an indirect chromogenic assay for rt-PA in which a synthetic substrate (S-2251) specific for plasmin is used to measure the plasmin generated upon incubation of rt-PA and plasminogen. The amount of plasmin generated in this test correlates with the plasminogen-activating potency of t-PA (115). The most biologically relevant in vitro potency assay for rt-PA is the in vitro clot lysis assay, which is based on measuring the time taken for a fixed amount of rt-PA to dissolve a fibrin clot. Typically, a fibrin clot is produced by combining fibrinogen and thrombin in the presence of plasminogen.

Then rt-PA is introduced to initiate the lytic reaction. If rt-PA is present in limiting amounts, the time for clot lysis is directly proportional to its concentration. The accuracy of the assay is dependent on the ability of the analyst to reproducibly measure the reaction endpoint. Several methods have been used, which include releasing entrapped air bubbles from the clot and dropping glass beads through the fibrin clot (116,117). In an automated version of the in vitro clot lysis assay, endpoint detection is based on turbidometric measurement performed by means of a commercially available microcentrifugal analyzer (118). Lysis of the clot is followed by measuring the decrease in absorbance at 340 nm. This automated version of the in vitro clot lysis assay is reliable and reproducible, with an accuracy of 99.5% and a precision of 5% (in the concentration range of 40–1200 ng/mL rt-PA) and can be performed at high throughput with minimal sample handling. The in vitro clot lysis assay has also been adapted to a microtiter plate format (119).

Another fibrinolytic assay system measures the lysis of clots prepared from human blood, then incubated in plasma in the presence of small amounts of added rt-PA (120). In this system, the fibrinolytic activity of rt-PA is determined by plotting the loss in blood clot weight against the concentration of rt-PA in the plasma sample. Since this so-called hanging-clot assay is very time consuming and not suited to automation, its application is limited to very specific investigations; it is not useful for routine quality control purposes.

General requirements for a stability test are high capacity and throughput of samples, as well as high accuracy and precision. Both criteria are fulfilled by the automated in vitro clot lysis assay for rt-PA. This assay is one of the stability-indicating assays for rt-PA and has been used to assign the expiration date (shelf life) of drug product lots.

In general, a functional, stability-indicating bioassay is required, with "functional" meaning relevant to the mechanism of action of the therapeutic protein. For example, for mABs, common cell-based potency assays include proliferation, inhibition of proliferation, and apoptosis. In cases where the cell-based bioassay is not sensitive, accurate, or precise enough, a choice for an in vitro potency assay can be a competitive specific binding assay. This competition assay can be designed as a sandwich assay, with the soluble antigen fixed to a microtiter plate. The displacement of a conjugated form of the mAB is measured for the nonconjugated test sample of the same mAB and for a defined reference material of the antibody on the same microtiter plate. From the displacement calibration curve for reference material, a relative binding potency for the sample can be determined. This assay format can be performed with high accuracy, precision, and capacity in sample throughput. However, a specific binding assay does not necessarily reflect the "true potency" of a protein drug. Therefore, care must be taken to correlate the data from in vitro binding assays with more relevant biological test systems. Most importantly, the stability-indicating properties of such an in vitro assay need to be established.

Product Purity

The determination of absolute as well as relative purity presents considerable analytical challenges, and the results are highly method dependent. Historically, the relative purity of a biological product has been expressed in terms of specific activity (units of biological activity per milligram of product), which is also highly method dependent. Consequently, both drug substance and drug products are assessed for purity by a combination of methods.

There is an inherent degree of structural heterogeneity in proteins, a result of the biosynthetic processes used by living organisms producing the protein. Therefore, the desired product can be a mixture of posttranslationally modified forms (e.g., glycoforms, as described above). These forms may be active, and their presence may have no deleterious effect on the safety and efficacy of the product. When variants of the desired product are formed during the manufacturing process and have properties comparable to the desired product, they are considered to be product-related substances, not impurities.

Biopharmaceuticals, composed of the desired product and multiple product-related substances, need to be tested for impurities, which may be either process related or product related.

Process-related impurities encompass those that are derived from the manufacturing process, classified in three major categories: cell-substrate derived, culture derived, and downstream derived. Impurities derived from cell substrates include host cell proteins, nucleic acid (host cell generic, vector, total DNA), lipids, polysaccharides, and viruses. For host cell proteins, a sensitive immunoassay capable of detecting a wide range of protein impurities is generally utilized. The polyclonal antibodies utilized in the test are generated from a crude preparation of a mock production organism (i.e., a production cell minus the product coding gene). Another common strategy in many companies is to develop a multiproduct enzyme-linked immunosorbent assay (ELISA) for all products derived from a particular cell type (proprietary assay), or to use a commercially available "generic" ELISA. The level of DNA from host cells can be detected by direct analysis of the product using DNA hybridization techniques or a threshold total DNA assay (121). DNA spiking experiments may be performed at laboratory scale in order to validate efficient removal of DNA. Internationally accepted specifications for impurities are not established, except for DNA, which has a WHO specification (122).

Safety Testing

Most biopharmaceutical drugs are delivered parenterally, which requires final products to be tested for mycoplasma, sterility, and pyrogenicity. Pyrogenicity testing may be replaced by limulus amoebocyte lysate testing for endotoxin according to the U.S. Pharmacopoeia (USP) or European Pharmacopoeia. The General Safety test is another safety test specific to biologics and is intended to detect any unexpected or unwanted biological reactivity with a product. The assay encompasses inoculation of guinea pigs and mice with the final formulated product. The final product lot will pass the test if no unforeseen reactions occur and no weight loss takes place during the test (123).

In addition to final product testing, good manufacturing practices are applied during processing to avoid any contamination with adventitiously introduced materials, not intended to be part of the manufacturing process, such as biochemical/chemical materials and/or microbial species (124). Special requirements are applied to products derived from mammalian cell cultures to avoid any contamination by viruses. These include a combination of testing and validation of the downstream purification process for removal and inactivation of adventitious viruses and viruses intrinsic to the cell line (125,126). Specific guidelines address these virus safety aspects for cell culture–derived products (127,128).

ESTABLISHING STABILITY-INDICATING ANALYTICAL METHODS FOR FORMULATION DEVELOPMENT AND STABILITY TESTING OF BIOPHARMACEUTICAL PRODUCTS

An extensive physicochemical characterization of the protein, as described above, forms the basis for the development of a stable formulation (129). The physicochemical properties of the protein and its behavior in different solutions, as well as the purpose and application of its in vivo use, will guide the choice of formulation. The strategy for development of protein formulations is discussed in more detail in Chapter 6. Analytical results from product characterization and first experiences concerning product stability obtained during process development provide a good basis for the selection of stability-indicating test methods. In many cases, these tests are product-specific assays such as those measuring potency or activity, or cover specific features unique to an individual protein. In addition, some general tests are performed and some general requirements for the scope of stability-indicating test methods for biopharmaceuticals can be defined.

Stability-indicating test methods should detect the most common degradation forms of biopharmaceuticals: inactive or denatured protein, soluble and insoluble aggregates, proteolytically truncated forms, and chemical modifications such as hydrolysis, deamidation, oxidation, disulfide exchange, β-elimination, and racemization. A more detailed discussion of degradation pathways of proteins is given in Chapters 2 and 3 (130). A summary of the most common degradation pathways and examples of methods used to detect degradation products is given in Table 1.

Table 1 Common Degradation Routes[a] and Methods[b] Applied to Detect
Degradation Products

Degradation rule	Region affected/ results	Major factors	Method
Aggregation	Whole protein; reversible or irreversible self-association	Shear, surface area, surfactants, pH, T, buff strength	Size exclusion chromatography, light scattering, analytical ultracentrifugation, field-flow fractionation
Deamidation	Asn or Gln; acidic product, isoform, or hydrolysis	pH, T, buffers, ionic strength	Isoelectric focusing, IEC, native electrophoresis, capillary electrophoresis, reversed-phase HPLC, mass spectrometry (following chemical cleavage or enzymatic digestion)
Cleavage	Asp-X; fragments (proteolysis also possible from trace proteases)	pH, T, buffers	N- and C-terminal sequencing, size exclusion chromatography, reversed-phase HPLC (peptide map), SDS-polyacrylamide gel electrophoresis, Isoelectric focusing, mass spectrometry
Oxidation	Met, Cys, Hid, Trp, Tyr; oxidized forms	Oxygen (ions, radicals, peroxide), light, pH, T, buffers, metals, (surfactants), free radical scavengers	Reversed-phase HPLC (peptide map), hydrophobic interaction chromatography, amino acid analysis, mass spectrometry (following chemical cleavage or enzymatic digestion)

(Continued)

Table 1 Common Degradation Routes[a] and Methods[b] Applied to Detect Degradation Products (*Continued*)

Degradation rule	Region affected/ results	Major factors	Method
Thiol disulfide exchange	Cys; mixed disulfides; intermolecular or intramolecular	pH, T, buffers, metal, thiol scavengers	Reversed-phase HPLC (±reduction), reversed-phase HPLC (peptide map), SDS-polyacrylaminade gel electrophoresis, mass spectrometry (following chemical cleavage or enzymatic digestion)
Altered secondary structure	Whole protein	Shear, surface area, surfactants, pH, T, buffers, ionic strength	Far-UV circular dichroism, Fourier transform infrared spectroscopy, Raman spectroscopy
Altered tertiary structure	Whole protein	Shear, surface area, surfactants, pH, T, buffers, ionic strength	Near-UV circular dichroism, UV absorption spectroscopy, fluorescence spectroscopy

[a]This table lists degradation pathways commonly observed for proteins and peptides. The listing is not comprehensive, however, and many of these degradation routes may occur independently or in combination with another.
[b]Methods that are frequently used for analysis of alteration.
Abbreviations: HPLC, high performance liquid chromatography; SDS, sodium dodecyl sulfate; UV, ultraviolet.

We now discuss these common degradation pathways, emphasizing the analytical techniques useful for monitoring changes in proteins during stability testing. These methods will be integral to the design and optimization of stable formulations.

Protein denaturation refers to a disruption of the higher-order structure, such as secondary and tertiary structure of a protein. Denaturation, which may be reversible or irreversible, can be caused by thermal stress, extremes of pH, and exposure to interfaces or denaturing chemicals. Denaturation typically involves unfolding of the protein. In some cases, the unfolded protein can be transformed back to its native state by using denaturants such as guanidine hydrochloride or urea, followed by dialysis (131). If the protein cannot easily recover its native state by refolding, denaturation is considered to be irreversible. In many cases, this leads to aggregation and precipitation phenomena.

Changes in the secondary and tertiary structure of proteins can be monitored by spectroscopic methods, such as UV-CD and FT-IR, as described earlier in this chapter. In some cases, unfolded proteins can be separated from their native forms based on differences in chromatographic behavior [e.g., separation by hydrophobic interaction chromatography (HIC) due to an altered hydrophobicity pattern of the protein]. Another useful technique for investigating changes in protein conformation as on exposure to specific environments is differential scanning calorimetry (DSC). As a protein is heated, the transition from the native state to the unfolded state is accompanied by the appearance of an endothermic peak on DSC.

The transition temperature, T_m, is analogous to the melting of a crystal and is affected by the environmental conditions (e.g., pH) and the presence of pharmaceutical excipients. Sugars such as glucose and sucrose and polyols such as sorbitol and glycerol have been found to increase the denaturation temperature of proteins (132,133). A pH dependency of the endothermic peaks in DSC has been shown for several proteins (83,134). For rt-PA, the melting temperature in phosphate buffer was found to be about 66°C. In the presence of arginine, which stabilized the protein in solution, the T_m shifted to 71°C (135). Figure 2 contrasts a DSC thermogram for rt-PA with a melting temperature of 69.87°C, with a scan for an rt-PA variant having a significantly lower T_m (64.20°C). The thermogram for the humanized mAB shown in Figure 3 exhibits multiple transition temperatures for the heavy and light chains in the antibody.

Aggregate formation is one of the most common forms of protein instability (136). Insoluble and soluble aggregates must be distinguished. Soluble aggregates can be detected and quantitated by gel filtration chromatography. They do not necessarily lead to opalescence or turbidity of the protein solution.

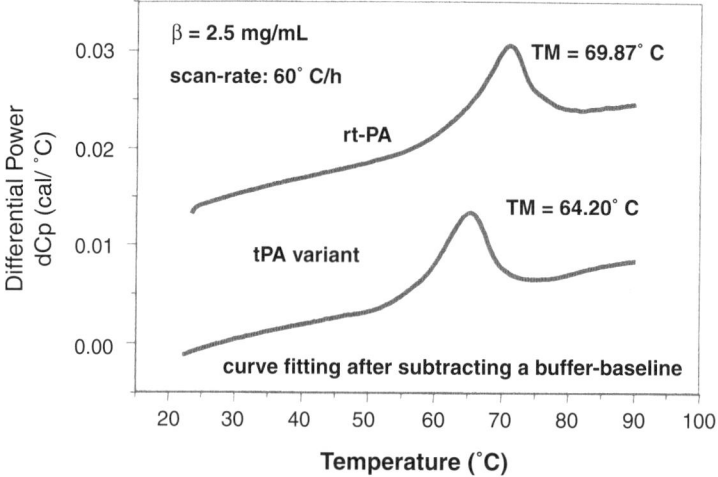

Figure 2   Differential scanning calorimetry of rt-PA and an rt-PA variant.

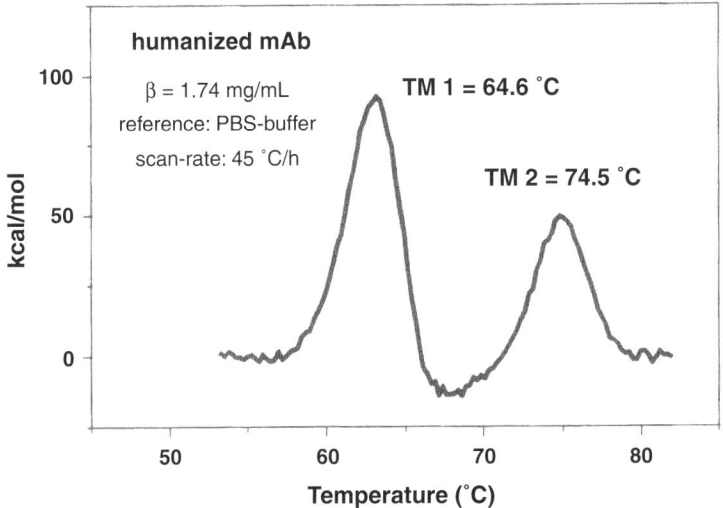

Figure 3   Differential scanning calorimetry of a humanized monoclonal antibody.

In contrast, insoluble aggregates consist of large protein particles that do not enter gel filtration columns and manifest themselves in the form of haziness or opalescence in a solution that is intended to be clear. Turbidity or opalescence in protein solutions can be caused by small amounts of insoluble aggregates, often less than 1%. All final product lots of biopharmaceuticals that are used as parenterals are inspected for appearance and clarity. A spectrophotometric determination of the opalescence of a protein solution in comparison with reference preparations as defined by the European Pharmacopoeia serves as an objective standard of evaluation for the presence of insoluble aggregates. To obtain accurate values when one is quantitating the concentration of a protein solution in the presence of opalescence, protein content measurements made by UV-absorption spectroscopy need to be corrected for increased absorbance, due to light scattering (137).

Insoluble protein aggregates can also be detected and quantitated as particles by light-scattering techniques. Particulates in injectables are assessed according to the USP Particulate Matter test and need to meet the acceptance criteria (<600 particles >25 μm and <6000 particles >10 μm). Photon correlation spectroscopy (PCS) is the most useful technique for particle size analysis of submicrometer particulates, having a range of application from a few nanometers (corresponding to the size of proteins in solution) to a few micrometers (138). PCS is a light-scattering technique for the measurement of the statistical intensity fluctuations in light scattered from the particles. These fluctuations are due to the random Brownian motion of the particles, which are size dependent. The diffusion coefficients of a protein can be determined from an autocorrelation function of the PCS, allowing particle size assignments. As with other light-scattering techniques, the analysis of PCS data becomes more complex when the sample particles are not monodisperse. However, PCS has been used to study the hydrodynamic size of proteins and their dependence on pH (139). The method is suitable for the quantitation of monomers and dimers and has been used to elucidate the refolding process of proteins (140). The measurement of monomers and dimers of proteins is disturbed by small amounts of aggregates, which can be detected at a very high sensitivity by PCS.

In addition to considering aggregation of the protein-active ingredient in a formulation, it is important to keep in mind that excipients in a formulation are also capable of aggregation. Proteins formulated at low concentrations may interact with container/closure systems, resulting in losses by absorption (141). In some cases, stabilizers such as human serum albumin (HSA) or surfactants are added to prevent such absorption effects. In one case described for an unstable HSA/dextrose solid formulation at elevated temperatures, however, a broadening and shift of the HSA peak together with a decrease in retention time in RP-HPLC was observed (Fig. 4). This was accompanied by a loss in mobility on SDS-PAGE gels (not shown) and an increase in the molecular weight of HSA as determined by MS analysis (Fig. 5). MS analysis of tryptic peptides of the HSA revealed a modification of lysine residues by a glycosylation of the ϵ amino group (142). These results strongly suggest an interaction between the HSA and dextrose excipients in the formulation at temperatures above 25°C. The development of HSA-free

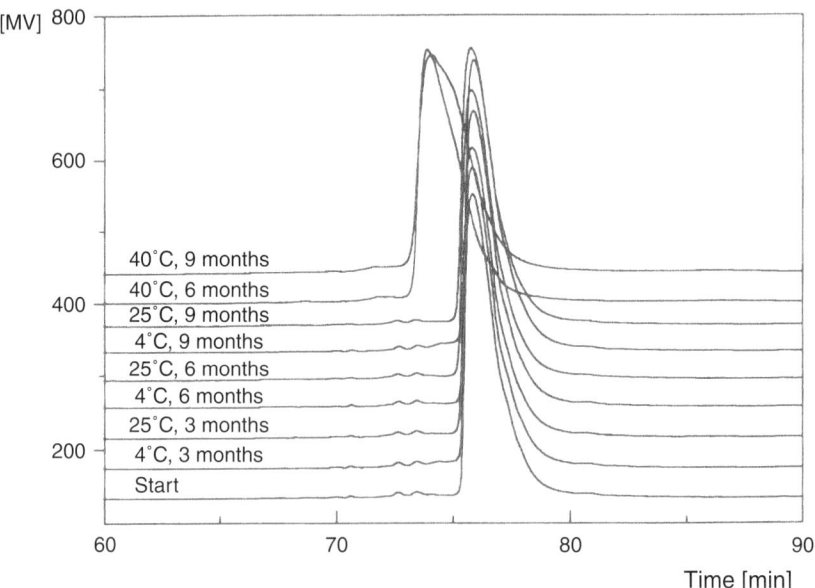

Figure 4 Stability profile of HSA in a solid dextrose formulation monitored by RP-HPLC.

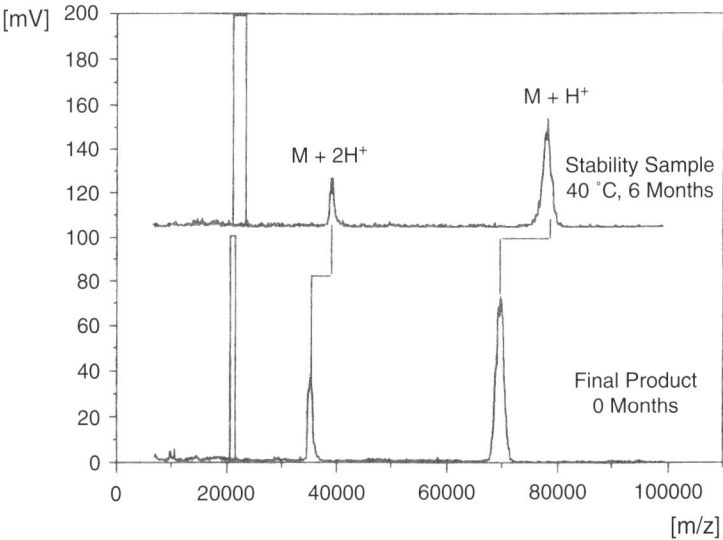

Figure 5   MALDI TOF mass spectroscopy of HSA in a dextrose formulation, before and after storage for 6 months at 40oC.

formulations is recommended even for low-dose biopharmaceuticals, however, because HSA is less well defined and less pure compared to recombinant proteins and additionally bears the potential risk of virus contamination due to the source of multidonor blood pools. Such protein excipients may also undergo the same degradation pathways as the therapeutic protein, introducing the need for stability testing here as well.

Another degradation pathway for proteins is hydrolysis through enzymatic and nonenzymatic routes, resulting in the cleavage of a peptide bond. Cleavage products can be visualized by SDS-PAGE using sensitive silver-staining techniques (143–145). Silver-staining techniques are mainly used when sensitivity is a major issue, such as in the detection of impurities in pharmaceutical protein preparations. However, the band intensities are not linearly proportional to the amount of protein loaded and the silver-binding properties of proteins (146). Thus, quantitation of silver-stained gels is problematic. When silver staining is used, the detection limit for proteolytic degradation products of proteins is in the range of 200 to 1000 parts per million for individual bands on a gel. Staining of SDS gels with Coomassie blue is 10 times less sensitive, but more suitable for quantitative measurement of degradation products by densitometric scanning of the gels.

Figures 6 and 7 illustrate the results of the quantitative measurement of degradation products during stability testing of an mAB by densitometric scanning of SDS gels. In Figure 6, the densitometric scan profiles from SDS-PAGE gels and main peaks of the heavy and light chains of a reduced antibody are shown, as well as peaks from degradation bands that increase during storage time. Figure 7 is the

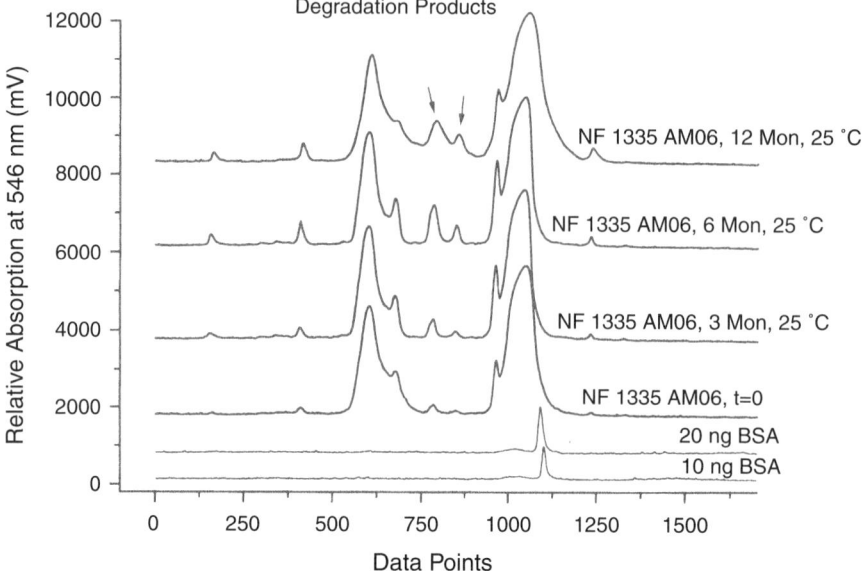

Figure 6 Stability testing of a monoclonal antibody preparation by SDS-PAGE and densitometric scanning of the gel.

graphic representation of the combined peak areas for the heavy and light chains and their changes as a function of storage time and temperature.

Proteolytic processing of proteins can also be measured by chromatographic techniques such as gel filtration chromatography. Rt-PA has a predominant proteolytic cleavage site at amino acid position 275–276 and is readily cleaved by

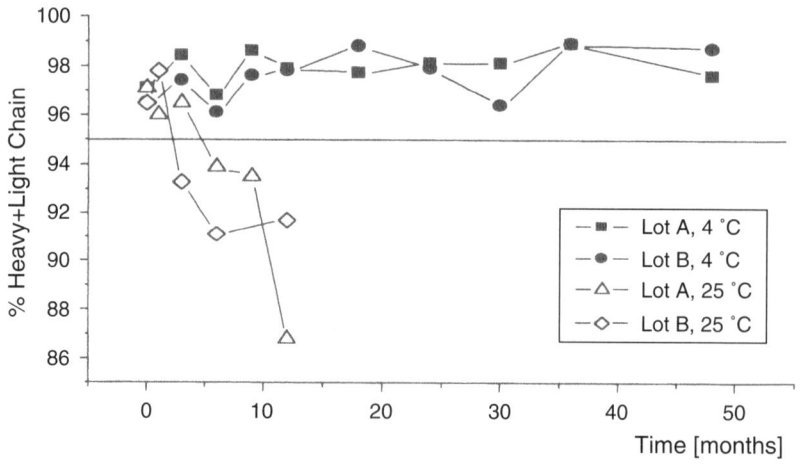

Figure 7 Densitometric scan evaluation of SDS-PAGE used for stability testing of monoclonal antibody preparations after storage at two different temperatures.

plasmin into two-chain rt-PA. The ratio of single-chain to two-chain t-PA can be determined by HPLC–gel permeation chromatography performed under reducing conditions. Rt-PA produced under serum-free conditions by mammalian cell culture exists predominantly in the single-chain form.

During long-term storage in liquid formulation, this molecule will be cleaved into the two-chain form, depending on the storage temperature and pH of the formulation buffer. Gel permeation chromatography is applied for stability testing of rt-PA for both quantitation of aggregates and for quantitation of the truncated two-chain form of the protein (Fig. 8).

In addition to proteolytic cleavage due to the presence of trace amounts of proteases, the peptide bond can undergo nonenzymatic hydrolysis, resulting in protein degradation. The Asp-Pro peptide bond is known to be the most susceptible to hydrolytic breakdown. The preferential hydrolysis of certain Asp-Pro bonds in a protein structure may be due to greater accessibility of the group (147) or to the location of these bonds adjacent to other amino acids that may influence the hydrolytic reaction (148). Degradation products or truncated forms of proteins generated by hydrolysis are analyzed by the same set of methods applied to truncated forms generated by proteolysis.

Oxidation is another of the major chemical degradation pathways of peptides and proteins, both in solution and in lyophilized formulations. Amino acids

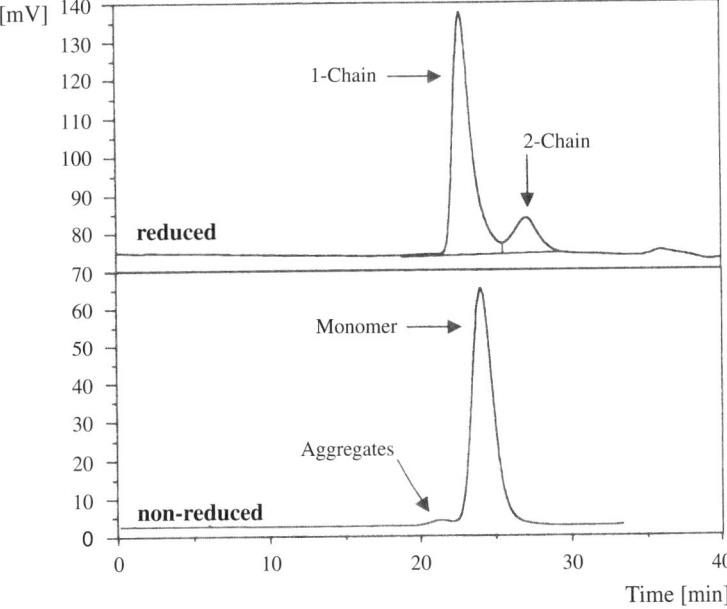

Figure 8   HPLC-SEC analysis of rt-PA under non-reduced and reduced conditions.

that may undergo oxidation include methionine, cysteine, histidine, tryptophan, and tyrosine. Most oxidation reactions commonly encountered in therapeutic proteins under normal storage conditions involve methionine and/or cysteine residues. Methionine is easily oxidized, even by atmospheric oxygen, to methionine sulfoxide. The oxidation of methionine is catalyzed by trace amounts of peroxide or metal ions, which may be present as contaminants from the manufacturing process or in trace amounts in excipients (54,131,149). The oxidation of methionine and cysteine is influenced by the three-dimensional structure of the protein. Residues that are buried in the interior of the protein are inaccessible to oxidation but can become reactive upon unfolding of the protein (113). Oxidized variants of proteins can be detected and characterized by liquid chromatographic techniques, such as RP-HPLC and HIC or by peptide mapping, followed by MS analysis. An example of an RP-HPLC separation of product variants including oxidized forms is shown in Figure 9.

Cysteine amino acid groups are also easily oxidized to yield cysteine disulfides. During long-term storage, free sulfhydryl groups may be oxidized to form intrachain or interchain disulfide linkages. Such interchain bonds between multiple protein molecules lead to protein aggregation (150). Under thermal stress, a protein will often undergo a destruction of the disulfide bonds by β-elimination from cysteine residues, resulting in free thiols that may contribute to other degradation pathways (131). The generation of free thiols by β-elimination may, in turn, catalyze disulfide interchange. Disulfide interchange may also result from the presence of unpaired cysteine residues. These cysteine residues can react at different sites to form new disulfide bridges, resulting in proteins with incorrect disulfide linkages and nonnative conformation (64,73). Confirmation of the correct linkage of disulfide bridges in a protein is a monumental analytical undertaking

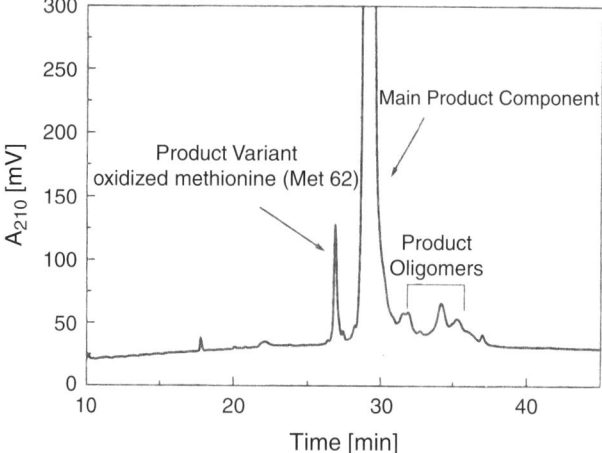

Figure 9   RP-HPLC separation of product variants.

and requires cleavage of the protein under reducing and nonreducing conditions and characterization of the isolated peptides by LC-MS to assign the cysteines involved in the disulfide bonds. Improperly folded proteins derived from disulfide interchange may be separated by high-resolution chromatographic techniques, such as RP-HPLC, HIC, size-exclusion chromatography (SEC), or IEC.

Nonenzymatic deamidation is a very common hydrolytic reaction responsible for degradation of peptides and proteins involving the amide group of asparaginyl or glutaminyl (Asn and Gln) residues. Asn and Gln residues are labile at extremes of pH and may be hydrolyzed easily to free carboxylic acids (Asp and Glu, respectively). Deamidation may have significant effects on protein bioactivity, half-life, conformation, aggregation, and/or immunogenicity. These effects must be evaluated on a case-by-case basis, since deamidation does not always affect bioactivity or the half-life of the product. However, deamidation results in a change in the original primary amino acid sequence of the protein, which may then be more susceptible to irreversible aggregation and more rapid clearance (136). Deamidated variants of proteins can be detected and quantitated by IEF followed by densitometric scanning or by IEC. Figure 10 illustrates the changes observed by IEF for an mAB stored for several months at 25°C. With increased storage time, there is a shift in the IEF pattern toward lower pH values, which would be consistent with a deamidation mechanism. IEC is also an especially simple and powerful technique to resolve and quantitate charge variants introduced by deamidation.

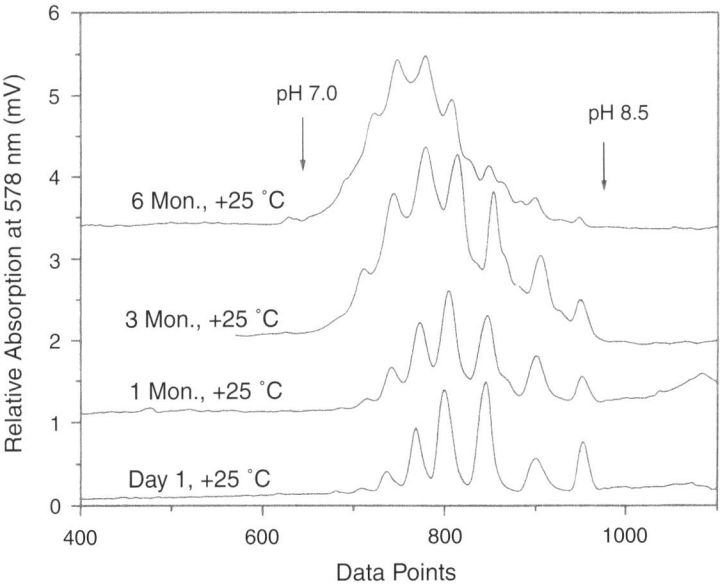

Figure 10 IEF profiles by densitometric scan analysis of a monoclonal antibody preparation, stored at 25°C in liquid formulation.

Once methods capable of detecting these common forms of protein degradation have been identified, they must be validated according to current guidelines for specificity, linearity, accuracy, recovery rate, sensitivity, precision, and limit of detection (151). The stability-indicating nature of each individual test method must be established for each individual protein. This process yields a specific stability profile for a given biopharmaceutical. We often determine whether a method is stability indicating by accelerating the degradation of a protein by exposure to elevated temperatures. Figure 11, for example, illustrates the temperature-dependent decrease in binding potency of an mAB in a competitive specific binding assay compared to binding of the reference standard.

One of the first goals of early formulation studies is to establish the degradation profile for a protein by accelerated stability testing, employing high temperatures, pH extremes, shear forces due to agitation, aeration, and other stresses. Based on the knowledge of all degradation pathways for a protein, strategies for optimization of the formulation can be applied. Figure 12 illustrates the results of accelerated stability testing for an mAB at 65°C and 70°C. At both temperatures, aggregation is first indicated by the occurrence of opalescence in the solution, followed by other parameters such as a decrease in monomer content (increase of soluble aggregates), decrease in binding activity and, only much later, decrease in protein content due to significant precipitation. This pattern is consistent with the observation that the most common degradation product for many proteins at elevated temperatures is the formation of aggregates. In a study of rt-PA, stability was assessed at 25°C for about one month in a 0.2 M arginine phosphate buffer

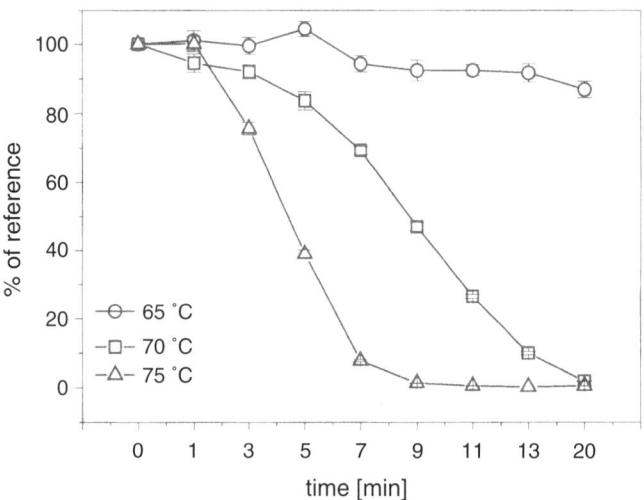

Figure 11 Accelerated stability testing of a monoclonal antibody preparation at different temperatures: competitive specific binding (% of reference) at three different temperatures.

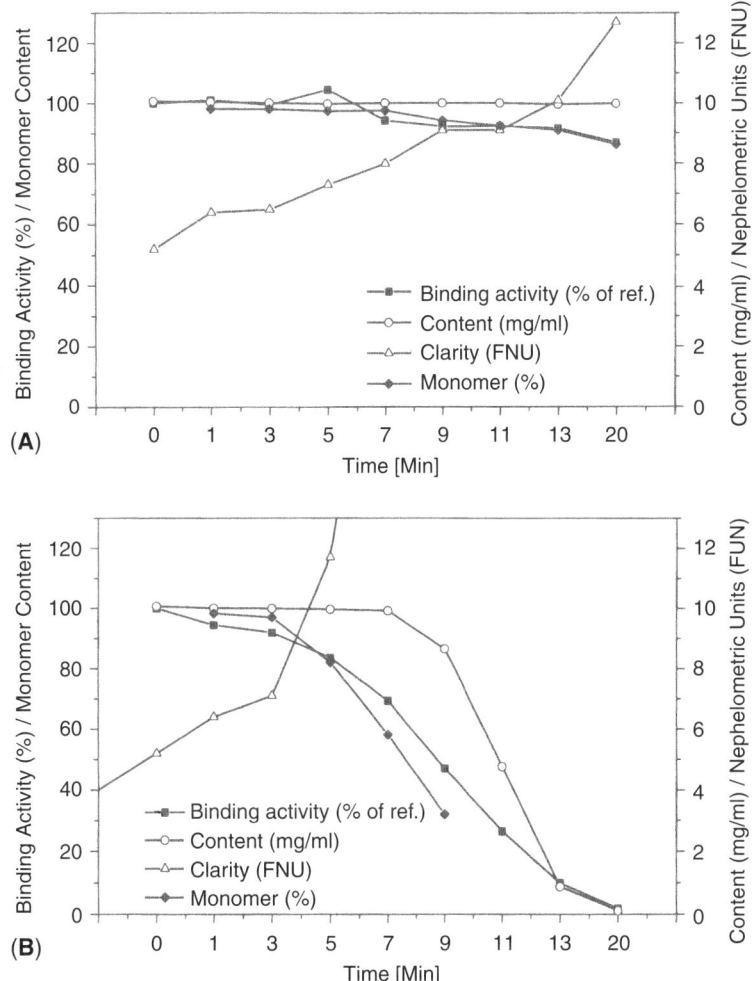

Figure 12 Accelerated stability testing of a monoclonal antibody at (**A**) 65°C and (**B**) 70°C.

system at various pH values. The stability-indicating test methods employed were HPLC-SEC for monomer content, HPLC-SEC with detergent in the running buffer for quantitation of two-chain rt-PA, and the clot lysis assay to measure potency. The pH optimum for rt-PA in this formulation was at pH of 6.0; a decrease in monomer content and in single-chain rt-PA occurred at higher pH values, whereas a decrease in clot lysis activity occurred at lower pH values (152).

Analytical methods are applied in the initial screening of formulation candidates that have been identified as indicative for the stability profile of a given protein pharmaceutical, reflecting the major degradation pathways discussed earlier. Based on screening studies with several formulation candidates

at accelerated temperatures and on further optimization of formulation compositions, one or two candidates are chosen for long-term real-time stability studies.
The overall design and scope of such a stability study as well as parameters to be
tested are described in the tripartite guideline of the International Conference on
Harmonization (153); an annex to this document deals with the specific requirements for biotechnological/biological products (154).

SUMMARY

This chapter describes the analytical methods used to characterize protein drugs
and support formulation development and stability testing. The presentation of
methods focuses on the criteria an analytical method must meet to be stability
indicating and useful in screening formulations and in formal stability programs.
A brief discussion of the manufacturing process is provided because the process
for production of a biopharmaceutical can have a strong influence on the resulting
characteristics and stability profile of the protein. The importance of producing a
consistent product by strict control and validation of fermentation and purification
processes is discussed. An extensive physicochemical characterization of the proteins forms the basis for choosing specific methods for formulation development,
as well as for product release and stability testing. The stability-indicating methods should detect the most common degradation forms of protein products. The
selection of methods used for an individual protein will be determined, at least
in part, by the properties of the protein and the design criteria for the formulation. The common degradation pathways of proteins are discussed, with emphasis
on analytical techniques useful for monitoring these changes in proteins during
stability testing. The understanding of the principles of these analytical methods,
the determination of factors controlling the major degradation pathways, and the
selection of appropriate strategies for minimizing such degradation are prerequisites for the development of stable formulations.

ACKNOWLEDGMENTS

The analytical data shown as examples in the figures were generated in the quality
control laboratories of Biopharmaceutical Manufacture, Boehringer Ingelheim,
Biberach, and were kindly provided by Dr. Michael Schlüter, Dr. Stefan Bassarab,
Dr. Jochen Wallach, and Dr. Kristina Kopp. I also thank my analytical colleagues
for many fruitful discussions on these topics.

REFERENCES

1. Lawrence S. Nat Biotechnol 2005; 23:1466.
2. Pavlou A, Reichert J. Nat Biotechnol 2004; 22:1513–1519.
3. Werner RG. Clinical Pharmacology of Biotechnology Products. Amsterdam:
    Elsevier Science Publishers, 1991:3–22.

4. Butler M. Appl Microbiol Biotechnol 2005; 68:283–291.
5. Chen M et al. Med Res Rev 2005; 25:343–369.
6. Jarvis D. Virology 2003; 310:1–7.
7. Gerngross TU. Nat Biotechnol 2004; 22:1409–1414.
8. Li H. Nat Biotechnol 2006; 24:210–215.
9. Kopp K, Schlüter M, Werner RG. In: Townsend RR, Hotch-kiss AT Jr, eds. Techniques in Glycobiology. New York: Marcel Dekker, 1997:475–489.
10. Note for Guidance on Biotechnological/Biological Products Subject to changes in their Manufacturing Process. CPMP/ICH/5721/03, CHMP, December 2004.
11. Guideline on Comparability of Medicinal Products containing Biotechnology-derived Proteins as Active Substance—Quality Issues CPMP/BWP/3207/00 Rev.1, CPMP, December 2003.
12. Allgaier H. Chimia 1994; 48:464–466.
13. Werner RG, Noe W. Arzneim-Forsch/Drug Res 1993; 43(11):1134–1139.
14. Werner RG, Noe W. Arzneim-Forsch/Drug Res 1993; 43(11):1242–1249.
15. Berthold W, Walter J. Biologicals 1994; 22:135–150.
16. The United States Pharmacopeia. Biotechnology-derived articles. Rockville, MD: U.S. Pharmacopeial Convention, Inc., 1994:1849–1859.
17. Chirino AJ, Mire-Sluis A. Nature Biotechnol 2004; 22:1383–1391.
18. Harris RJ et al. J Chrom B 2001; 752:233–245.
19. Jones AJS. Adv Drug Delivery Rev 1993; 10:29–90.
20. Kopp K, Schlüter M, Werner RG. In: Techniques in Glycobiology. Townsend RR, Hotchkiss AT Jr, eds. New York: Marcel Dekker, 1997:475–489.
21. Hancock WS, Bishop CA, Hearn MTW. Anal Biochem 1979; 92:170–173.
22. Nguyen DN et al. Protein mass spectrometry: applications to analytical biotechnology. J Chromatogr 1995; A705:21–45.
23. Perkins M et al. Determination of the origin of charge heterogeneity in a murine monoclonal antibody. Pharm Res 2000; 17:1110–1117.
24. Wang L et al. Structural characterization of the maytansinoid–monoclonal antibody immunoconjugate, huN901–DM1, by mass spectrometry. Protein Sci 2005; 14:2436–2446.
25. Chloupek RC, Harris RJ, Leonard CK, et al. J Chromatogr 1989; 463:375–396.
26. Dougherty JJ, Snyder LM, Sinclair RL, Robbins RH. Anal Biochem 1990; 190:7–20.
27. Edman P, Begg C. Eur J Biochem 1967; 1:80–91.
28. Jones AJS, Garnick RL. In: Lubniecki AS, ed. Large-Scale Mammalian Cell Culture Technology. New York: Marcel Dekker, 1990:543–566.
29. Gibson BW, Yu Z, Gillece-Castro B, Aberth W, Walls FC, Burlingame AL. In: Hugli TE, ed. Techniques in Protein Chemistry. Vol. 3. San Diego: Academic Press, 1989:135–151.
30. Hernandez P, Müller M, Appel RD. Mass Spectrometry Rev 2006; 25:235–254.
31. Havel HA, Chao RS, Haskel RJ, Thamann TJ. Anal Chem 1989; 61:642–650.
32. Strickland EH. CRC Crit Rev Biochem 1974; 2:113–175.
33. Chen YH, Yang JT, Chau KH. Biochemistry 1974; 13:3350–3359.
34. Johnson WC. Annu Rev Biophys Chem 1988; 17:145–166.
35. Pearlman R, Nguyen TH. Biochemistry 1991; 10:824–830.
36. Bewley TA. Recent Prog Hormone Res 1979; 35:155–213.
37. Jones AJS, O'Connor JV. In: Gueriguian JL, Bransome ED Jr, Outschoorn AS, eds. Hormone Drugs. Rockville, MD: U.S. Pharmacopeia, 1982:335–351.

38. Davio SR, Hageman MJ. In: Wang JY, Pearlman R, eds. Stability and Characterization of Protein and Peptide Drugs: Case Histories. New York: Plenum Press, 1993:59–90.
39. Susi H, Byler DM. Methods Enzymol 1986; 130:290–311.
40. Surewicz WK, Mantsch HH, Chapman D. Biochemistry 1993; 32:389–394.
41. Arrondo JL, Muga A, Castresana J, Goni FM. Prog Biophys Mol Biol 1993; 59:23–56.
42. Prestrelski SJ, Arakawa T, Carpenter JF. Arch Biochem Biophys 1993; 303:465–473.
43. Van der Weert M et al. Fourier transform infrared spectrometric analysis of protein conformation: effect of sampling method and stress factors. Anal Biochem 2001; 297:160–169.
44. Remmele RL Jr et al. Differential scanning calorimetry: a practical tool for elucidating stability in liquid pharmaceuticals. Biopharm Eur 2000; 56:58–60.
45. Kronman MJ, Holmes L.G. The fluorescence of native, denatured, and reduced-denatured proteins. Photocem. Photobiol 1971; 14:113–134.
46. Burstein EA, Vedenkina NS, Ivkova MN. Photchem Photobiol 1973; 18:263–279.
47. Montreuil J. Pure Appl Chem 1975; 42:431–477.
48. Montreuil J. Adv Carbohydr Chem Biochem 1980; 37:157–223.
49. Tsuda E, Kawanishi G, Ueda M, Masuda S, Sasaki R. Eur J Biochem 1990; 188:405–411.
50. Wolle J, Jansen H, Smith LC, Chan L. J Lipid Res 1993; 34:2169–2176.
51. Parekh RB, Dwek RA, Rudd PM, et al. Biochemistry 1989; 28:7670–7679.
52. Elbein AD. Trends Biotechnol 1991; 9:346–352.
53. Berman PW, Lasky LA. Trends Biotechnol 1985; 3:51–53.
54. Steer CF, Ashwell G. Prog Liver Dis 1986; 8:99–123.
55. Takeuchi M, Inoue N, Strickland TW, et al. Proc Natl Acad Sci USA 1989; 86:7819–7822.
56. Wright A, Morrison SL. Effect of C2-associated carbohydrate structure on IgG effector function: studies with chimeric mouse-human iGG1 antibodies in glycosylation mutants of Chinese hamster ovary cells. J Immunol 1998; 160:3393–3402.
57. Shields RL et al. Lack of fucose on human IgG1 N-Linked oligosaccharide improves binding to human FcgRIII and antibody-dependent cellular toxicity. J Biol Chem 2002; 277:26733–26740.
58. Jefferis R. Glycosylation of human IgG antibodies: relevance to therapeutic applications. Biopharm 2001; 14:19–26.
59. Goochee CF, Monica T. Biotechnology 1990; 8:421–427.
60. Goochee CF, Gramer MJ, Anderson DC, Bahr JB, Rasmussen JR. Biotechnology 1991; 9:1347–1355.
61. Gawlitzek M, Conradt HS, Wagner R. Biotechnol Bioeng 1995; 467:536–544.
62. Borys MC, Linzer DIH, Papoutsakis ET. Biotechnol Bioeng 1994; 43:505–514.
63. Borys MC, Linzer DIH, Papoutsakis ET. Biotechnol 1993; 11:720–724.
64. Gramer MJ, Goochee CF. Biotechnol Prog 1993; 9:366–377.
65. Gramer MJ, Schaffer DV, Sliwkowski MB, Goochee C. Glycobiol 1994; 4:611–616.
66. Hardy MR, Townsend RR. Proc Natl Acad Sci USA 1988; 85:3289–3293.
67. Townsend RR, Hardy MR. Glycobiol 1991; 1:139–147.
68. Mechref Y, Novotny MV. Structural investigations of glycoconjugates at high sensitivity. Chem Rev 2002; 102:321–369.
69. Sheeley DM et al. Characterization of monoclonal antibody glycosylation: comparison of expression systems and identification of terminal a-linked galactose. Anal Biochem 1997; 247:102–110.

70. Lowry OH, Rosebrough NJ, Farr AL, Randall RJ. J Biol Chem 1951; 193:265–275.
71. Bradford MM. Anal Biochem 1976; 72:248–254.
72. Wetlaufer DB. Adv Protein Chem 1962; 17:303–390.
73. Wang YJ. In: Avis KE, Lieberman HA, Lachman L, eds. Pharmaceutical Dosage Forms: Parenteral Medications. New York: Marcel Dekker, 1992:283–319.
74. Winder AF, Gent WLG. Biopolymers 1971; 10:1243–1252.
75. Gill SC, Hippel PH. Anal Biochem 1989; 182:319–326.
76. Bewly TA. Anal Biochem 1982; 123:55–65.
77. Harnes BD, Rickwood D, eds. Gel Electrophoresis of Proteins: A Practical Approach. Oxford: IRL Press, 1981.
78. Kleparnik K, Boocek P. J Chromatogr 1991; 569:3–42.
79. Righetti PB. Isoelectric Focusing: Theory, Methodology, and Applications. Amsterdam: Elsevier, 1983.
80. Righetti PG. In: Creighton TE, ed. Protein Structure: A Practical Approach. Oxford: IRL Press, 1989:23–63.
81. Righetti PG. Immobilized pH Gradients: Theory and Methodology. Amsterdam: Elsevier, 1990.
82. Vesterberg O, Svensson H. Acta Chem Scand 1966; 20:820–834.
83. Nguyen TH, Ward C. In: Wang JY, Pearlman R, eds. Stability and Characterization of Protein and Peptide Drugs: Case Histories. New York: Plenum Press, 1993:91–134.
84. Canova-Davis E, Teshima GM, Kessler TJ, Lee PJ, Guzzetta AW, Hancock WS. Am Chem Soc Symp 1990; 434:90–112.
85. Pearlman R, Nguyen TH. In: Marshak D, Liu D, eds. Therapeutic Peptides and Proteins: Formulation, Delivery, and Targeting. Plainview, NY: Cold Spring Harbor Laboratory, 1989:23–30.
86. Good DL. Capillary electrophoresis of proteins in a quality control environment. Capillary Electrophoresis of Proteins and Peptides, Methods in Molecular Biology. New Jersey: Human Press Inc, 2004:276.
87. Wehr T, Rodriguez-Diaz R, Zhu M. Capillary Electrophoresis of Proteins, Chroatography Science Series. New York: Marcel Dekker, Inc., 1999:80.
88. Henry MP. In: Oliver RWA, ed. HPLC of Macromolecules: A Practical Approach. Oxford: IRL Press, 1989:9–125.
89. Sluyterman LA Ae, Elgersma O. J Chromatogr 1978; 150:17–30.
90. Sluyterman LA Ae. Trends Biochem Sci 1982; 7:168–170.
91. Reynolds JA, Tanford C. J Biol Chem 1970; 245:5161–5165.
92. See YP, Jackowski G. In: Creighton TE, ed. Protein Structure: A Practical Approach, Oxford: IRL Press, 1989:1–21.
93. Shapiro AL, Vinuela E, Maizel JV. Biochem Biophys Res Commun 1967; 28:815–820.
94. Weber K, Osborne M. J Biol Chem 1969; 244:4406–4412.
95. Laemmli UK. Nature 1970; 27:680–685.
96. Neville DM Jr. J Biol Chem 1971; 246:6328–6334.
97. Harnes BD. In: Harnes BD, Rickwood D, eds. Gel Electrophoresis of Proteins: A Practical Approach. Oxford: IRL Press, 1981:1–91.
98. Gooding KM, Regnier FE. In: Gooding KM, Regnier FE, eds. HPLC of Biological Macromolecules. New York: Marcel Dekker, 1990:47–75.
99. Andrews P. Methods Biochem. Anal 1970; 18:1–53.
100. Whitaker R, Determination of molecular weights of proteins bygel. Anal Chera, 1963 35:1950–53.

101. Van Liedekerke BM, Nelis HJ, Kint JA, Vanneste FW, De Leenheer AP. J Pharm Sci 1991; 80:11–16.
102. Townsend MW, DeLuca PP. J Pharm Sci 1991; 80:63–66.
103. Smith RD, Loo JA, Loo RRO, Busman M, Udseth HR. Mass Spectrom Rev 1991; 10:359–451.
104. Bristow AWT. Mass Spectrometry Rev 2006; 25:99–111.
105. Maquin F, Schoot BM, Devaux PG, Green BN. Rapid Commun Mass Spectrom 1991; 5:299–302.
106. Hillenkamp F, Karas M. Methods Enzymol 1990; 193:280–294.
107. Karas M, Bahr U, Giessmann U. Mass Spectrom Rev 1991; 10:335–357.
108. Beavis RC, Chait BT. Proc Natl Acad Sci USA 1990; 87:6873–6877.
109. Groesbeck MD, Parlow AF. Endocrinology 1987; 120:2582.
110. Gery I, Gershon RK, Waksman BH. J Exp Med 1972; 136:128–142.
111. Gery I, Davies P, Derr J, Krett N, Barranger JA. Cell Immunol 1981; 64:293–303.
112. Gu L, Fausnaugh J. Pharmaceutical biotechnology. In: Wang YJ, Pearlman R, eds. Stability and Characterization of Proteins and Peptid Drugs: Case Histories. Vol. 5. New York: Plenum Press, 1993:221–248.
113. Stewart WE II. The Interferon System. Vienna: Springer Verlag, 1981:13–25.
114. Finter NB. Interferon assays and standards. In: Finger NB, ed. Interferons. Amsterdam: North Holland, 1966:87.
115. Ranby M. Biochim Biophys Acta 1982; 704:461–469.
116. Collen D, Tytgat G, Verstaete M. J Clin Pathol 1968; 21:705–707.
117. Gaffney PJ, Curtis AD. Thromb Haemostasis 1985; 53:134–136.
118. Carlson RH, Garnick RL, Jones AJS, Meunier AN. Anal Biochem 1988; 168:28–435.
119. Jones AJ, Meunier AM. Thromb Haemostasis 1990; 64:455–463.
120. Werner RG, Bassarab S, Hoffman H, Schlüter M. Arzneim Forsch/Drug Res 1991; 41(11):1196–1200.
121. Kung VT, Panfili PR, Sheldon FL, et al. Anal Biochem 1990; 187:220.
122. WHO Epidemiological Records. 1997; 72(20):143–145.
123. Schiff LJ, Moore WA, Brown J, Wisher H. BioPharm 1992; 5(5):36–39.
124. Hill D, Beatrice M. BioPharm 1989; 2(10):28–32.
125. Werner RG, Walter J. BioEngineering 1990; 5(90):14–19.
126. Walter J, Allgaier H. In: Hauser H, Wagner R, eds. Mammalian Cell Biotechnology in Protein Production. Berlin, New York: Walter de Gruyter, 1997; 451(1)–483.
127. CPMP/ICH/295/95. Q5A Viral Safety Evaluation of Biotechnology Products Derived from Cell Lines of Human or Animal Origin, 1997.
128. CPMP/ICH/294/95. Q5D Quality of Biotechnological/Biological Products: Derivation and Characterization of Cell Substrates Used for Production of Biotechnological/Biological Products, 1998.
129. Jones AJS. In: Cleland JL, Langer R, eds. Formulation and Delivery of Proteins and Peptides. Washington, DC: American Chemical Society, 1994:22–45.
130. Banga AK. Therapeutic Peptides and Proteins: Formulation, Processing and Delivery Systems. Lancaster, PA: Technomic Pub, 1995:61–80.
131. Manning MC, Patel K, Borchardt RT. Pharm Res 1989; 6:903–918.
132. Chang BS, Randall CS, Lee YS. Pharm Res 1993; 10:1478–1483.
133. Back JF, Oakenfull D, Smith MB. Biochemistry 1979; 18:5191–5196.
134. Vermuri S, Beylin I, Sluzky V, Stratton P, Eberlein G, Wang YJ. J Pharm Pharmacol 1994; 46:481–486.

135. Pearlman R, Nguyen T. J Pharm Pharmacol 1992; 44:178–185.
136. Cleland JL, Powell MF, Shire SJ. Crit Rev Thera Drug Carrier Syst 1993; 10:307–377.
137. Winder AF, Gent WLG. Correction of light-scattering errors in spectro-photometric protein determinations. Biopoly 1971; 10:1243–51.
138. Washington C. Particle Size Analysis in Pharmaceutics and Other Industries: Theory and Practice. New York: Ellis Horwood 1992:Chap. 7.
139. Nicoli DF, Benedek GB. Biopolymers 1976; 15:2421–2437.
140. Cleland JL, Wang DIC. Biochemistry 1990; 29:11072–11078.
141. Geigert J. J Parenter Sci Technol 1989; 43(5):220–224.
142. Wallach J, Schlüter M, Werner RG. Stable formulations for biopharmaceuticals. In: Duchene D, ed. Recent Advances in Peptide and Protein Delivery. Minutes of the 8th International Pharmaceutical Technology Symposium, 1997.
143. Oakley BR, Kirsch DR, Morris NR. Anal Biochem 1980; 105:361–363.
144. Merril CR, Goldman D, Sedman SA, Ebert MH. Science 1981; 211:1437–1438.
145. Morrissey JH. Anal Biochem 1981; 117:307–310.
146. Poehling HM, Neuhoff V. Electrophoresis 1981; 2:141–147.
147. Schrier JA, Kenley RA, Williams R, et al. Pharm Res 1993; 10:933–944.
148. Kenley RA, Warne NW. Pharm Res 1994; 11(1):72–76.
149. Powell MF. In: Cleland JL, Langer R, eds. Formulation and Delivery of Proteins and Peptides. Washington, DC: American Chemical Society, 1994:100–117.
150. Arakawa T, Prestrelski SJ, Kenney WC, Carpenter JF. Adv Drug Delivery Rev 1993; 10:1–28.
151. ICH Guidance Validation of Analytical Procedures. Target and Methodology, CPMP/ICH/381/95–1CHQ2 (Res).
152. Hoffmann H, Bassarab S, Schlüter M, Werz W, Werner RG. Stability testing of biopharmaceuticals. Paperback APV, Band 32, Stability Testing in the EC, Japan and the USA. Stuttgart: Wissenschaftliche Verlag, 1993:245–272.
153. International Conference on Harmonization. www.emea.europa.eu/htms/human/ich/ichquality.htm. CPMP/ICH/2736/99. Aug. 2003.
154. International Conference on Harmonization. Quality of biotechnological products: stability testing of biotechnological/biological products (Topic Q5C). CPMP/ICH/138/95. July 1996.

5

Preformulation Development of Protein Drugs

Frank K. Bedu-Addo

Integritas Drug Development Consulting, LLC, Cincinnati, Ohio, U.S.A.

INTRODUCTION

Preformulation is typically the first step in characterizing and developing a drug product. Very often, a protein identified through genomics or proteomics may go into preclinical development prior to having a full understanding of its mechanism of action. Due to the various potential degradation pathways of proteins, understanding and overcoming their inherent physical and chemical instabilities becomes one of the major challenges faced by the formulation scientist. The various studies and drug development activities performed with the goal of overcoming the protein's inherent instabilities constitute the formulation development process. Developing a product that is stable, safe, efficacious, and marketable will be the focus of the formulation scientist's efforts.

A successful formulation process typically begins with the preformulation step. In preformulation development, the main goal is to perform the basic studies that provide a good understanding of the physico–chemical characteristics of the protein molecule. Such a study will also involve understanding the effects of various pH conditions and excipients on both the physical and the chemical stability of the molecule, as well as their effects on solubility. The goal of the formulation scientist at this stage of the development process should be to gain a good feel of what formulation conditions would be most suitable to maintain the stability and biological activity of the protein, and what conditions should be avoided. This study upon completion should also provide a good feel for what excipients and stabilizers might require further evaluation in the next stage of development, which is the actual formulation study.

Many proteins are only biologically active in their native conformations. Change in conformation can lead not only to changes in biological activity, but also to changes in the stability of the molecule. A protein that undergoes unfolding could expose previously hidden hydrophobic residues, thereby leading to hydrophobic interactions with other molecules or surfaces, resulting in both aggregation and surface adsorption respectively. It therefore becomes important for the formulation scientist to also obtain an understanding of the impact of various conditions on the biophysical characteristics of the protein and the resulting potential impact on both stability and biological activity. The use of biophysical characterization tools in preformulation development will be discussed further.

One of the key decisions to be made in a drug development program is very often the timing of preformulation studies. For the formulation development scientist, initiation of formal preformulation studies will typically begin upon the availability of purified material typical of that to be used in biodistribution, toxicology, and early-stage clinical studies. Very often, even earlier-stage preformulation studies can be very useful in developing the cell culture and purification processes. This early preformulation will typically be performed using protein obtained from an experimental process, and the data then fed back to the process development groups for use in developing a more robust process.

Separation and purification of recombinant proteins is a critical element of modern bioprocessing, and currently represents the major manufacturing costs (1). Advances and cost reduction in recombinant proteins will depend on several factors including advances in molecular biology and proteomics. Such advances will also depend heavily on innovations in the optimization of purification processes, and the participation of the preformulation scientist in determining and understanding basic protein behavior. pH–solubility and pH–stability relationships and the specific effects of these factors on protein conformation, stability, and even adsorption behavior should be well understood prior to developing the process. In addition to separation processes, material handling and transfer as well as in-process hold steps can significantly impact product quality and yield. Selection of the right process conditions therefore becomes very important. The need for deterministic models that combine the chromatographic behavior of proteins with their biophysical and structural behavior is very important in designing efficient purification processes (1). An example of how preformulation studies utilizing biophysical characterization were applied to efficient purification process development will be presented in the case studies section of the chapter.

During the preformulation stage, it is important for the formulation scientist to anticipate the intended mode of delivery of the protein as well as the intended dosage form, concentration, and administration dose. For example, is the protein being encapsulated in a delivery system? In such a case, it may have to be exposed to organic solvents, and the effects on stability should be evaluated. Is it going to be lyophilized? In such a case, the effects of freeze–thaw, various stabilizers, and concentration should be evaluated. For products that have to be formulated and administered at high concentrations, e.g., some antibodies, understanding the solubility profile and other characteristics of the molecule that could impact the quality attributes of the formulation, such as the tendency towards self-association, become critical at the early preformulation stage.

Statistical design is a very useful tool that is underutilized in the formulation of biopharmaceutical products. Utilizing statistical design tools during the preformulation stage provides the formulation scientist with the ability to elucidate interactions that may occur between the parameters being evaluated. Very often, understanding these interactions between factors under investigation, for example, pH and ionic strength, may be the key to identifying the true effect of the various parameters on the stability, conformation, and biological activity of a protein. The statistical design approach, as opposed to the typical one-factor-at-a-time (OFAT) approach provides contrast of averages and therefore provides statistical power towards estimating the effects of factors being investigated (2). OFAT methods require replicating tests or studies to provide equivalent statistical power. As the number of factors under investigation increases, the number of replicates required with the typical approach increases significantly, thus impacting drug development time. OFAT approaches are not suitable for understanding interactions. The statistical design approach also provides a larger area or volume of space from which inferences regarding the product can be made. An example of how statistical design was used in preformulation studies of a Pegylated recombinant protein will be discussed in the case studies section of the chapter.

It is important to remember that the essential goal of the preformulation study is to determine the inherent characteristics of the protein molecule, and therefore enable the development of a rational strategy for formulating a safe, biologically active, and stable product, coupled with an efficient and economical manufacturing process. For this to be achieved, the formulation scientist, the analytical chemist, and the process engineer should begin to collaborate at a very early stage in product development.

UNDERSTANDING THE CRITICAL QUALITY ATTRIBUTES OF THE PROTEIN

Proteins are susceptible to degradation by several potential degradation pathways, e.g., oxidation, deamidation, deglycosylation, aggregation, disulfide scrambling, cleavage, adsorption, and denaturation. As a result, a number of analytical methods are usually required to obtain a full understanding of protein stability behavior. At the beginning of the preformulation study, the following characteristics of the protein should be known: molecular weight, primary structure or primary amino acid sequence, and protein charge. It is also important to understand at this stage, the native aggregation and biologically active state of the protein, which could be monomeric, dimeric, or oligomeric. Protein primary structure will also provide some insight into the protein's susceptibility to chemical changes such as oxidation and deamidation, and disulfide bond formation (3).

The complex nature of protein structure and behavior also requires that any changes in the secondary and tertiary structure of the protein under the various conditions be well understood. The secondary structure of a protein refers to the arrangement of the individual amino acids along the protein backbone, often resulting in very specific and well-defined structures such as α-helices and β-sheets. Tertiary structure is the three-dimensional arrangement of the protein molecule. The tertiary structure is formed as a result of the manner in which the specific secondary structural elements interact with each other and side chains to form stable domains.

ANALYTICAL AND BIOPHYSICAL METHODS IN PREFORMULATION

In order for the preformulation study to be adequately performed, it is critical that the right analytical tools be in place to enable physical and chemical changes in the molecule to be detected and characterized. At this early preclinical development stage, it is, however, not necessary for the analytical methods to be fully validated according to the International Conference Harmonization guidelines. However, tests should be performed to ensure that the analytical and preformulation scientists are confident in the data being generated by the particular method. For example, a chromatographic method being developed to quantify the protein should be evaluated to determine the suitable linearity range. A method being developed to separate and identify a particular degradation product should be evaluated to determine the specificity of the particular method, and the ability to efficiently separate the intact from the degraded product. This is often referred to as a qualification of the method.

Protein Charge and Isoelectric pH

Protein surface charge may be altered as a result of chemical modifications occurring during storage or processing. The tendency of the molecule to undergo changes in surface charge should be evaluated as a part of the preformulation study. Causes of such changes may be chemical degradation or changes in protein glycosylation patterns, for example.

Electrophoretic Method

Isoelectric focusing (IEF) is a very useful electrophoretic technique that separates proteins by differences in surface charge under a pH gradient (4). In IEF, by applying a voltage, a pH gradient is established across a gel matrix by using a mixture of amphoteric substances dissolved in the gel (5). Low-pH values migrate towards the anode and high-pH values to the cathode. When proteins are subjected to the IEF system, they will migrate until a state of charge neutrality is attained. This is the isoelectric pH (pI) of the protein (4–6). The pH values across the gel are usually determined by running pI markers as standards in one lane of the gel.

Chromatographic Methods

The most common chromatographic methods for detecting changes in protein surface charge are reverse phase chromatography and ion exchange chromatography, both of which are discussed below.

Protein Size and Aggregation

Electrophoretic Method

Sodium dodecyl sulfate–polyacrylamide gel electrophoresis (SDS-PAGE) is a rapid method for the estimation of molecular weight, and for characterization of aggregation as well as cleavage or clipped products (6,7). After denaturing the protein with SDS by heating, the electrophoretic separation is carried out on polyacrylamide gel in a buffer containing SDS. The mobility of the protein is dependent on its molecular weight. The molecular weight of the protein is estimated by calibrating the gel with a standard, containing proteins of known size. Often, a reducing agent such as β-mercaptoethanol or dithiothreitol can be added to cleave disulfide bonds. The protein, once separated, can be visualized by various staining methods. Coomassie blue and silver staining are the most common (8–11). Within a set concentration range, Coomassie blue binds to proteins in a stoichiometric manner (4). As a result, laser densitometry can often be utilized to estimate the amount of protein in each band, thereby providing a quantitative analysis of protein aggregation or breakdown. By running both reduced and nonreduced gels, the formulator is able to determine whether or not observed aggregates are due to covalent bond formation.

Chromatographic Method

High-performance size exclusion chromatography (SEC), sometimes referred to as gel filtration or gel permeation chromatography, separates molecules in an aqueous solution based on size. This method allows for the characterization of both aggregates and breakdown products in the formulation. The principle of macromolecular separation in SEC is based on the fact that different-sized molecules diffuse into the column matrix to differing extents during their migration through the column (12,13). Smaller molecules enter the pores of the stationary phase or matrix more readily, and as a result, elute more slowly from the column than larger molecules. Various stationary phases utilized in such columns permit the diffusion of molecules into the matrix up to a specified molecular weight cutoff. In liquid chromatography, proteins eluting from the column are typically monitored using ultraviolet (UV) spectroscopy at two or more wavelengths. This practice of utilizing multiple wavelengths is useful in ensuring that the eluting peak is due to protein absorption rather than light scattering by insoluble aggregates.

Deamidation

Deamidation of proteins is a common chemical change that can lead to changes in the biological activity of the protein as well as its physical parameters (3,14).

The biological half-lives of proteins in certain instances have been correlated with their deamidation rates (14). In deamidation, the amide side chains of glutamine and asparagines deamidate to glutamate and aspartate residues. The rate of protein deamidation is known to be very dependent on the amino acid sequence next to the asparagines and glutamine side chains (3).

It is important, as part of the preformulation study, that the formulation scientist evaluates the effect of various potential formulation conditions such as pH, ionic strength, excipients, temperature, etc., on the susceptibility of the molecule towards deamidation. To do so, analytical methods that are capable of detecting the chemical change in the molecule must be applied.

Ion Exchange Chromatography

Deamidation of proteins is accompanied by a change in charge and increase of one unit in mass. A number of chromatographic and electrophoretic techniques can therefore be exploited in order to separate the deamidated fraction from the intact protein for characterization and quantification. Ion exchange chromatography is the most widely applied chromatographic technique for the isolation of deamidated proteins. Ion exchange chromatography separates molecules on the basis of charge. Hydrophobic and polar interactions may also contribute to the resulting retention or separation behavior. In ion exchange chromatography, the molecules to be separated are reversibly bound to the stationary phase of the column or ion exchanger by means of electrostatic interactions. The column is then eluted with buffers of varying pH or ionic strength. Elution of the bound molecule then occurs as a result of preferential binding of the buffer ions to the molecular binding sites of the ion exchanger. The ion exchange method, being selective for changes in net charge, is therefore optimal for the resolution of asparagine and deamidated products (aspartate and isoaspartate) (15,16), as well as for distinguishing aspartate from succinimide. (17). In ion exchange, separation is performed under nondenaturing conditions. The protein's native conformation can therefore be preserved, allowing for the evaluation of bioactivity and antigenic properties of the eluting proteins. A disadvantage of ion exchange in detecting deamidation is the fact that the formation of either a succinimide from an asparagine or an isoaspartate from aspartate does not result in a change in net charge and therefore resolution by ion exchange will not be possible.

Reverse-Phase Chromatography

Reverse-phase high-performance liquid chromatography (RP-HPLC) has been used successfully in certain cases to separate and resolve the deamidated protein from the intact molecule. Reverse phase chromatography discriminates among molecules mainly on the basis of their hydrophobic character (18,19). Hydrophobicity of the component amino acids is the key factor determining retention and separation behavior. In the presence of a polar mobile phase, protein retention on the reverse phase column is based on the strength of interactions occurring between the hydrophobic portion of the molecule, the surface of the hydrophobic stationary phase, and components of the eluting phase. With asparagine deamidation for example, it is possible in several cases to distinguish asparagine from succinimide as well as aspartate from isoaspartate by exploiting their subtle differences in hydrophobicity, leading to differences in retention behavior associated with the structural differences. In reverse phase chromatography, the "contact region," i.e., the amino acid residues interacting with the stationary phase, can be manipulated by varying the mobile phase conditions, usually the buffer pH as well as type and amount of organic modifier. Typical RP-HPLC conditions will normally result in partial to complete protein denaturation.

Hydrophobic Interaction Chromatography

Hydrophobic interaction chromatography (HIC) can be used to resolve deamidated and nondeamidated proteins. HIC is based on the adsorption of proteins onto a hydrophobic stationary phase in the presence of high concentrations of antichaotropic salts such as ammonium sulfate and sodium sulfate. Such salts tend to order the structure of water (20,21). As result of decreasing the interactions between the water and protein, these salts cause an increased interaction between the hydrophobic residues of the protein and the stationary phase. Protein elution is achieved by utilizing a gradient of decreasing salt concentration. HIC is carried out under fully aqueous conditions in contrast to RP-HPLC. Strongly hydrophobic proteins will sometimes require the addition of small amounts of organic solvent to the mobile phase. Fausnaugh et al. compared HIC with RP-HPLC of proteins (22). In HIC, the conditions are therefore much less denaturing compared to RP-HPLC, due to the lack of organic modifiers in the mobile phase and the use of lower density, less nonpolar stationary phases. Sample loading capacity is also relatively high.

Protein Oxidation

Proteins that contain free cysteine are very susceptible to oxidation. The thiol (SH) group of the cysteine is extremely reactive at neutral pH. These thiol groups can also undergo spontaneous oxidation to form disulfides (–S–S–) (3,23). The resultant oxidation product of two cysteine residues is known as cystine. Proteins containing both cysteine and cystine may undergo disulfide scrambling, resulting in a number of covalent isomers. The sulfur atom of methionine is also very susceptible to alkylation or reversible oxidation to a sulfoxide, which can be further oxidized irreversibly to a sulfone. Other less susceptible groups are the carboxyl groups of aspartic acid, glutamic acid, and the indole ring of tryptophan. Covalent aggregates resulting from intermolecular disulfide bonds can be observed and distinguished by using simple techniques such as a combination of reduced and nonreduced SDS-PAGE. Intramolecular disulfides that do not result in aggregates or conformational changes resulting in lack of altered surface charge are more efficiently characterized using sophisticated methods such as peptide mapping and mass spectrometry. Oxidation of methionine and other residues is most commonly characterized by means of reverse phase chromatography.

Hydrolysis and Proteolysis

Hydrolysis of peptide bonds within a protein may occur when the protein is subjected to harsh conditions such as extremes of pH or temperature (24). Hydrolysis can clearly result in reduced biological activity. Hydrolysis typically results in the formation of clipped or cleaved species with reduced molecular weight. Hydrolytic products can therefore be detected by the use of techniques that separate compounds based on molecular weight, such as SDS-PAGE and SEC. The identification of conditions that promote hydrolysis should be evaluated as part of the preformulation study.

The presence of proteolytic enzymes in the formulation may also lead to hydrolysis. Such enzymes could be introduced via bacterial contamination or during the recovery of recombinant proteins through the copurification from cell extracts. Manipulation of the purification conditions or the addition of protease inhibitors will be necessary to minimize such a problem.

Protein Concentration

UV spectroscopy is a rapid and convenient means of determining protein concentration during preformulation studies (25). The near-UV absorption spectrum of proteins has a maximum around 280 nm, with tyrosine and tryptophan being the major contributors to the spectrum. Tyrosine and tryptophan have absorptivities of 5700 and 1300/M/ cm respectively. Therefore the absorptivity of the protein can be calculated from the number of these residues in the protein sequence (26). Absorbance at 280 nm (A_{280}) = molar absorptivity (ε) × molar concentration (c) × light path length in centimeters. Beer's Law states that molar absorptivity is constant for a given substance dissolved in a given solute and measured at a given wavelength (27). As a result, molar absorptivities are known as molar absorption coefficients or molar extinction coefficients. Standard laboratory spectrophotometers are typically fitted for use with 1 cm path length cuvettes; hence the path length is usually assumed to be equal to 1. Therefore $c = A/\varepsilon$.

Protein concentration can also be evaluated using any of the quantitative liquid chromatography methods described above.

Biophysical Characterization Techniques

Biophysical characterization tools that can be used very effectively in the preformulation study will be discussed in this section, and further in the case studies. Light scattering, circular dichroism (CD), fluorescence spectroscopy, and microcalorimetry [high-sensitivity differential scanning calorimetry (DSC)] are very useful tools that can aid the formulation scientist in understanding the effect of potential formulation conditions on protein conformation and also on the conformational stability of the protein. The conformational stability refers to the propensity of the protein to undergo structural changes such as unfolding. Protein unfolding is usually the first step in the protein denaturation process leading to aggregation. Other methods that can also be used effectively are Fourier transform infrared spectroscopy (FT-IR) (28), to evaluate secondary structure, and ultracentrifugation, to determine the protein's aggregation state in solution. Ultracentrifugation is especially useful for high-concentration formulations in which reversible self-association could occur at high concentrations, but which can, however, not be detected under the low-concentration conditions that may be required for chromatographic analysis (29,30).

Dynamic Light Scattering

Most folded proteins assume a compact globular conformation. When a solution of such molecules is illuminated with a light source, the amount of light scattered by the molecule can be measured (31). This is done at a light wavelength at which the protein does not absorb. In dynamic light scattering, the dependence of the average intensity of the scattered light on solute concentration and the scattering angle is exploited to determine the molecular weight and dimensions of the solute. Powerful laser light sources currently available allow small changes in molecular size of a few nanometers to be accurately recorded.

Spectroscopy is a quick and useful tool that can be easily used by the formulation scientist during preformulation to evaluate the formation of insoluble aggregates in the formulation. Scattering of light by particulate matter in the visible range where the protein does not absorb will result in an absorbance reading and therefore signify the presence of particulate matter.

Circular Dichroism

CD measured in the far-UV region reflects protein secondary structure (protein backbone), while the spectrum in the near-UV region is related to the tertiary structure or the environment around the aromatic side chains. CD is a very useful tool in monitoring the properties of both the aromatic and the peptide chromophores of a protein during folding and unfolding events (32–34). The resulting circular dichroic absorbance or ellipticity is reflective of the extent of folding or unfolding of the protein. Folding or unfolding can therefore be monitored by observing changes in the ellipticity of the spectrum at specific wavelengths. A typical α-helical spectrum is characterized by two pronounced minima at 209 and 222 nm, and a maximum around 190 nm. In the case of β-sheet proteins, a peak is observed at 194 nm and a minimum at 217 nm. A typical random coil conformation exhibits a broad positive band at 218 nm and a negative band with a minimum at 197 nm.

Fluorescence Spectroscopy

Fluorescence spectroscopy is a rapid and useful tool for detecting conformational changes occurring within a protein (35). The aromatic side chains of tryptophan and tyrosine are strong fluorophores when exposed to UV-light wavelength in the vicinity of their absorption maxima, which exist between 280 and 290 nm. The emission spectra of tryptophan and tyrosine exhibit maxima around 250 and 303 nm respectively. Changes in the environment immediately surrounding these amino acids lead to significant shifts in both the spectra and the emitted radiation. In general, increasing the polarity of the environment surrounding the tryptophan and tyrosine, as will typically occur during unfolding, results in a decrease in fluorescence intensity. This increase in polarity will also typically result in a shift toward radiation of higher wavelengths.

High-Sensitivity DSC

High-sensitivity DSC (HSDSC) or microcalorimetry can be used to measure the energy changes that occur during protein unfolding under controlled increase in temperature (36–38). Proteins in solution exist in equilibrium between their native folded state and the denatured or unfolded state. Conditions resulting in a higher unfolding transition midpoint (T_m) when 50% of the biomolecules are unfolded, indicate a more conformationally stable molecule, and thus a more stable formulation. Therefore, a buffer or pH condition that results in a higher T_m could be said to impart greater conformational stability to the protein, suggesting that under those conditions, the protein would be more resistant to unfolding and hence denaturation. Comparison of T_m values under different conditions therefore provides a useful screening tool for the preformulation study. DSC is also used to determine the enthalpy (ΔH) of denaturation. In a DSC experiment, the protein solutions are typically heated from about 10°C to approximately 100°C. Parameters such as the peak widths of the unfolding transitions and also the peak shapes provide very useful information relating to the overall conformation of the protein (39). Heating the protein above the unfolding transition, cooling, and rescanning provides useful information on the reversibility of the unfolding transition under the particular formulation conditions.

Biological Activity

Biological activity is the most important property of a protein drug product. The method of measuring biological activity will depend on the protein and its intended therapeutic activity. These may be enzymatic assays, immunological assays, or actual in vivo tests.

At the preformulation stage, biological activity screening will not typically be performed on all samples screened. However, the formulation scientist should select a very limited number of representative samples (e.g., significantly different conformations, specific degradation product) in order to start generating some useful information on the relationship between protein conformation and stability and also between conformation and biological activity if possible.

FORCED DEGRADATION STUDIES: UNDERSTANDING POTENTIAL DEGRADATION MECHANISMS AND KEY DEGRADATION PRODUCTS

Forced degradation studies are typically performed as a part of the overall preformulation strategy to achieve two important goals. The first goal is to ensure that the analytical methods being used in the preformulation study can actually detect the various specific degradation products. The second goal is to obtain additional information on the potential degradation pathways of the particular protein.

The stress tests that are often utilized in the forced degradation studies and the conditions applied to achieve the specific physical or chemical changes are shown in Table 1.

At this stage, the various stressed samples should be tested using the developed chromatographic and spectroscopic methods, in order to ensure that the intended changes are actually observed. Identifying the chromatographic retention times and characteristics of the intact protein and the degradation products formed under the different stress conditions will allow for effective identification of degradation products that may be formed during the preformulation studies or early processing and storage. The analytical chemist should be very closely involved at this stage, and mass spectrometric/peptide map analysis of the samples should also be performed to ensure that the intended physical and chemical changes do actually occur. It is important for the developed or selected analytical methods to be effective in detecting the specific degradation products. Such methods, which effectively detect the various degradation products, can be described as being "stability indicating."

Peptide Mapping

Most proteins of biological interest, due to their large size, are enzymatically digested in order to cleave the polypeptide chains into smaller peptide fragments, which can then be separated chromatographically and efficiently analyzed by mass spectrometry (39,43–45). In order to facilitate digestion, the molecule is typically denatured by the addition of urea or guanidine hydrochloride. By evaluating the digests under both reduced and nonreduced conditions, the presence of covalent bonds such as disulfides and the exact location within the primary structure of the protein can be elucidated. The resulting peptides are then separated chromatographically. Trypsin and Lys C are probably the more specific of the most commonly used enzymes. Trypsin catalyzes the hydrolysis of lysl and arginyl peptide bonds except for lys-pro or arg-pro. Lys C breaks down the protein at the C terminus of Lys residues. The most common separation methods are

Table 1 Typical Stress Conditions Utilized in Preformulation Development

Physical/chemical change	Stress conditions applied to protein
Oxidation	3–4% hydrogen peroxide for 4 hr at room temperature (40)
Deamidation	pH 9–11 at room temperature for 24–48 hr (41,42)
Acid hydrolysis	pH 2.5 for 1–12 hr at room temperature (24)
Aggregation/denaturation	Expose to five freeze–thaw cycles
	Agitation for 72 hr

reverse phase and ion exchange chromatography. Peptide mapping will result in a "fingerprint" pattern of peptides unique to the particular protein. Tandem HPLC utilizing both reverse phase and ion exchange HPLC has proven to be a very efficient separation method for large proteins, which result in a large number of peptides upon digestion (46). The individual peptide peaks can then be efficiently identified using mass spectrometry.

PERFORMING THE PREFORMULATION STUDY

Once the formulation scientist has some representative protein material and analytical methods in place, the preformulation study can be performed. An important goal during this stage of the study is often to utilize as little material as possible.

Solubility Studies

The typical pH range over which protein solubility is usually evaluated is the physiologically relevant pH of 4 to 9. Understanding protein solubility behavior under various pH and ionic strength conditions is critical for both the formulator and the process development engineer. Most proteins have the lowest solubility at the pI, with solubility increasing both above and below the pI. This profile can however be altered in the presence of certain salts, due to interactions with the salt ions leading to a masking of protein charge (39). Protein conformational changes resulting in the exposure of previously hidden residues could also impact protein solubility. Ultrafiltration is a quick method for estimating protein solubility under various buffer and excipient conditions. With this approach, a membrane having the appropriate molecular weight cutoff is used to concentrate the formulation until some precipitation begins to occur. The formulation is then filtered and the protein concentration rapidly determined by means of UV spectroscopy at 280 nm if the extinction coefficient is known, or by chromatography. The disadvantage of ultrafiltration is the fact that in certain instances, super saturation may occur, resulting in an overestimation of protein solubility. Another commonly used approach is to concentrate the protein to the solubility limit under a known high-solubility condition, followed by dialysis into the desired buffer and excipients. The resulting formulation is then filtered and analyzed for concentration.

Stability Screening Studies

At the preformulation stage, the goal of the study is not to determine the exact degradation mechanisms, but rather to evaluate whether or not the drug will exhibit certain instabilities under selected formulation, storage, and potential processing conditions. The use of selected stability-indicating assays in this study as described above will provide useful information on the types of degradation products that are observed and could therefore potentially be observed during normal processing and handling of the product.

Protein supply may be severely limited at this stage. If so, the formulation scientist should perform a rational selection of the pH, ionic strength, and excipient conditions to be evaluated in this screening study. Selection of the conditions will be based on solubility of the protein, information from forced degradation, potential processing conditions, and additional information on the properties of the protein such as charge and pI. The concentration of protein selected should be based on the desired dosage if known, and also the detection limits of the analytical and biophysical methods to be used. Again, protein supply may be low, requiring these studies to be performed at very low protein concentration. Should sufficient material be available, more detailed studies including a statistical design approach may be utilized to gain further insight into the effects of various parameters on stability of the formulation.

Once the conditions to be analyzed have been determined, the selected formulations can be screened to evaluate their effects on the biophysical and structural characteristics of the protein, by means of DSC, CD, or fluorimetry.

The selected formulations will usually be screened by placing on accelerated stability for two to four weeks, and analyzed by analytical methods. The accelerated temperature condition should be carefully selected. DSC studies provide very useful information on the unfolding transition temperatures of the protein. Studies should normally be conducted at temperatures below the unfolding transition of the protein, preferably below even the onset of the unfolding transition. Higher temperatures will lead to changes in the overall conformation of the protein, which could significantly alter the degradation mechanisms. It is usually a good idea to select two to three temperature conditions under which to perform these studies. This approach enables an efficient screening process to be utilized. Formulations that do not show any difference in degradation at the low temperature may undergo further but differing extents of degradation at the higher temperature. Conversely, formulations that undergo excessive degradation at the higher temperatures may prevent effective discrimination between the formulations. Degradation may be less extensive at the lower temperatures, thus facilitating discrimination between the formulations.

The above-described studies will shed some light on the relationship between conformation and stability of the protein. If a quick biological activity assay is available, additional information on how these changes and conditions impact biological activity can also be evaluated.

Additional compatibility studies may be desirable, depending on the intended use or type of formulation. For example, it may be necessary to evaluate both the stability and the solubility behavior of the protein in the presence of various mixtures of aqueous and organic solvents if the protein will be exposed to similar conditions during processing. It may also be necessary to evaluate compatibility of the protein with potential intravenous solutions.

CASE STUDIES

Case Study 1

Preformulation Development of a Heavy Chain Fragment of Botulinum Serotype B: Identification of Suitable Purification Conditions

Purpose: The purpose of this preformulation study was to investigate the physico–chemical and structural characteristics of recombinant Botulinum serotype B under various conditions, and to utilize the information in developing robust purification process conditions (39).

Preformulation studies: The solubility of the protein was evaluated at pH 4, 5, 6, 7.5, 8, and 9 in the presence and absence of 150 mM NaCl (effect of ionic strength). This was done in order to understand the impact of protein exposure to various pH and ionic strength conditions. This information would then be utilized to select separation processes and conditions that would maximize protein integrity and stability, resulting in optimal protein yield and biological activity. Secondary structure of the protein under the above-stated pH and ionic strength conditions were evaluated using CD. Conformational stability of the protein was monitored using HSDSC. HIC, SEC-HPLC, SDS-PAGE, peptide mapping, and UV spectroscopy were used to monitor protein stability under the various pH and excipient conditions. Protein concentration was evaluated by UV spectroscopy at 280 nm, and the formation of insoluble aggregates was monitored by light scattering at 360 nm wavelength.

Results:

Solubility. Ionic strength was observed to have a significant impact on solubility. At high-pH conditions, the pH of minimum solubility was shifted to higher pH values. This phenomenon suggested a possible interaction between cations and the negatively charged protein (above the pI), possibly resulting in a masking of the surface electrostatic charges. In the absence of NaCl, solubility was as expected, having the lowest solubility at the pI of about 7.5. UV_{360} also indicated the presence of significant amounts of insoluble particulate matter in the high-pH, high-salt formulations upon storing at 30°C for one week. Negligible insoluble aggregate matter was observed in non–NaCl containing formulations.

Differential scanning calorimetry. The effect of pH and NaCl on the unfolding transition temperature T_m is shown (Fig. 1). The effect of pH and NaCl on the enthalpy (ΔH) of the unfolding transition was similar to the effect on T_m. T_m and ΔH both decreased steadily as pH increased, suggesting that the protein conformation became less stable as pH increased. This implied that the protein would therefore be more susceptible to unfolding and physical degradation. At a pH of 7.5 and greater, 150 mM NaCl led to a decrease in T_m, with the decrease being very significant at pH 9. Above pH 7.5, a similar distinct decrease in ΔH was observed (Fig. 2). This observed decrease in both the unfolding transition temperature and the energy associated with the unfolding transition suggested that

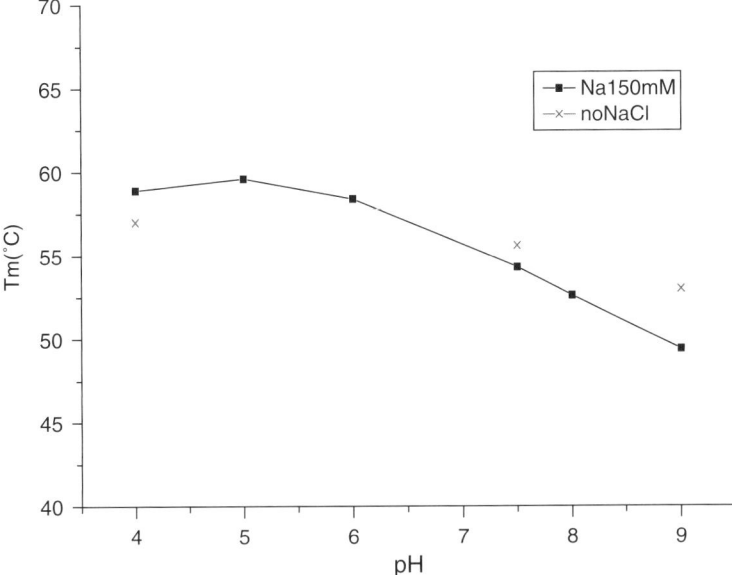

Figure 1 Plot of unfolding transition temperature T_m of the heavy chain fragment of the Botulinum serotype B protein versus pH.

at high-pH conditions, the presence of NaCl significantly destabilized the protein. The broadened peak widths at high pH also signified a more loosely packed (more unfolded) protein structure.

Circular dichroism. CD studies (Fig. 3) indicated that the protein consisted predominantly of β-sheets. The classical pattern typical of β-sheet structures with a single minimum at 216 to 218 nm wavelength (32–34) was observed. Spectral analysis by CD revealed identical secondary structures between formulations at pH 4 to 7.5. At pH 8 and 9, a reduction in the positive peak at 230 nm (aromatic side chains) and the negative peak at 216 nm (peptide bonds) was observed, signifying some unfolding of the protein backbone structure at high pH. By evaluating the secondary structure over a period of one week, protein conformation when stored under high-pH conditions was shown to change with time (unfolding occurred), contrary to the effect observed at low pH.

SDS-PAGE. The results of SDS-PAGE after one week at 30°C are shown (Fig. 4). A single band, indicating a homogenous population of the monomeric protein, was observed at pH 4 to 6 (not shown). At pH 7.5 and higher, several higher molecular weight bands were observed in the nonreduced gels. These high-molecular-weight bands were converted to monomer bands in the reduced gels, signifying that these aggregates were formed as a result of intermolecular disulfide bonds.

Size exclusion chromatography. SEC was performed only on the pH 4 and pH 9 formulations to complement the SDS-PAGE studies. The protein eluted

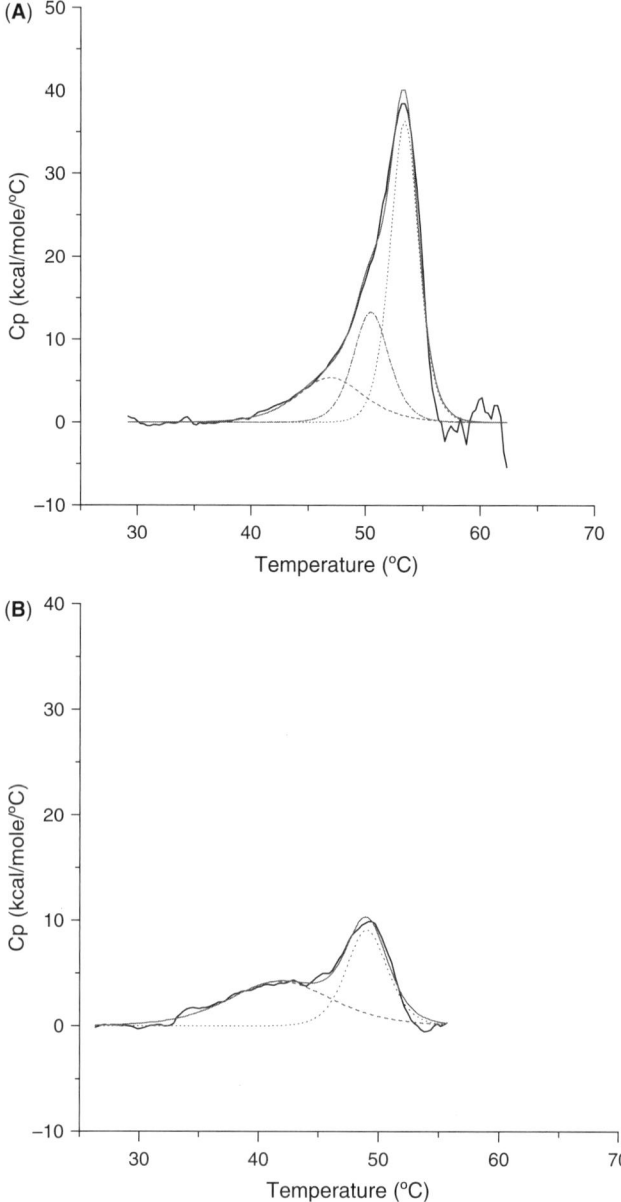

Figure 2 (**A**) Unfolding transition below the pI. (**B**) Unfolding transition above the pI with 150 mM NaCl. The solid lines represent the overall unfolding transitions obtained from the DSC scan, whereas the dotted lines are results of the deconvolution representing separate theoretical unfolding domains. *Abbreviations*: DSC, differential scanning calorimetry; pI, iso-electric pH.

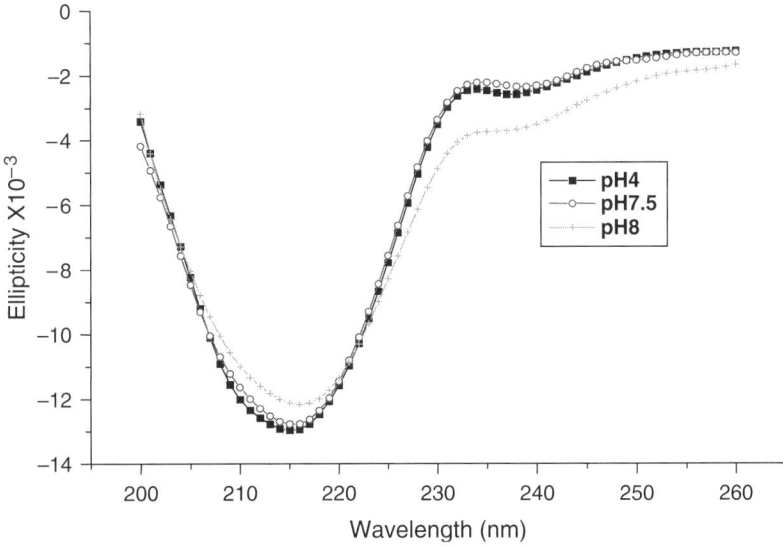

Figure 3 Effect of pH on secondary structure of the heavy chain fragment of the botulinum serotype B protein.

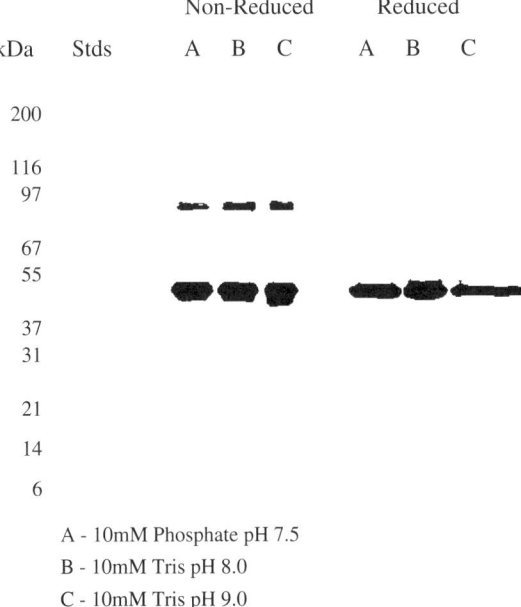

Figure 4 Reduced and nonreduced SDS-PAGE gels of the heavy chain fragment of the Botulinum serotype B protein at high pH. *Abbreviation*: SDS-PAGE, sodium dodecyl sulfate-polyacrylamide gel electrophoresis.

from the SEC column with an apparent molecular weight of 16 kDa. However, calculation of molecular weight using light scattering confirmed an eluting product of molecular weight 50 kDa. After storage at 30°C for one week, a single peak with a retention time of 18.8 minutes was maintained for the pH 4 formulation, confirming that degradation had not occurred. The pH 9 condition, however, provided evidence of a higher molecular weight fraction eluting at approximately 10 minutes.

Peptide mapping. The reverse phase chromatogram obtained by peptide mapping analysis of the protein is shown (Fig. 5). Lys C was used to digest the protein, resulting in 43 peptides. The four cysteine-containing peptides were identified. By utilizing mass spectrometry, peaks corresponding to three deamidated peptides, which showed a 1 Da mass increase, were also identified. In high-pH formulations, peptides L35 and L9 containing free cysteines showed slight decrease in intensity while new peaks corresponding to disulfide bonded peptides L35 to 42 (S2, 56.4 minutes), L9 to 38 (S4, 58.9 minutes) and L9 to 35 (S3, 57.9 minutes) appeared.

Conclusions: The preformulation study provided a very good understanding of the effect of pH on the solubility, as well as on the physical and chemical stability of the protein. Chemical and physical stability were found to be optimal at low pH. The preformulation studies suggested that long-term exposure to

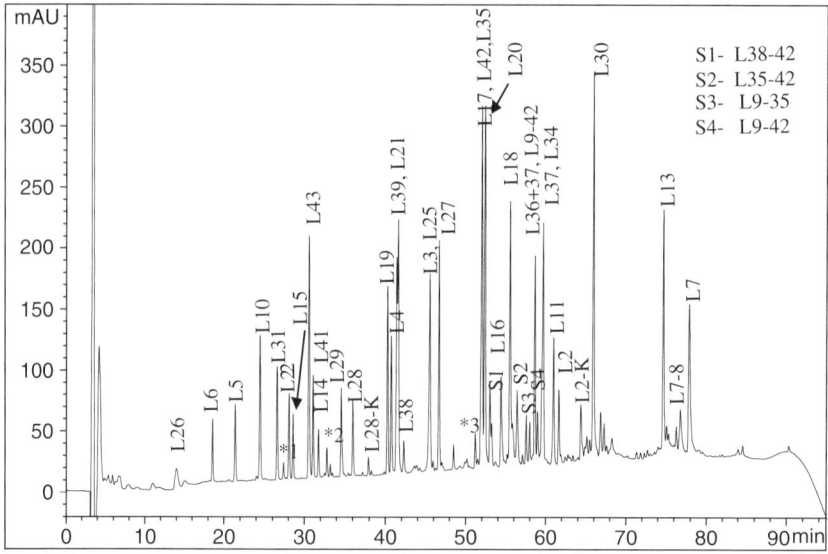

Figure 5 Peptide map showing resulting peptides of the digested heavy chain fragment of the Botulinum serotype B protein separated by reverse-phase HPLC chromatography. *Abbreviations*: SDS-PAGE, sodium dodecyl sulfate-polyacrylamide gel electrophoresis; HPLC, high-performance liquid chromatography.

high-pH conditions during processing would lead to a decrease in product yields due to potential losses from covalent and noncovalent aggregation. Increased deamidation would also be favored above pH 7.5. Conformational changes at high pH could also potentially negatively impact biological activity of the product, and it was also important to avoid high-pH, high-salt conditions. Biological activity of the protein was shown to be dependent on conformation. Any hold step during the process required the protein to be stored under low-pH conditions.

This approach of utilizing biophysical studies in addition to the traditionally utilized analytical methods was very useful in understanding how various conditions could affect the protein during the purification process. This preformulation study would subsequently help in the development of a robust purification process.

Case Study 2

Preformulation Development of Recombinant Pegylated Staphylokinase SY161 using Statistical Design

Purpose: The purpose of the study was to perform preformulation development of SY161 in order to understand the effects of three basic formulation factors influencing protein stability using a statistical design approach: pH (5–9), buffer strength (10–100 mM), and ionic strength (0–250 mM) (37). NaCl, which was used to evaluate ionic strength, also acts as a protein stabilizer by the mechanism of preferential hydration (47,48) of the solution. The pH range selected focused on the physiologically relevant pH. A wide enough range for buffer and salt concentrations was selected, such that the statistical design would provide an understanding of protein behavior within the entire selected range.

Preformulation studies: A central composite two-level factorial design was performed. Protein secondary structure was evaluated using CD. Stability towards protein unfolding was evaluated by means of HSDSC. Depegylation, aggregation (most significant forms of observed degradation), and protein loss were evaluated by storing the various formulations at 40°C for two weeks and evaluating by SEC with online light scattering. A central composite statistical design was utilized. The center point conditions were replicated six times in order to get an estimate of experimental or pure error. The design resulted in 20 formulations/runs with the various combinations of pH, buffer strength, and NaCl concentration. The statistical design and data analysis were performed using Stat-Ease software (2).

Results: The resulting CD spectra showed small differences in ellipticity, suggesting some differences in the extents of protein backbone folding. Evaluation of a single formulation (pH 7, 150 mM NaCl, 20 mM phosphate buffer) by FT-IR indicated that the protein secondary structure contains 0% α-helix, 50% β-sheet, 12% bend, 12% turns, and 21% random coil.

pH was observed to have the most significant impact on SY161 conformational and solution stability. The SEC chromatogram (Fig. 6) shows effective

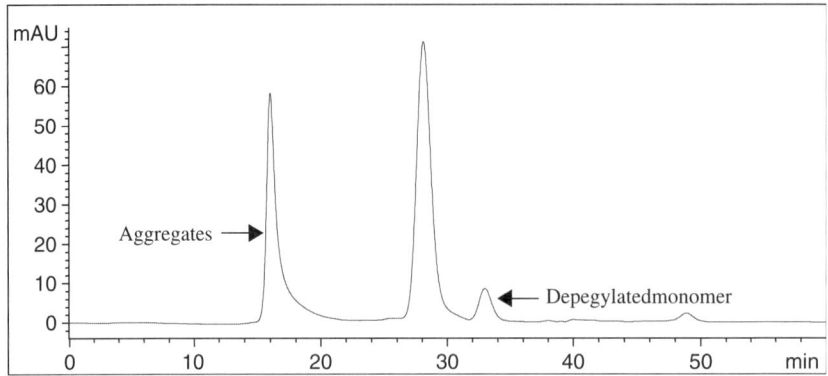

Figure 6 SEC chromatogram showing the Pegylated monomer separated from aggregates and depegylated monomer. *Abbreviation*: SEC, size exclusion chromatography.

separation of SY161, aggregates, and the depegylated monomer for SY161 stored at pH 8 in 77.5 mM Tris buffer and 187.5 mM NaCl for two weeks at 40°C. Molecular weights of the various separated fractions were confirmed by online light scattering. A significant interaction between pH and ionic strength on T_m was observed (Fig. 7). Negligible effect of ionic strength on T_m was observed at low pH in contrast to

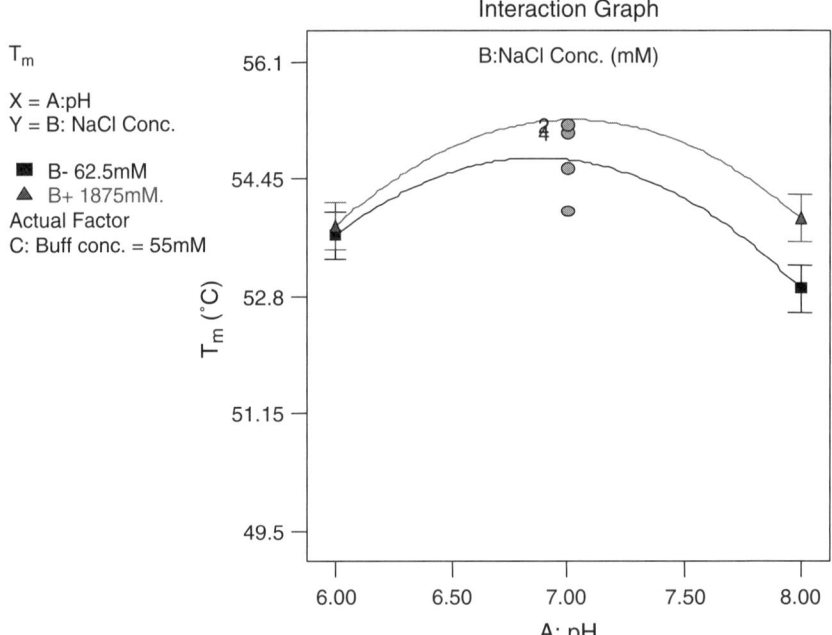

Figure 7 Interaction plot showing the interaction between pH and NaCl concentration on the conformational stability of SY161.

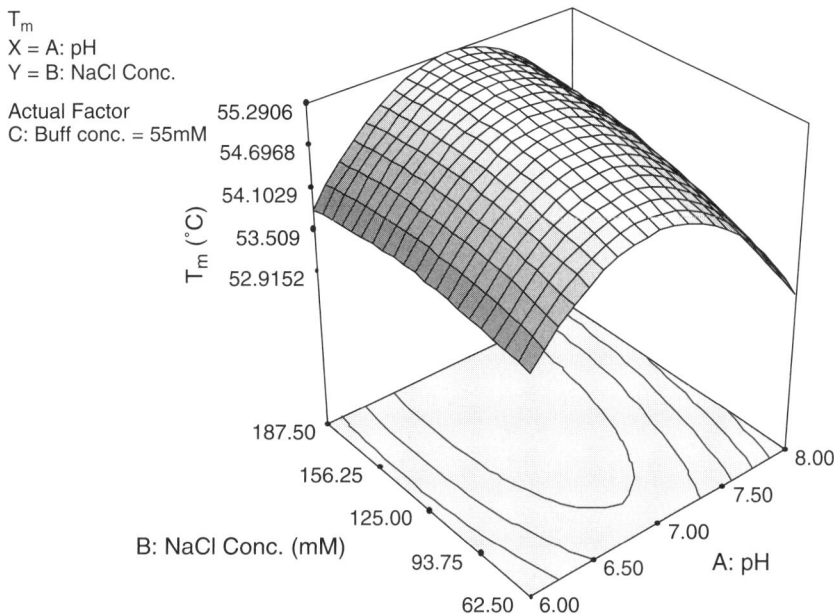

Figure 8 Three-dimensional response surface plot showing the effects of pH and NaCl concentration on conformational stability of SY161.

the effect of ionic strength on T_m at high pH. A response surface plot (Fig. 8) shows how T_m changes as a function of both pH and NaCl concentration. The statistical model obtained for the percentage of SY161 remaining in solution by SEC (solution stability) was virtually identical to that obtained for T_m (conformational stability). Under the conditions of the study, the conformational stability of SY161 was therefore seen to accurately predict the solution stability of the protein.

Conclusions: The statistical design approach enabled a full understanding of the effects of pH, NaCl, and buffer strength on SY161. Accurate models describing the effects of the various formulation factors on SY161 were also obtained. This study approach provided a wider inductive basis from which inferences regarding the effects of the various factors on SY161 could be drawn. SY161 was observed to be most stable at a pI of 7. At this pH, both buffer and NaCl concentration effects were significant. Stability towards unfolding increased as both NaCl and buffer concentrations were increased, resulting in optimal solution stability at pH 7.

REFERENCES

1. Hearn MTW, Anspach B. Chemical, physical and biochemical concepts in isolation and purification of proteins. In: Asenjo JA, ed. Separation Processes in Biotechnology. New York: Marcel Dekker, Inc., 1990:17–64.

2. Anderson MJ, Witcomb PJ. DOE Simplified: Practical Tools for Effective Experimentation. Portland, Orgeon: Productivity, Inc., 2000.
3. Oeswein JQ, Shire SJ. Physical biochemistry of protein drugs. In: Lee VHL, ed. Peptide and Protein Drug Delivery. New York: Marcel Dekker, Inc., 1991:167–202.
4. Righetti PG. Isolectric focusing: Theory Methodology and Applications. Amsterdam: Elsevier, 1983.
5. Righetti PG, Gianzza E, Bosisio AB. Biochemical and clinical application of isoelectric focusing, in recent developments. In: Frigerio A, Renoz L, eds. Chromatography and Electrophoresis. Amsterdam: Elsevier Biomedical Press, 1979:1–36.
6. Pearlman R, Nguyen TH. Analysis of protein drugs. In: Lee VHL, ed. Peptide and Protein Drug Delivery. New York: Marcel Dekker, Inc., 1991:247–302.
7. Laemmli UK. Cleavage of structural proteins during the assembly of the head of bacteriophage T4. Nature 1970; 227:680.
8. Morrisey JH. Silver stain for proteins in polyacrylamide. Gels: a modified procedure with enhanced uniform sensitivity. Anal Biochem 1981; 117:307.
9. Oakley BL, Kirsch DR, Morris NR. A simplified ultra sensitive silver stain for detecting proteins in polyacrylamide gels. Anal Biochem 1980; 105:361.
10. Wesrermier R. Electrophoresis in Practice. Germany: VCH Press, 1997.
11. Neuhoff V, Stamm R, Ebil H. Clear background and highly sensitive protein staining with Coomassie Blue dyes in polyacrylamide gels: a systematic analysis. Electrophoresis 1985; 6:427.
12. Ackers GK. Analytical gel chromatography of proteins. Adv Protein Chem 1970; 24:343.
13. Regnier FE. High-performance liquid-chromatography of bio-polymers. Science 1983; 222:245.
14. Robinson AB, McKerrow JH, Cary P. Controlled deamidation of peptides and proteins: an experimental hazard and a possible biological timer. Proc Natl Acad Sci USA 1970; 66(3):753.
15. Teshima, G. Porter J, Yim K, et al. Deamidation of soluble CD4 at asparagine-52 results in reduced binding capacity for the hiv-1 envelope glycoprotein gp120. Biochemistry 1991; 30:3916.
16. Cacia J, Quan CP, Vasser M, et al. Protein sorting by high-performance liquid chromatography. I: Biomimetic interaction chromatography of recombinant human deoxyribonuclease I on polyionic stationary phases. J Chromatogr 1993; 634:229.
17. Teshima G, Stults JT, Ling V, et al. Isolation and characterization of a succinimide variant of methionyl human growth hormone. J Biol Chem 1991; 266(21):13544.
18. Hancock WS, Sparrow JT. HPLC Analysis of Biological Compounds: A Laboratory Guide. New York: Marcel Dekker, 1984.
19. O'hare MJ, Nice EC. Hydrophobic high-performance liquid chromatography of hormonal polypeptides and proteins on alkylsilane-bonded silica. J Chromatogr 1979; 171:209.
20. Shaltiel S. Hydrophobic chromatography. Methods Enzymol 1984; 104:69.
21. Melander WR, Corradini D, Horvath C. Salt-mediated retention of proteins in hydrophobic-interaction chromatography. J Chromatogr 1984; 317:67.
22. Fausnaugh JL, Kennedy LA, Regnier FE. Comparison of hydrophobic-interaction and reversed-phase chromatography of proteins. J Chrmatogr 1984; 317:141.
23. Creighton TE. Proteins: Structures and Molecular Properties. New York: W.H. Freeman and Company, 1983.

24. Klibanov AM. Stabilization of enzymes against thermal inactivation. Adv Appl Microbiol 1983; 29:1.
25. Stoscheck CM. Quantitation of protein. Methods Enzymol 1990; 182:50.
26. Gill SC, von Hippel PH. Calculation of protein extinction coefficients from amino acid sequence data. Anal Biochem 1989; 182:319.
27. Weast RC. Handbook of Chemistry and Physics. 56th ed. Cleveland: CRC Press, 1975.
28. Surewicz WK, Mantsch HH. Infrared absorption methods for examining protein structure. In: Havel HA, ed. Spectroscopic Methods for Determining Protein Structure in Solution. New York: VCH Publishers, 1996:135–162.
29. Perez Sanchez H, Tatarenko K, Nigen M, et al. Organization of Human Interferon gamma-heparin complexes from solution properties and hydrodynamics. Biochemistry 2006; 45(44):13227.
30. Berkowitz SA. Role of analytical ultracentrifugation in assessing the aggregation of protein biopharmaceuticals. AAPS J 2006; 8(3):E590.
31. Berne BJ, Pecora R. Dynamic light scattering. USA: John Wiley, 1975.
32. Manning MCW, Woody R. Theoretical determination of the CD of proteins containing closely packed antiparallel β-sheets. Biopolymers 1987; 26:1731.
33. Woody RW. Circular dichroism of peptides. In: Hruby V, ed. The Peptides, Analysis Synthesis and Biology. New York: Academic Press, 1985:15–104.
34. Mulkerrin MG. Protein structure analysis using circular dichroism. In: Havel HA, ed. Spectroscopic Methods for Determining Protein Structure in Solution. New York: VCH Publishers, 1996:5–27.
35. Bell JE. Spectroscopy in Biochemistry. Boca Raton, Florida: CRC Press, 1981.
36. Wang W. Instability, stabilization and formulation of liquid protein pharmaceuticals. Int J Pharm 1999; 185:129.
37. Bedu-Addo F, Moreadith R, Advant S. Preformulation development of recombinant pegylated staphylokinase SY161 using statistical design. AAPS Pharm Sci 2002; 4(4):19.
38. Tsai PK, Volkin DB, Dabora JM, et al. Formulation design of acidic fibroblast growth factor. Pharm Res 1993; 10:649.
39. Bedu-Addo F, Johnson C, Jeyarajah S, et al. Preformulation development of a heavy chain fragment of Botulinum serotype b: identification of suitable purification conditions. Pharm Res 2004; 21(8):1353.
40. Nguyen TH, Burnier J, Meng W. The kinetics of relaxin oxidation by hydrogen peroxide. Pharm Res 1993; 10:1563.
41. Xie M, Shahrokh Z, Kadkhodayan M, et al. Asparagine deamidation in recombinant human lymphotoxin: hindrance by three-dimensional structures. J Pharm Sci 2003; 92(4):869.
42. Stratton LP, Kelly RM, Rowe J, et al. Controlling deamidation rates in a model peptide: effects of temperature, peptide concentration, and additives. J Pharm Sci 2001; 90(12):2141.
43. Gadgil HS, Bondarenko PV, Pipes GD, et al. Identification of cysteinylation of a free cysteine in the Fab region of a recombinant monoclonal IgG1 antibody using Lys-C limited proteolysis coupled with LC/MS analysis. Anal Biochem 2006; 355(2):165.
44. Houde D, Kauppinen P, Mhatre R, Lyubarskaya Y. Determination of protein oxidation by mass spectrometry and method transfer to quality control. J Chromatogr 2006; 1123(2):189.

45. Xie M, Shahrokh Z, Kadkhodayan M, et al. Asparagine deamidation in recombinant human lymphotoxin: hindrance by three-dimensional structures. J Pharm Sci 2003; 92(4):869.
46. Takahashi N, Ishioka N, Takahashi Y, Putnam FW. Automated tandem HPLC system for separation of extremely complex peptide mixtures. J Chromatogr 1985; 326:407.
47. Arakawa T, Timasheff SN. Mechanism of poly(ethylene glycol) interaction with proteins. Biochemistry 1985; 24:6756.
48. Arakawa T, Timasheff SN. The stabilization of proteins by osmolytes. Biophys J 1985; 47:411.

6

Solution Formulation of Proteins/Peptides

Paul McGoff

ZymoGenetics, Seattle, Washington, U.S.A.

David S. Scher

Alkermes, Inc., Cambridge, Massachusetts, U.S.A.

INTRODUCTION

The simplest and most economical way to formulate a protein and/or peptide is to develop a solution formulation. Lack of adequate protein stability in solution may ultimately require development of an alternative, more complex formulation such as a lyophilizate. A well-designed formulation study will allow the formulation scientist to determine whether a solution formulation will be acceptable for a given protein.

The first requirement for a formulation study is the availability of significant quantities of purified protein and/or peptide. Along with this protein or peptide should come a vast amount of practical knowledge obtained during various stages of process development. This knowledge will not only be anecdotal in nature but will also provide some specific physicochemical characteristics of the molecule.

Purification in-process analytical tests should be available to assess basic parameters such as purity, concentration, and some measure of activity. Experiences during early stages of process development can contribute to this initial knowledge base. For example, problems such as poor solubility or aggregation that may occur during fermentation, harvesting, and purification of the molecule can result in a reduced recovery yield. Knowledge of the conditions and circumstances under which these problems occurred can aid in focusing and prioritizing the initial formulation studies. For example, a particular pH or mix of buffer components used during processing may reduce solubility, induce aggregation, or promote specific degradation pathways.

Even physical processing steps like diafiltration can expose the protein to shear and denaturing liquid–solid or liquid–air interfaces, which can ultimately cause protein aggregation as well. Alternatively, investigating the effect of freezing might be advisable, since freezing is often used as a convenient process-hold step. Feedback from these experiences can be used to design formulation experiments to gain insight into tactics to avoid, minimize, or explain process problems. Conversely, the results from these formulation experiments can aid in refining the purification process as well as in guiding future formulation development to address specific potential stability problems.

The second prerequisite for starting protein solution formulation studies is availability of analytical test procedures for characterizing the physicochemical properties of the protein. Methodologies for the characterization, beyond the basic ones used to assess purity during stages of purification, are necessary. As part of preformulation studies, specific stability indicating analytical methods will need to be developed. These methods must be capable of detecting changes in protein samples that have been altered by physical or chemical stress.

Typical properties monitored include changes in charge [studied by means of isoelectric focusing (IEF), native-polyacrylamide gel electrophoresis (native-PAGE), and ion exchange chromatography], conformation [size exclusion chromatography (SEC), circular dichroism, fluorescence spectroscopy, native-PAGE, and capillary zone electrophoresis], size [SEC, sodium dodecyl sulfate-polyacrylamide gel electrophoresis (SDS-PAGE), laser light scattering, matrix-assisted laser desorption ionization time-of-flight mass spectrometry, native-PAGE, and capillary electrophoresis], hydrophobicity (hydrophobic interaction chromatography, reversed phase chromatography, and micellar electrokinetic chromatography), and biological activity (cell-based assay, enzymatic assay, and enzyme-linked immunosorbent assays).

Additionally, knowledge of a protein's unique structural and functional characteristics can often indicate which analytical tests will be most useful in assessing stability. For example, when one is probing for changes that effect hydrophobicity, such as denaturation or methionine oxidation, hydrophobic interaction chromatography is generally more useful for large proteins (immunoglobulins), whereas reversed phase techniques are more appropriate for small proteins (molecular weight < 40 kDa) and peptides.

Many glycosylated proteins contain negatively charged sialic acid residues, which create surface charge heterogeneity and the potential for complicated IEF gel patterns and poorly resolved native-PAGE gels. Such complexity might mask charge changes that one might expect due to deamidation, for example. Extensive glycosylation can also result in size heterogeneity that may contribute to zone broadening in SDS-PAGE or SEC. Calculations based on amino acid composition or sequence can be used to estimate charge as a function of pH (including isoelectric point) for a particular protein (1). This is useful in choosing electrophoretic or ion exchange conditions and may predict solubility behavior, since solubility can be strongly influenced by charge.

Coupling specific structural knowledge of a protein with general knowledge of protein degradation will help in choosing the most appropriate analytical methods for a particular protein. In addition, it will give one an idea of what results to expect when a particular analytical method is applied.

Consideration of all the aforementioned prerequisites will allow for some prioritization and flexibility in designing solution formulation studies. In the end, formulation studies should address solubility as a function of pH and salt and the influence of increased temperature, solution pH, buffer ion, salt, protein concentration, and other excipients and preservatives (when necessary) on stability. Other studies to include are photostability, cavitation/shaking, and freeze–thaw cycling. Finally, material compatibility studies should be performed with any storage containers or medical device the molecule/formulation may contact.

SOLUTION FORMULATION

Solubility Studies

Typically, the relevant pH range over which solubility needs to be determined, for protein formulation studies, is 4 to 9. Although processing may expose the protein or peptide to pH values varying from 2.5 to 11, extremes in pH are avoided in formulations, since many degradative processes become more prevalent below pH 5 (acid hydrolysis) and above pH 7 (deamidation).

To assess the minimum solubility required to formulate and deliver a protein drug, it is useful to consider the maximum dosing limit. A survey of the Physician's Desk Reference (PDR) reveals that for direct injection intravenous (IV) formulations, the upper limits on the volume and dose for currently marketed protein therapeutics, are 10 to 20 mL and 10 µg to 100 mg of protein, respectively. Therefore, this means that one may need to achieve a minimum solubility in the range of 0.1 to 5 mg/mL. If the solubility is less than 0.1 mg/mL in this pH 4 to 9 range, additional excipients or sodium chloride may need to be examined as a means of increasing solubility.

If lyophilized, excipient-free protein is not readily available or obtainable owing to specific protein chemistry considerations, an alternative (and more laborious) way to perform solubility studies is to diafilter a protein stock solution into the desired buffer and then concentrate it as high as is practically possible. A variety of small-volume, centrifuge-driven, pressure-driven, and vacuum-driven devices are commercially available for this purpose. The concentrated stock solution can then be diluted with the appropriate series of buffers to be used in the solubility studies. An example of a solubility study protocol is as follows:

1. A ready supply of either lyophilized, excipient-free protein or diafiltration equipment is needed.
2. Start with protein in dry form and incrementally add weighed amounts to a small volume (typically ≤1 mL) of test buffer.

3. When the solubility limit is exceeded (i.e., when precipitation occurs), the undissolved protein pellet is centrifuged and the supernatant is analyzed for concentration.

4. Limitations on the amount of protein to add are discretionary as long as acceptable solubility is achieved.

Typically, if protein solubility exceeds 50 mg/mL at the pH of interest, no further solubility studies are necessary. The solubility desired should be minimally two-fold and maximally 10-fold higher than the maximum conceivable dose.

If protein solubility is limited at relevant formulation pH values, additional measures will need to be taken to increase solubility. Some of the ways in which solubility can be increased include increasing/decreasing sodium chloride, adding other salts, varying the buffer species at a given pH, or including glycerol, lipids, polymers, cyclodextrins, or surfactants. Generally, one or the other of these approaches has successfully been applied to increasing solubility in protein solution formulations (2,3).

Isotonicity Considerations

Since isotonicity alone may have a tremendous influence on solubility, the basic approach should be to adjust the sodium chloride concentration before other excipients are added to the formulation. From a physicochemical standpoint, it is best to have as simple a formulation as possible, with fewer opportunities for potential complications due to undesired excipient-protein interactions. Furthermore, when problems do arise, sorting out what is happening is much easier.

When considering salt concentration, our recommendation is to formulate at or near isotonic conditions. Although hypotonicity is not typically a problem for injectables, nasals, and topical products, hypertonic solutions may cause undesirable local tissue irritation or a burning sensation upon administration.

Solution Stability Studies

Generally, a formulation pH stability study should be limited to the physiologically relevant pH range of 4 to 9. Preformulation studies will typically address conditions and degradation pathways outside this pH range. The choice of buffer species should be physiologically compatible and listed as generally regarded as safe (GRAS) by the U.S. Food and Drug Administration (FDA). A useful source of excipients used in formulations is the FDA guide to inactive ingredients (4). Often at a given target pH, where multiple buffer choices are available, there will be differences between specific buffer ions with respect to effect on certain degradative pathways or rates of degradation.

Specific Buffer Ion Effects

To illustrate the importance of considering the specific ion effects of the buffer chosen, we discuss an example for a nine amino acid peptide (RMP-7). One clinical formulation for RMP-7 drug product consisted of a pH 4.0 unbuffered solution in normal saline (0.9% sodium chloride, pH adjusted with acid/base). This product was reformulated into a buffered system at the same pH to minimize pH drift due to leaching of hydroxyl ions from the surface of the glass vials of the drug product.

The two choices for buffer species were citrate (pK 2.5, 4.5, 6.0) and acetate (pK 4.5), both of which possess good buffering capacity near the target pH of 4.0. The study consisted of formulating RMP-7 drug substance in a series of citrate buffers ranging in pH from 2.5 to 6.0 and in acetate buffers at pH 4.0, 4.5, and 5.0. Solutions were stored for up to one month at 60°C in a well-controlled temperature stability cabinet.

Potency and purity of RMP-7 solution formulations were assessed as a function of time using a validated reversed-phase high-performance liquid chromatography (HPLC) assay. Figure 1 graphically illustrates the effect of the two buffer species on the stability of RMP-7, as measured by drug purity as a function of solution pH verses the unbuffered formulation. In this case, the use of citrate ion caused more rapid peptide degradation than the acetate formulations over the pH range of 4.0 to 5.0. The acetate formulations were equivalent to the unbuffered formulation at pH 4.0.

This was an important finding in the formulation development of RMP-7 and illustrates the importance of choosing the appropriate buffer species for a formulation. This choice will always be dependent on the characteristics of the individual drug substance and cannot easily be predicted in advance nor generalized to other drugs. Once the likely pH of the final product formulation has been determined

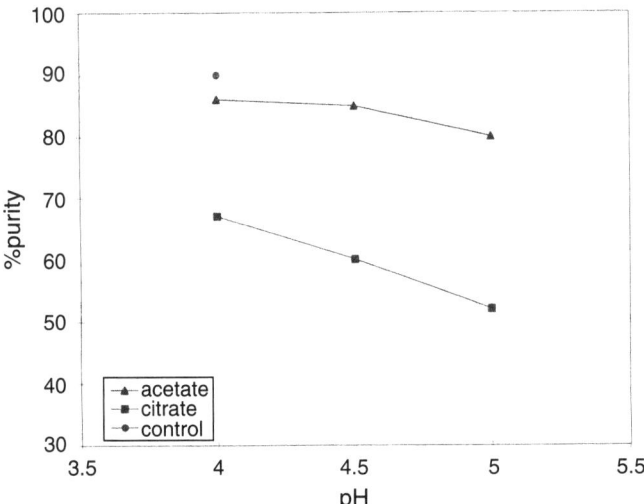

Figure 1 Degradation of the peptide RMP-7 as a function of pH in two types of equivalent molarity buffer. The control solution was unbuffered with pH adjusted to 4 by addition of NaOH and HCl. All solutions were adjusted with NaCl to isotonicity and stored for one month at 60°C in molded glass vials. The difference in degradation rate demonstrates the buffer ion effect on RMP-7 stability.

(from earlier preformulation studies), testing more than one buffer species at multiple concentrations at the chosen pH is advisable.

To measure the buffer ion effect on a degradative pathway, the buffer ion concentration in a formulation is varied and the tonicity is kept constant by addition of sodium chloride. A plot of the observed degradative rate constant versus buffer concentration will have a slope proportional to the buffer ion activity. Extremes in temperature and pH will accelerate degradation. Study design will require frequent sampling (i.e., hourly).

Once the pH range has been narrowed to where major degradation processes are slowed, if degradation is still unacceptable (projected shelf life at 4°C <2 years), the effects of adding other potentially stabilizing excipients should be studied. However, if projected degradation rate is acceptable within this narrowed range, formulation choice becomes more discretionary and can be based upon experience and other dosing or delivery considerations.

With proteins, one should be cautious about projecting 5°C (2–8°C) stability from accelerated temperature studies. At elevated temperatures, other pathways/ mechanisms can become more prevalent, often confusing analytical results and resulting in erroneous shelf-life predictions. Some degradation processes have nonlinear plots of rate as a function temperature or may have plateaus, complicating projection (5–9).

Excipients

The inclusion of stabilizing excipients leads to the most interesting phase of formulation development studies. Here exists a vast sea of folklore and anecdotal information alongside excellent case history literature (3). Personal preferences often dictate excipient choices based on past experiences with different proteins (not always with positive results). Every protein has unique physicochemical characteristics and the potential to behave differently stability wise in the presence of a particular excipient.

For example, the common usage of surfactants to stabilize small recombinant proteins does not automatically warrant this inclusion in monoclonal antibodies formulations. Excipients, if necessary, should be chosen for their known functionality in slowing or arresting specific degradative pathways. Various amino acids have been employed as specific stabilizers of proteins. Another example of specific excipient choice based on known functionality is the use of hydroxypropyl-β-cyclodextrin in stabilizing interleukin-2 aggregation (2,10).

Another important consideration is the quality of the excipient chosen for early formulation studies. In some instances, even very low levels of impurities (<100 ppm) in excipients can cause degradation of the protein in the final drug-product formulation

In some protein drug product formulations, it is necessary to include chelating agents, e.g., ethylenediamine tetraacetic acid (EDTA) to stabilize the protein and/or to prevent protease clipping. Unfortunately, there are potential clinical consequences to the usage of chelating agents.

Temperature Stability

Normally, proteins in solution are formulated for storage at 5°C (2–8°C) with a minimum target shelf-life of two years. As stated in the International Conference Harmonization (ICH) stability guideline (11),

> Since most biotechnological/biological products need precisely defined storage temperatures, the storage conditions for the real-time/real temperature stability studies may be confined to the proposed storage temperature. However, it is strongly suggested that studies are conducted on the drug substance and drug product under accelerated and stressed conditions. Studies under stress conditions may be useful in determining whether accidental exposure to conditions other than those proposed (e.g. during transportation) are deleterious to the product. Conditions should be carefully selected on a case-by-case basis.

Most solutions of proteins are not thermally stable above room temperature for prolonged periods of time; therefore, accelerated studies are typically conducted at 25°C, 30°C, and/or 40°C, and generally no higher. However, in the case of peptides, accelerated studies at higher temperatures are often permissible because their inherent temperature stability is greater. Conducting a temperature rate study of degradation may allow not only a projection of stability at 5°C (2–8°C) but also an estimation of the activation energy of a particular process (12).

Protein Concentration

Another important formulation parameter that needs to be studied is the effect of protein concentration on stability. Of course, early on in development, availability of sufficient amounts of protein active may limit the extent to which high concentration studies can be conducted. However, such studies should be performed because rates of degradation typically vary as a function of protein concentration. Based on overall clinical experience with protein drugs and their maximum inherent solubilities, we recommend starting in the 1 to 10 mg/mL protein concentration range. This assumes that the final drug product will be administered by IV injection in a volume of 1 to 10 mL.

Photostability

Another ICH guideline states, "The intrinsic photostability of new drug substances and drug products should be evaluated to demonstrate that, as appropriate, light exposure does not result in unacceptable change. Normally, photostability testing is carried out on a single batch of material" (13). This particular test will be most relevant for products formulated as solutions where there is potential for storage underlighted conditions.

Since most biologicals are stored refrigerated (the light goes out when you close the door), the assessment of photodegradation is a minor consideration, but nonetheless, a regulatory requirement. After controlled light exposure of RMP-7 drug product, the nine amino acid peptide discussed earlier, an increase in one impurity was observed, but the product still met acceptance criteria of impurity of 1% or below. This is in contrast to RMP-7 drug substance, which exhibited no photodegradation when exposed to the same conditions.

Preservatives

General Considerations

If the protein formulation is intended for multidose use, a preservative will need to be included to prevent microbial growth. However, preservatives are toxic, and there are maximum limits described for injectables (14). Table 1, assembled from

Table 1 Currently Used Preservatives for Proteins and Peptides from the PDR

Product, manufacturer; drug	Route of administration	Container	pH	Buffer system	Preservative concentration
Proteins					
Alferon N, Hemispher X Biopharma, Inc.; interferon alfa-n3	IL	1 mL vial	7.4	Sodium and potassium phosphate	Phenol: 3.3 mg/mL
Epogen, Amgen; recombinant erythropoietin (epoetin alfa)	IV or SC	1 and 2 mL vials	6.1	Sodium citrate	Benzyl alcohol: 1%
Follistim AQ Cartridge, Organon; follitropin beta (hFSH)	SC	0.21, 0.42, and 0.78 mL cartridges	7	Sodium citrate	Benzyl alcohol: 1%
Gonal-f RFF PEN, Serono; follitropin alpha	SC	0.5, 0.75, and 1.5 mL		Sodium phosphate	3.0 mg/mL *m*-cresol
Intron A, Schering; recombinant interferon alfa-2b	IM, SC, IV, or IL	Multidose pens and vials		Sodium phosphate	1.5 mg/mL *m*-cresol, 0.1 mg/mL EDTA
Nutropin AQ, Genentech; somatropin	SC	2 mL vial and 2 mL pen cartridge	6.0	Sodium citrate	2.5 mg/mL phenol

(Continued)

Table 1 Currently Used Preservatives for Proteins and Peptides from the PDR (*Continued*)

Product, manufacturer; drug	Route of administration	Container	pH	Buffer system	Preservative concentration
Procrit, Ortho Biotech; recombinant erythropoietin (epoetin alfa)	IV or SC	1 and 2 mL vials	6.1	Sodium citrate	Benzyl alcohol: 1%
Roferon-A, Roche; recombinant interferon alfa-2a	SC or IM	0.5 mL prefilled syringes	NA	ammonium acetate	Benzyl alcohol: 1%
Peptides					
Byetta, Amylin; exenatide	SC	Pen-injector	4.5	Sodium acetate	Metacresol: 2.2 mg/mL
Calcimar, Rhone Poulenc Rorer; calcitonin	SC or IM	2 mL vial	NA	Sodium acetate	Phenol: 5 mg/mL
Miacalcin, Novartis; calcitonin (salmon)	SC or IM	2 mL vial	NA	Sodium acetate	Phenol: 2.25 mg/mL
Miacalcin, Novartis; calcitonin	IN	3.7 mL glass bottle	NA	No additional	Benzalkonium chloride: 0.1 mg/mL
Pitocin, King; oxytocin	IV	1 mL vials and ampoules	NA	Acetic acid	Chlorbutanol: 0.5%
Sandostatin, Novartis; octreotide acetate	SC or IV	5 mL vials	4.2	Lactic acid, sodium bicarbonate	Phenol: 5 mg/mL
Stimate nasal spray, ZLB Berhing; desmopressin acetate	IN	Nasal pump	NA	Citric acid, sodium phosphate	Chlorbutanol; 5 mg/mL
Synarel, Searle; nafarelin acetate	IN	8 mL bottle	NA	Acetic acid	Benzalkonium chloride: 0.1%

Abbreviations: PDR, Physician's Desk Reference; IN, intranasal; IV, intravenous; IL, intralesional; SC, subcutaneous. NA, not available.

a search of the PDR, lists preservatives used in some approved protein and peptide pharmaceuticals. Selection of an appropriate preservative is not trivial and requires the consideration of many factors, which complicates formulation.

Focus initially will be to choose several potential preservative candidates based on current use and route of drug delivery, ideal solution pH of formulation, and compatibility with other formulation components. Once a selection has been made, accelerated and real-time stability testing of the protein in the presence of these selected preservative candidates should be performed.

Chapter 4 on preformulation contains a discussion on the use of differential scanning calorimetry to screen preservatives with respect to protein interactions. In addition, an article on the effect of benzyl alcohol on the stability of γ-interferon is available (15). These authors do not identify the mechanism of interaction but do show that by choosing the correct buffer salt and minimizing the amount of preservative, a stable formulation can be developed.

Timing for Inclusion in a Formulation

Performing extensive stability testing of the protein without a preservative will provide one with an idea of the buffer system and pH on which to focus stability testing. If the objective of the initial phase I clinical trial is proof of concept, the simplest approach is formulation without a preservative. However, this approach will mean repeating the phase I trial with a second preserved formulation, and additional stability studies will be required. If it is known that the final drug product will be used in a multidose form, a preserved formulation must be developed at some point. The timing of the decision to develop and test the preserved formulation will depend heavily on a cost-risk benefit assessment.

A decision to develop a preserved formulation will require developing methods to assay both protein and preservative. Processing equipment and container compatibility with the preservative, as with the protein, will need to be addressed. Finally, preservative efficacy in the protein formulation will need to be performed on the preservative candidates.

Quick elimination of a particular candidate formulation due to stability problems will reduce the amount of samples that need to be tested by preservative efficacy.

Choosing a Preservative

There are a number of prerequisites for selecting a potential preservative candidate. Is it GRAS and pharmaceutically acceptable (GRAS-listed)?

The intended route of administration is also an important consideration. A preservative appropriate for IV administration may not be appropriate for nasal administration. Some preservatives that are acceptable for IV administration may have a strong odor or taste (phenolics); others may dry nasal mucosa or be cilia static or toxic (16). Sensitivity of target patient population also may be an important consideration. For example, asthmatics have been shown to be more sensitive to certain types of preservative. Benzalkonium chloride and EDTA from nebulized solutions have both been reported to induce dose-related bronchoconstriction in asthmatics (17).

Formulation pH is another important consideration in choosing a preservative. Preservatives are most effective against microbes within specific pH ranges. When formulated outside these ranges, many preservatives change ionization state and lose efficacy. The candidate should have a wide spectrum of activity, preventing the growth of fungi, as well as exhibiting cidal activity against various types of gram-positive and -negative bacteria. Preservatives can often be used in combination to achieve synergistic effects (18,19).

Compatibility with other formulation components is another important consideration. Certain preservatives (e.g., mercurial preservatives are incompatible with metals; others, e.g., phenolics, quaternary ammonium, paraben preservatives), are incompatible with surfactants. Choice of preservative concentration to use can initially be based upon current use in the industry (PDR) and specific literature references for preservative safety and efficacy against various types of microbe (19). Eventually, preservative efficacy studies performed with the protein will determine the optimum preservative concentration.

Analytical Methods

Many preservatives have characteristics that interfere with analytical methods that work extremely well for the protein in their absence. For example, many preservatives exhibit high ultraviolet adsorption, making it necessary to have analytical techniques capable of separating the protein active from the preservative. Often the preservative needs to be removed before cell-based activity assays are performed, since the preservatives are toxic to the cells.

Ideally, it would be nice to have quantitative separation methods capable of resolving the preservative(s) from the protein with no interfering interactions from any of the components. We have seen that paraben preservatives interact with hydrophilic SEC media, complicating results. In contrast, a reversed-phase method that worked well to separate and quantitate parabens irreversibly adsorbed the protein and thus modified column performance. Developing analytical techniques to get around these problems can be a challenge.

Compatibility

In addition, container considerations may change when a preservative is added to a protein formulation. Some preservatives are sensitive to light and/or air and may require storage in brown glass containers and/or airtight or purged containers. Phenolics, benzyl alcohol, chlorobutanol, parabens, and mercurial preservatives are examples.

Container-preservative compatibility can be a problem. Parabens and benzalkonium chloride adsorb to and chlorobutanol diffuses through plastic containers and delivery devices, reducing preservative strength (20,21). Furthermore, rubber stoppers adsorb many preservatives. Additionally, adsorption/interaction with processing equipment that contacts preservative, such as vessels, tubing, and filters, can be potentially problematic. Adsorption studies should be performed to address these potential pitfalls.

Preservative Efficacy

Preservative efficacy studies are performed on preserved formulations as described under the following general test chapter of the U.S. Pharmacopeia: *Antimicrobial Preservatives—Effectiveness* 51® (USP 30, p. 79–81) and are useful in identifying the most appropriate preservative for a particular protein at the optimum preservative concentration to maintain a growth-free solution. The assay involves challenging the formulation with bacteria of three types (*Staphylococcus aureus, Escherichia coli,* and *Pseudomonas aeruginosa*) and fungi of two types (*Candida albicans* and *Aspergillus niger*). The effective concentration range of the preservative is determined by adding various levels to the formulation and performing USP <51>. For example, if the intended formulated preservative concentration of phenol is 0.3% w/v, the effectiveness of this concentration is tested at this level and several lower levels (e.g., 0.2%, 0.15%, 0.1% w/v). The lower concentration levels are tested to ensure a margin of safety, since preservatives can become less effective owing to either degradation or adsorption over time.

Once preservative efficacy has been demonstrated over the lifetime of the drug product, it is possible and highly recommended to submit an application to the regulatory agency to replace the accelarated exposure test, USP <51> testing with a chemical analysis method.

Cavitation/Shaking

Proteins in solution denature at air–liquid interfaces. Under significant stress—for example, during filtration or physical shaking—a portion of the soluble protein will often denature and irreversibly aggregate. A typical result of this denaturation process is a turbid, cloudy solution, or a solution containing insoluble protein particles. Therefore, it is a good idea to test the ruggedness of a formulation to withstand shaking or agitation that might occur during handling.

To test a formulation's ruggedness, ultimately, the final dosage form (in its final container) should be shaken or cavitated. A mechanical shaker is used to cavitate the protein solution for days and up to a week at room temperature. If the solution becomes cloudy or SEC reveals soluble aggregates, adding volume to fill container often reduces the problem. Another approach would be to supplement the formulation by adding glycerol, polymer, or a surfactant, but this ultimately depends on the sensitivity of the protein. If the solution does become cloudy, consider adding the instruction "Do Not Shake" to the container label.

Freezing Studies

Freezing is another handling consideration. At various stages in processing, it may be necessary to freeze bulk protein solutions. Although the storage condition of final drug product is generally 2°C to 8°C, there is always the possibility of inadvertent exposure to freezing temperatures during shipping and storage. Freezing might occur during air shipments of protein solutions, since packages may end up in uncontrolled-temperature cargo bays or in refrigerated storage units having colder areas that are below freezing. Thus, it is important to do freeze–thaw cycling studies to determine the ruggedness of the protein solution. In the processing of proteins, the availability of containers of many types and sizes makes it a challenge to the design a relevant scalable freeze–thaw simulation at the small scale.

Variations in the process of freezing can radically affect the rate of freezing and ultimately the integrity of the protein. The temperature of the freezer (−20°C or −80°C), the type of freezer, the method of freezing (contact of shelf, liquid, or air), the location within the freezer, the volume of solution, and the container geometry all affect the way in which the solution freezes. Details of how these parameters affect protein integrity should be carefully considered in freeze–thaw cycling experimental design. One approach is the use of mini-tanks to assess the potential damage caused by repeated freeze–thaw cycles in a scaled-down version of process tanks. Studies of these types have been used to support the freezing and subsequent thawing of bulk drag substances stored in large-scale tanks. Protein solutions are analyzed both visually and by SEC for aggregation.

For final drug product, such studies are somewhat easier, because the container is a limiting constraint. Freeze–thaw studies using the final drug product are considered relevant because product may encounter hostile environmental conditions en route to its destination or accidentally in the hands of health care workers (doctors, nurses, and pharmacists).

Excipient choice can also impact stability to freezing. For example, freezing sodium phosphate buffered solutions can induce acidic pH shifts, which impact protein integrity (22–24).

Materials Adsorption

Materials adsorption is mainly a consideration for low-concentration protein/peptide formulations, generally at concentrations less than 1 mg/mL. Increasing surface area, contact time, and temperature all can enhance adsorption. Adsorption has the potential to occur during processing and during storage in final drug product container. Materials that contact the protein solution during processing stages and in the final container will determine what compatibility testing needs to be done. Table 2 shows what types of material may need to be tested and at the stage during processing at which contact occurs.

If the active drug is formulated at a low concentration and adsorbs to the container of choice, there are several options. One option is to fill the container to maximum volume. Filling a container to near capacity can reduce surface area contact between protein/peptide solution and the container, thus minimizing potential adsorption. The second and more difficult option is to change container material

Table 2 Material Contact at Various Processing Stages

Surface material	Drug substances	Formulated bulk	Final drug product
Plastics/glass	√	√	√
Membranes/filters	√	√	√
Tubing	√	√	
IV bags/catheters			√
Syringes			√
Stoppers			√

Abbreviation: IV, intravenous.

type. As a last resort, one may consider the addition of another excipient (e.g., human serum albumin, polymers, or surfactants) to the formulation, to inhibit absorption.

Placebo Considerations

In developing placebos for protein-solution formulations, one must consider that the properties of the placebo solution will differ slightly from those of the active substance. There will be a difference in the solution viscosity, and a shaken protein solution will exhibit a certain degree of foaming, which may allow a clever clinician the opportunity to unblind a trial. As far as stability is concerned, there should be no difference between the placebo and active solutions. To demonstrate this, it is prudent to manufacture and monitor the stability of the placebo product prior to initiating clinical trials. This will ensure that placebo of an acceptable quality and adequate blinding properties is available.

MANUFACTURING OF DRUG PRODUCT: ASEPTIC PROCESS CONSIDERATIONS

Terminal Sterilization versus Aseptic Processing

Drugs administered parenterally must be either terminally sterilized or subjected to aseptic processing. Terminal sterilization by heat or radiation exposure is generally simpler and cheaper to validate and implement. Unfortunately, most proteins and peptides will not tolerate terminal sterilization; therefore, they are processed aseptically. The FDA has provided aseptic processing guidance for the industry (25). In the case of RMP-7 drug product, several vials of product were subjected to normal autoclave cycles to determine whether aseptic processing could be supported. The results of this study indicated that an unacceptable level of degradation occurred during even reduced autoclave cycle times (1, 3, 10, 15, 20, and 30 minutes). This study demonstrated that aseptic processing was the only acceptable means of sterilizing RMP-7 final drug product.

Process Equipment Leachates

In filling low-concentration (2 µg/mL) products, highly sensitive HPLC methods are capable of detecting and quantitating very-low-level process/formulation contaminants. Guidelines for impurities in peptide drug products allow for 1% or less of single impurities (ICH guideline). There is the potential during processing and formulation stages to leach into final drug-product formulations low-level contaminants that can be detected in final filled drug product as impurities. One should take care in testing all potential product contact equipment/containers for potential leachates.

It is not always obvious which individual pieces of process/formulation equipment come in contact with drug product during filling operations. Caution should be exercised in determining which pieces of process equipment will contact drug formulations. One should thoroughly investigate this potential problem area before embarking on the formulation of low-concentration drug products. All process and formulation equipment should be thoroughly cleaned and rinsed, and the rinsate solutions analyzed for leachates, before active drug product is processed. This precaution could save a great deal of time, money, and lost product, in addition to avoiding a lengthy investigation.

Incompatibilities Between Process and Product

In addition to monitoring the product during the autoclaving process, it is important to bear in mind that the container closure system can be adversely affected by the terminal sterilization process. Several publications have addressed the issue of extractables from container closure systems and the associated problems. In one example, leachates in a final product were implicated in inducing an immune response to the therapeutic protein in patients (26). In another example, leaching of zinc salts from various rubber closures has been reported to result in precipitation, discoloration, and contamination of products (27). Such leaching processes can be accelerated at elevated temperature, as would be encountered during the sterilization process.

An example of this was encountered during the aseptic filling and terminal sterilization of a phosphate buffered saline solution that was being manufactured for use as a placebo. During visual inspection of the autoclaved product vials, insoluble particulates were observed in the solution and on the gray butyl rubber vial closure. Upon isolation and examination of the particulates by X-ray microanalysis, the insoluble particulates were identified as zinc phosphate.

A series of experimental compatibility studies was performed and the source of the particulates was identified as the gray butyl rubber stoppers encountered during autoclaving. These stoppers contained zinc in the form of zinc oxide, and the hypothesis was that zinc was leaching out of the stopper and in the presence of the phosphate buffer yielded water-insoluble zinc phosphate crystals. When Teflon-coated gray butyl rubber stoppers were used in the manufacture and autoclaving of the phosphate buffer, no particulates were generated.

LINK OF FORMULATION BACK TO MANUFACTURING

Well-designed formulation studies yield much knowledge of the inherent degradation pathways to which a particular protein (or peptide) molecule is susceptible. They also reveal under which conditions these degradation processes proceed most rapidly. This in-depth understanding of the molecule's physicochemical stability can aid in explaining problems that might have occurred during purification or how subtle changes in pH or storage and handling conditions might affect the final recovery and integrity of the protein.

For example, knowledge of how freeze–thaw cycling affects the protein might aid in making decisions on storage of bulk drug substance at various in-process stages. A protein molecule known (from formulation cavitation studies) to be susceptible to air–water interface denaturation may need to be treated more gently, or a protective excipient may need to be added to the formulation during filtration. A change in the process that reduces air–water interface exposure might be another option in handling the sensitive molecule.

In our experience, changes in the large-scale cell culture process have resulted in changes in the IEF pattern of the purified and formulated protein. Knowledge of deamidation rates under cell culture conditions and exposure times to the media components in the cell culture process allowed prediction of the IEF banding pattern when the process was changed. Such projections are analogous to shelf-life projections performed for final products. Modeling of this deamidation process based on solution formulation studies allowed extension of this model to the actual cell-culture process conditions. Ultimately, the model was used to explain that the changes seen in the purified molecule were only minor charge changes resulting from the extended exposure time of the protein in the modified cell culture process.

Another concern with highly glycosylated protein is the change in glyco (or sialo-glyco) form distribution with changes in cell-culture parameters. Process scale up, changes in fermentation media components, changes in cell-culture process conditions (times, temperatures, DO_2, etc.) all have the potential to produce major changes in the glyco-form distribution of the drug product. Glyco-forms (or sialo-glyco forms) generally do not impart major physicochemical changes to the drug product that affect solubility, pH stability, or temperature stability but do create interesting analytical challenges.

CONCLUSION

The following step-by-step, how-to protocol should enable the formulation scientist to proceed through a protein solution formulation development study. Although it is important to address all aspects of protein solution formulation, there is a natural sequence to the order in which individual parameters are investigated. Ideally, the protein concentration and solubility are determined first to establish a working range for the initial pH stability study. It is also advisable to assess freeze–thaw stability early in the plan. Once these parameters have been investigated, the initiation of a simple range-finding stability study, as outlined below, is recommended.

Step 1 Protocol

1. Protein concentration: 1 and 10 mg/mL
2. pH range in 1° increments from 4.0 to 9.0
3. Temperature: 2°C to 8°C only
4. Additional excipients: NaCl to near isotonic
5. Time frame: maximum three months
6. Sampling interval: weekly at pH extremes; monthly at remainder

Step 2 Protocol

The results of the initial range-finding studies will enable one to refine and narrow the scope of future experiments. The next phase of experiments should focus on a narrower pH and concentration range and should study higher temperatures, possibly adding stabilizing excipients and extending the timeframe of the study to at least one year, sampling at monthly intervals. Specific buffer ion effects could be investigated during this phase.

The final phase, step 3, would involve a series of individual studies to address the remainder of the solution formulation parameters.

Step 3 Protocol

1. Autoclave study
2. Photostability study
3. Preservative study (if multidose product)
4. Cavitation/shaking
5. Materials adsorption
6. More in-depth freeze–thaw studies

ACKNOWLEDGMENTS

The authors gratefully acknowledge the efforts of Antonio Pinho and Gil Olson at Alkermes Inc. In addition, we thank Gerry Bell, Christine Hall, Alison Mares, and Dr. Eugene McNally at Boehringer-Ingelheim Pharmaceuticals Inc.

REFERENCES

1. Sillero A, Ribeiro JM. Isoelectric points of proteins: theoretical determination. Anal. Biochem 1989; 179:319–325.
2. Brewster ME, Hora MS, Simpkins JW, Bodor N. Use of 2-hydroxy-propyl-β-cyclodextrin as a solubilizing and stabilizing excipient for protein drugs. Pharm Res 1991; 8:792.

3. John Wang Y, Rodney Pearlman, eds. Stability and Characterization of Protein and Peptide Drugs: Case Histories. New York: Plenum Press, 1993.
4. Inactive Ingredient Guide. Washington, D.C. Division of Drug Information Resources, U.S. Food and Drug Administration Center for Drug Evaluation and Research, Office of Management, 1996.
5. Jossang T, Feder J, Rosenquist E. Heat aggregation kinetics of human IgG. J Chem Phys 1985; 82(1):574.
6. Rosenquist E, Jossang T, Feder J, Harbitz O. Characterization of a heat-stable fraction of human IgG. J Protein Chem 1986; 5(5):323.
7. Uemura Y. Dissociation of aggregated IgG and denaturation of monomeric IgG by acid treatment. Tohoku J Exp Med 1983; 141:337.
8. Cheng-Der Yu, Niek Roosdorp, Shamim Pushpala. Physical stability of a recombinant α_1,-antitrypsin injection. Pharm Res 1988; 5(12):800.
9. Rosenquist E, Jossang T, Feder J. Thermal properties of human IgG. Mol Immunol 1987; 24(5):495.
10. Hora MS, Rana RK, Wilcox CL, et al. Development of a lyophilized formulation of interleukin-2. Dev Biol Standard 1992; 74:295.
11. ICH Harmonized Tripartite Guideline, Quality of Biotechnological Products: Stability Testing of Biotechnological/Biological Products, recommended for adoption at Step 4 of the ICH Process, ICH Steering Committee, International Conference Harmonization, Nov 30, 1995.
12. Pearlman R, Nguyen T. Pharmaceutics of protein drugs. J Pharm Pharmacol 1992; 44:178–185.
13. ICH Harmonized Tripartite Guideline, Stability Testing: Photostability Testing of New Drug Substances and Products, Recommended for adoption at Step 4 of the ICH, International Conference Harmonization, ICH: Steering Committee, Nov 6, 1996.
14. The United States Pharmacopeia 30. (1) Injections. Vol. 1. Rockville, MD: The United States Pharmacopeial Convention, Inc., 2007:33–36.
15. Lam XM, Patapoff TW, Nguyen TH. The effect of benzyl alcohol on recombinant human interferon-gamma. Pharm Res 1997; 14:725–728.
16. Batts AH, Marriott C, Martin GP, Wood CF, Bond SW. The effect of some preservatives used in nasal preparations on the mucus and ciliary components of mucociliary clearance. J Pharm Pharmacol 1990; 42(3):145.
17. Beasley CRW, Rafferty P, Holgate ST. Bronchoconstrictive properties of preservatives in ipratropium bromide (Atrovent) nebuliser solution. Br Med J 1987; 294:1197.
18. Dabbah R, Chang W, Cooper MS. The use of preservatives in compendial articles. Pharm Forum 1996; 22(4):2696–2704.
19. Raymond C Rowe, Paul J Sheskey, Paul J Weller, eds. Handbook of Pharmaceutical Excipients. Washington DC: American Pharmaceutical Association; London: Pharmaceutical Press, Royal Pharmaceutical Society of Great Britain, 2003.
20. Kakemi K, Sezaki H, Arakawa E, Kimura K, Ideda K. Interaction of parabens and other pharmaceutical adjuvants with plastics containers. J Chem Pharm Bull 1971; 19(12):2523.
21. Akers MJ. Considerations in selecting antimicrobial preservative agents for parental product development. Pharm Technol 1984; 8(5):36–44.
22. Anchordoquy TJ, Carpenter JF. Polymers protect lactate dehydrogenase during freeze-drying by inhibiting dissociation in the frozen state. Arch Biochem Biophys 1996; 332(2):231–238.

23. Pikal-Cleland KA, Rodrguez-Hornedo N. Protein denaturation during freezing and thawing in phosphate buffer systems: monomeric and tetrameric—galactosidase. Arch Biochem Biophys 2000; 384(2):398–406.

24. Pikal-Cleland KA, Cleland JL, Anchordoquy TJ, Carpenter JF. Effect of glycine on pH changes and protein stability during freeze-thawing in phosphate buffer systems. J Pharm Sci 2002; 91(9):1969–1979.

25. Guidance for Industry: Sterile Drug Products Produced by Aseptic Processing—Current Good Manufacturing Practice. U.S. Department of Health and Human Services, Food and Drug Administration, Center for Drug Evaluation and Research (CDER), Center for Biologics Evaluation and Research (CBER), Office of Regulatory affairs (ORA), September 2004.

26. Sharma B, Bader F, Templeman T, Lisi P, Ryan M, Heavner GA. Technical investigations into the cause of the increased incidence of antibody-mediated pure red cell aplasia associated with EPREX. Industry Science Nr 2004; 5:86–91.

27. Wang YJ Chien YW. Sterile pharmaceuticals: Packaging, compatibility and stability. Technical Report No. 5, Bethesda, MD: Parenteral Drug Association, 1984.

7

Formulation of Leuprolide at High Concentration for Delivery from a One-Year Duration Implant

Cynthia L. Stevenson

Nektar Therapeutics, San Carlos, California, U.S.A.

INTRODUCTION

A primary driving force for the creation and development of novel delivery systems was the advent of biotechnologically derived pharmaceuticals. Most protein and peptide pharmaceuticals are formulated for intravenous or subcutaneous injection; however, many patients have an adverse reaction to needles, and the desire for alternate dosage forms sets new expectations for the industry. Furthermore, these biomolecules often have limited solubility and/or aggregate with loss of activity. In many cases, aggregation, gelation, and precipitation occur, resulting in irreversible denaturation, precipitation, and stability issues. Reversible aggregation, usually marked by increased viscosity, may be acceptable if manufacturing filtration steps and syringability issues are alleviated (1). These formulation challenges are further accentuated by the need for an efficacious dose in a small volume, and thus, formulation strategies at relatively high protein concentrations, where aggregation is a bimolecular degradation pathway and concentration dependent, become increasingly difficult.

However, examples of formulation and manufacturing success are present in the literature. Biopharmaceuticals are routinely purified, precipitated, and crystallized under conditions requiring high protein concentrations (60 mg/mL) (2,3). Furthermore, individual formulation processing steps for parenteral injections and controlled release depot injections require protein concentrations in excess of 100 mg/mL (1,4–6). For example, bovine serum albumin has been spray-freeze-dried at 20 to 100 mg/mL with little loss of monomer (5). Human growth hormone (hGH) was formulated with trehalose or mannitol, lyophilized, and reconstituted at 200 and 400 mg/mL hGH prior to encapsulation in polylactic-coglycolic acid (PLGA) (4). Similarly, γ-interferon (γ-INF) was formulated at 137 mg/mL prior to encapsulation with good results (4).

Parenteral injections and ambulatory external pumps offer improved stability and controlled delivery (insulin); however, patient compliance and convenience are not optimal. Implantable drug delivery systems offer benefits over the repetitive administration of conventional drug therapy by providing unattended continuous delivery within the therapeutic window. Controlled implantable drug delivery avoids the highly variable peak and trough drug concentrations often seen after immediate release dosing (tablets and injections). Alleviation of this peak and trough serum profile by maintaining a continuous drug concentration can result in enhanced drug efficacy, minimized side effects, and increased patient compliance. Osmotic delivery systems are usually unaffected by in vivo variables and exhibit excellent correlation between in vivo and in vitro release (7–11). They offer precise delivery rates and are not dependent on the chemical or physical properties of the drug formulation. Therefore, these systems can be designed to provide a variety of release profiles for targeted or systemic delivery (12,13).

The Duros implant was designed after positive feedback from pharmaceutical researchers using the Alzet osmotic pumps in laboratory animals (preclinical studies) and veterinary applications (breeding endangered species and somatotropin delivery in pigs and cows) (1,2). The Viadur leuprolide acetate implant was designed to provide an alternative to periodic depot injections of leuprolide for the palliative treatment of prostate cancer. The implant delivers a highly concentrated solution formulation of leuprolide (~400 mg/mL) continuously over one year at ~120 µg/day (0.4 µL/day) from a 150 µL drug reservoir. Continuous administration of leuprolide results in the saturation and downregulation of pituitary receptors, resulting in decreased serum testosterone to castrate levels and retardation of tumor growth (14–16). The progress of prostate cancer is dependent on circulating androgen levels.

SOLUTION FORMULATION

Drug Substance

Leuprolide (pGlu-His-Trp-Ser-Tyr-D-Leu-Leu-Arg-Pro-NHEt) is a potent, luteinizing hormone–releasing hormone agonist. The N-terminal residues, pGlu-His-Trp, appear to be responsible for activity (17). The active conformer has been proposed to consist of two β-turns: a Type II′ β-turn centered at Tyr^5-Arg^8, and a Type III β-turn perpendicular to the first, centered at $pGlu^1$-Ser^4 (18,19). Leuprolide was supplied as an acetate salt (pH 5.0) and is highly soluble in a variety of solvents, making high concentration formulation more achievable.

Solubility

The solubility of pharmaceutically active proteins under nonaqueous conditions has been well characterized (4,20–23). Most of the solvents utilized in the literature are not pharmaceutically acceptable; however, the rationale for selecting a solvent is applicable, where the physical characteristics of organic solvents that are useful in selecting a solvent include hydrophobicity (log *P*), dielectric constant (ε), dipole moment (μ), and the Hildebrand solubility parameter (δ). The solubility of lysozyme in 34 solvent systems was determined to correlate well with hydrophobicity, but little to no correlation was observed with dielectric constant, the solubility parameter, or dipole moment (24). However, others have observed enhanced dissolution in solvents with a high dielectric constant (20,24). Formulation screens in organic solutions have also been used to correlate successful stability of PLGA-encapsulated hGH and γ-INF (4).

One should keep in mind that little theoretical meaning can be attributed to a nonaqueous pH reading, and the definition of pH in cosolvent solutions may be better described as an apparent pH. Under aqueous conditions, the concentration of hydrogen ions in solution is usually expressed in terms of the hydrogen ion concentration or activity, or in terms of pH units. The pH of a nonaqueous solution or cosolvent solution is more difficult to measure, since pH only relates to purely aqueous conditions. The pH of a water-miscible solution can be measured with an electrode; however, interpretation of the pH value must be done with care.

A parameter affected by the solvent is the liquid junction error, where the glass electrode may be off by as much as a pH unit due to liquid junction potential (25). For example, a 52% solution of ethanol (EtOH) has a liquid junction potential error of −0.46 pH units (25). Furthermore, the addition of a cosolvent with varying polarity will affect the solubility of the solute. Addition of EtOH increases the solubility of the unionized species, by decreasing the polarity of the solvent. The alcohol also decreases the dissociation of the solute, where solubility decreases as the pK_a increases and the dissociation constant decreases (26). Additionally, the dissociation constants of many organic solvents are not the same as that of water. For example, an aqueous solution is neutral at pH 7.0, but methanol (MeOH) and EtOH are neutral at 8.42 and 9.55, respectively (25). This means that alcohols are less dissociated than water at a pH reading of 7.0. Finally, the pH scales for solvents are of different range and breadth. The pH scale for water ranges from 0 to 14, while the pH scale for EtOH ranges from −4.2 to 14.9. In addition to the dissociation constants and the range of the pH scales, the effect shielding of the solvent will vary. For example, MeOH has a dielectric constant of 32.7 compared with 78.3 for water at 25°C, indicating that MeOH will not shield separated charges as well as water. Many cosolvent mixtures will have a combination of characteristics; however, these effects may not be linear with the titration of organic content. Small ratios of MeOH in water will not affect the pH dramatically, but small ratios of water in MeOH will significantly change the pH (27). For EtOH, the required adjustment in the pH value is approximately

0.25 units at 70% EtOH. Some tables for adjusting pH readings are published, but are not comprehensive (25,27). The pH measurements in the literature include a range of appropriately adjusted pH measurements, uncorrected pH readings, and pH values pertaining to the "pH memory" of the lyophilized material prior to dissolution.

Solubility is a function of pH, where more charge on the protein surface correlates to increased solubility. The isoelectric point (pI) of the protein should be used to determine if the pH of the solution will have a large or small effect on the overall charge on the molecule. If the pH of the solution is far from the pI, the difference in the apparent pH and the actual pH may not have a large effect on the stability or conformation of the protein. Furthermore, solubility may be enhanced by lyophilizing the protein at a pH away from its pI (21,24,28) prior to dissolution in a relatively polar organic vehicle. Lysozyme (pI 11.0) showed a 1000-fold decrease in solubility in 56% acetonitrile/44% dimethyl sulfoxide (DMSO) when reconstituted from pH 10.0, as compared with pH 2.0 (24). Similarly, insulin (pI 5.3) solubility increased from 160 to 1100 μg/mL as the pH decreased from 7.4 to 3.0 (29). These findings suggest that pH values equidistantly above and below the pI should not be assumed to afford equivalent and/or linear increases in solubility, as the distribution of amino acid pK_as exposed to the surface will dictate these characteristics. The resultant increase in solubility can be generally attributed to increased charge repulsion, decreased aggregation, and water retention on the protein (20).

Preliminary leuprolide formulations were characterized in aqueous and nonaqueous conditions, where the criteria for success were high solution solubility and good physical (lack of gelation or precipitation) and chemical stability. Leuprolide solubility was determined to be in excess of 400 mg/mL in water, propylene glycol (PG), and DMSO at 25°C. Specifically, leuprolide was most readily soluble in DMSO, where the saturation solubility in DMSO was determined to be 580 mg/mL (30). Leuprolide solubility in DMSO, at 400 mg/mL, was determined to be relatively consistent over a wide temperature range (–20°C to 50°C), and only decreased slightly (10–15%) with increasing moisture (0–15% water) and with increasing acetate (10–70 mg/mL ammonium acetate).

Aggregation and Gelation

Three leuprolide formulations (solubilized in water, DMSO, and PG) were placed on stability at 25°C and 37°C, where both the water and PG formulations were observed to gel over time (31). The gelled formulations were assessed by Fourier transform infrared (FTIR). The ungelled water formulation showed two distinct bands at 1615/cm and 1630/cm, corresponding to aggregate and β-sheet structure, respectively (31). The lack of β-turn bands was attributed to the extremely high peptide concentration, where it was hypothesized that increasing peptide concentration would transition the intramolecular hydrogen bonds stabilizing the two β-turn structures to intermolecular hydrogen bonds stabilizing the β-sheet structure.

The 1630/cm band has been characterized in the literature as an extended inter-molecular antiparallel β-sheet structure (32–34). Furthermore, FTIR analysis of the gelled water formulation revealed that the aggregate band at 1615/cm grew in intensity, and could be correlated to the onset of gelation.

Similar FTIR results were obtained for the PG formulation. The ungelled PG formulation FTIR spectra exhibited a β-turn band at 1690/cm and β-sheet bands at 1642/cm and 1630/cm. The FTIR band shift between water and PG was attributed to the solvent being less polar. Upon gelation of the PG formulation, the bands at 1690/cm and 1642/cm were observed to increase, suggesting aggregation of leuprolide molecules in an antiparallel conformation. Other researchers concluded that a combination of 1630/cm and 1680/cm were indicative of an anti-parallel β-sheet structure for melittin (35).

A common approach to enhancing the solubility in aqueous and in nonpolar organic solvents can be achieved by the addition of salts (36,37). An increase in protein solubility with the addition of salt can be explained by the Debye–Huckel theory. Binding between protein and weakly hydrated ions results in decreased protein electrostatic free energy, increased solvent activity, and increased solubility. Conversely, the addition of salts can also result in gelation and/or precipitation. Interfacial effects between the protein and strongly hydrated ions essentially remove water molecules and desolvate the surface (38,39). For example, the solubility of human insulin–like growth factor in 140 mM benzyl alcohol and 145 mM NaCl increased aggregation and decreased solubility (38). In organic solutions, increasing ionic strength can mask charged groups on the protein with salt ions, increasing protein solubility in less polar or nonpolar solvents.

The addition of $CaCl_2$ to aqueous leuprolide solutions resulted in faster onset of gelation and the production of firmer gels, and therefore in no improvement of solubility. Increasing salt concentration increased gelation. Furthermore, the addition of divalent anions instead of monovalent anions produced firmer gels ($SO_4^{-2} > H_2PO_4^- > HCO_3 > Cl^-$), consistent with the Hofmeister series (31,40). The Hofmeister lyotropic series has been used to select an anionic or cationic species, where the precipitation ability is related to the hydration of the ion and its ability to separate water molecules from the hydrophilic regions of the molecule. These findings were similar to previous work on dilute aqueous solutions of detirelix (41,42).

Both the aqueous and PG gelled formulations were observed to be birefringent under a polarized light microscope. The appearance of birefringence was consistent with those observed for nafarelin and detirelix, indicative of lyotropic liquid crystal formation (41–44). Detirelix and nafarelin (4–8 mg/mL) formed birefringent nematic liquid crystals at much lower concentrations than leuprolide (42). Aqueous solutions of detirelix (4 mg/mL) rapidly formed nematic liquid crystals of undulose extinction that birefringe less than 0.001 where the onset of liquid crystal formation and critical melting temperature for detirelix were determined (41). Liquid crystals are defined as lyotropic and thermotropic (45,46). Thermotropic liquid crystals are induced by a change in temperature and are substantially solvent free. Lyotropic liquid crystals are induced by the presence of solvent, and are usually formed by amphophilic molecules.

A structure for the leuprolide liquid crystal has been proposed (31). For aqueous leuprolide gels prepared at pH 5, both His[2] and Arg[8] are protonated. The charged residues appear on the same face of the leuprolide β-sheet structure and can be rotated to the solvent-exposed interface, allowing dimers to form along the hydrophobic faces. Association of the dimers to allow ring stacking for the hydrophobic interiors was proposed as a feasible two-dimensional liquid crystal structure (Fig. 1). Addition of salts allows anions to form weak salt bridges between the charged residues in the β-sheet–rich dimers to further stabilize the structure. If ionic interactions between anions and protonated His and Arg residues occur, then divalent anions would be more effective than monovalent anions, consistent with the studies on detirelix.

Finally, the leuprolide formulation dissolved in DMSO did not gel over time and revealed a radically different FTIR spectra. The bands were wide, indicative of a rapidly changing structure, and comprised α-helix at 1658/cm, random coil at 1648/cm, and minor β-sheet and aggregate bands at 1632/cm

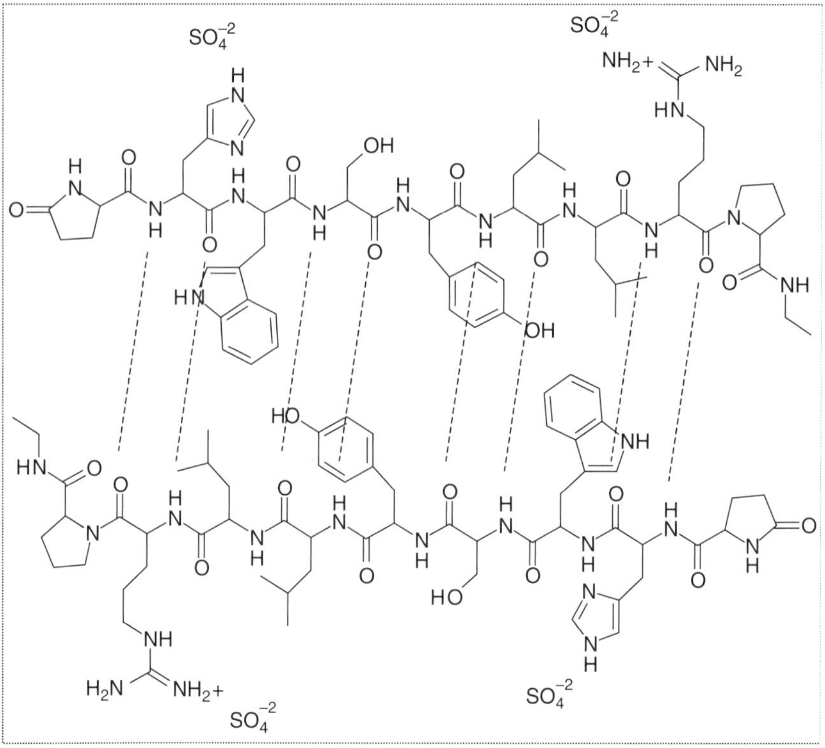

Figure 1 Proposed structure of leuprolide liquid crystal in the presence of divalent anions.

and 1615/cm, respectively. Therefore, this formulation in DMSO became the lead candidate.

Solution Stability

Leuprolide formulations in water, PG, and DMSO were placed on stability for three years at 37°C, and assayed by reversed phase-high performance liquid chromatography (RP-HPLC) and size exclusion chromatography (SEC) (Fig. 2) (8,47,48). The DMSO formulation provided the best chemical stability, where leuprolide stored for two years at 37°C showed that 75%, 82%, and 93% leuprolide remaining in water (pH 5.0), PG, and DMSO, respectively (30,48). Therefore, the DMSO formulation was utilized in clinical studies because it was gelation free and provided the best stability. However, for the purposes of this chapter, the DMSO and water formulations were analyzed in order to compare and contrast their chemical stability.

The stability data was fit to pseudo-first-order kinetics and revealed a linear Arrhenius plot, where an $E_a = 20.5 \pm 2.0$ kcal/mol was calculated for the aqueous formulation (48). A pseudo-first-order fit was linear and utilized, even though several of the degradation pathways may be second order. The major degradation products were identified by liquid chromatography/mass spectrometry (LC/MS) and grouped into four major degradation pathways (48). The proportion of leuprolide degradation products in water, at 37°C, were hydrolysis > aggregation > isomerization > oxidation, where the prevalence of the pathway was not observed to change with increasing temperature.

Similarly, the DMSO formulation also revealed a linear Arrhenius plot, where $E_a = 22.6 \pm 1.2$ kcal/mol (48). However, the proportions of leuprolide degradation products were different in DMSO, at 37°C, and were aggregation > oxidation > hydrolysis > isomerization. Hence, one of the major benefits of formulating under nonaqueous conditions is that hydrolytic and isomerization degradation pathways can be minimized. Furthermore, the use of an aprotic solvent limits the hydrogen source for initiation of degradation.

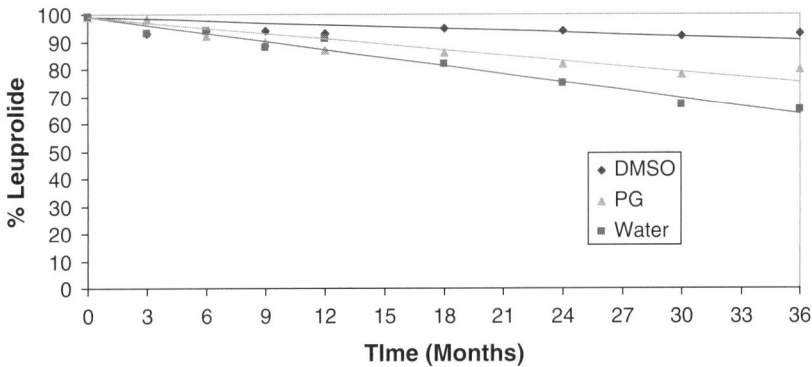

Figure 2 Leuprolide stability in implants at 37°C ($n = 3$) for three years.

The majority of degradation products in DMSO were similar to the water formulation, where the hydrolysis pathway was primarily backbone cleavage C-terminal to Trp^3, Ser^4, Tyr^5, Leu^6, and Leu^7 (30,49). Isomers of leuprolide were attributed to His^2, Trp^3, and Ser^4 (49–51), and three oxidation products of Trp^3 were identified (30,48,52–57). A few degradation products were more prevalent in a specific formulation. For example, N-terminal acetylation and cyclo(His-Trp) were most abundant in the water formulation (48). Conversely, β-elimination at Ser^4, followed by hydrolytic cleavage between the α-carbon and amide nitrogen of the resulting dehydroalanine residue, resulting in the C-terminal fragment HO-Dha-Tyr-Leu-Leu-Arg-Pro-NHEt, was observed in DMSO (30,48,58–60).

Leuprolide in DMSO showed increased stability with increasing peptide concentration (50–400 mg/mL), where 73% and 80% leuprolide remained in 50 and 400 mg/mL formulations stored at 80°C for two months (30). This suggests that leuprolide may be self stabilizing at high concentrations. Overall, the proportion of degradation products at 80°C did not change between high and low leuprolide concentration. Some individual hydrolytic and isomerization degradation products were observed to decrease with increasing leuprolide concentration, but this effect was minor (30).

The effect of temperature on leuprolide stability in DMSO was examined at 37°C, 50°C, 65°C, and 80°C for up to three years (30). Leuprolide degradation at 50°C, 65°C, and 80°C showed accelerated exponential decay with time; however, the rate of degradation at 37°C appeared to plateau after six months. This may be attributed to the consumption of residual water, other reactive impurities, the kinetics relating to the specific degradation pathway, and changes to the major degradation pathway with temperature. For example, the proportion of leuprolide degradation products changed from aggregation > oxidation > hydrolysis > isomerization at 37°C to aggregation > isomerization > hydrolysis > oxidation at 80°C (30). Therefore, the relative importance of oxidation from temperature-accelerated conditions would have been underestimated.

The effect of residual moisture on leuprolide stability in DMSO was also characterized (48). A lyophilized leuprolide acetate drug substance containing 5% to 8% moisture results in a leuprolide formulation containing 2% to 3% moisture. Residual water in an aprotic solvent may remain preferentially bound to charged peptide side chains or it may become equilibrated in the highly polar DMSO (μ = 4.0 D). Either way, the leuprolide molecules are effectively exposed to equivalent or less moisture than the lyophilized drug substance. Leuprolide solutions in DMSO were spiked to obtain 2% to 15% water and placed on stability for six months at 50°C (48). As expected, increased moisture content resulted in accelerated leuprolide degradation by 4% with 15% water. Therefore, moisture content in the formulation was minimized through manufacturing and process steps including a dry nitrogen headspace (61). It should also be mentioned that under these "dry" conditions (four to six water molecules per leuprolide molecule), hydrolytic degradation pathways may no longer fit first-order criteria.

The effects of leuprolide concentration, temperature, and moisture may be further accentuated by solvent issues such as dielectric constant, dissociation constant, and oxygen solubility. First, as the dielectric constant decreases (ε_{water} = 80.0, ε_{DMSO} = 47.2 at 20°C) the degradation rate decreases, consistent with studies on the rate of deamidation in organic cosolvents (62–64). Specifically, the rate of asparagine deamidation for Val-Tyr-Pro-Asn-Gly-Ala (pH 7.4) in water, glycerol, EtOH, and dioxane decreased with decreasing dielectric constant (63). Theoretically, the decrease in the rate of deamidation may be due to destabilization of the deprotonated nitrogen anion in the peptide backbone responsible for attack on the asparagine side chain and formation of the succinimide intermediate. Furthermore, a lower dielectric constant for leuprolide in DMSO was consistent with a shift in proportion of degradation products, resulting in less hydrolytic/isomerization.

Second, as the temperature increases, the dissociation constant of residual water increases from 2.57×10^{-14} at 37°C to 23.4×10^{-14} at 80°C (26). The dielectric constant of DMSO will also affect the dissociation of residual water, where it may be dissociated to a lesser extent in DMSO. An increase in the dissociation of water with increasing temperature would accelerate hydrolytic degradation, consistent with the data. The limited source of residual water would also be consumed faster.

Third, oxygen solubility decreases with increasing temperature, and will therefore result in less-apparent temperature sensitivity (52,65). As temperature decreases, oxygen solubility and the concentration of reactive species increase, resulting in increased oxidation products.

Sterilization

Early process and scale-up studies explored the terminal sterilization of the Viadur leuprolide acetate implant (61). When the formulation was subjected to γ-irradiation (25–35 kGy), the formulation stability resulted in a 2% loss of leuprolide, followed by a slightly increased rate of degradation on stability. Therefore, the leuprolide formulation in DMSO exhibited no appreciable change in viscosity or gelation and was sterile filtered prior to aseptic fill.

SYSTEM DESIGN

The Duros implant releases a therapeutic agent at a predetermined, typically zero-order, delivery rate based on the principles of osmosis (8,47,66,67). Osmosis is the natural movement of a solvent through a semipermeable membrane into a solution of higher solute concentration, leading to equal concentrations of solute on both sides of the membrane.

The implant is a miniature (4 mm diameter by 45 mm length), osmotically driven, drug delivery system designed for the long-term, parenteral, zero-order delivery of potent therapeutic agents in humans. These single-use implants are

sterile, nonpyogenic, and nonbiodegradable. The implant consists of an impermeable titanium alloy cylinder capped on one end by a rate-controlling, semipermeable membrane and on the other end by an orifice (diffusion moderator) for drug
delivery (Fig. 3). The titanium alloy reservoir can withstand impact at the implant
site and was designed to contain the semipermeable membrane when pressure is
exerted by a swelling engine compartment. The titanium reservoir is impermeable
to water, ensuring drug stability and continual zero-order release.

The semipermeable membrane is permeable to water, but impermeable to
ionic or higher molecular weight compounds. Therefore, the membrane is not
influenced by extracellular fluid and remains chemically stable under physiological
conditions. The membrane can be constructed of cellulosic esters, polyamides, or
polyurethanes, and can be modified by varying the polymer chemistry and process
variables (8). For example, water content varies linearly, and hydraulic permeability varies exponentially with acetyl content in cellulosic ester membranes. Typical
membrane permeabilities range from 10^{-7} to 10^{-6} g/cm/sec (7,8). The Viadur leuprolide acetate implant utilizes a polyurethane membrane. In the implant, the surface area and thickness of the membrane are controlled by design specification
tolerances and the manufacturing process. The permeability of the polyurethane
membrane has been shown to be constant over time, both in vitro and in vivo.

The orifice was designed with a small inner diameter and a suitable length
so that the diffusional contribution to the release rate is minimized at low delivery
rates. For example, the orifice design is small enough to minimize diffusional
fluxes and large enough to prevent pressure buildup. If the effective cross-sectional
area of the orifice is too large and the orifice length is too short, then the diffusional fluxes can be on the same order of magnitude as the osmotic delivery rate
(66). The orifice is designed to prevent back diffusion of extracellular components
when the system is implanted, preventing exposure of the drug formulation to the
surrounding tissue prior to delivery. The orifice can be constructed of biocompatible polymers or titanium.

The interior of the implant contains a polymeric piston that separates the
osmotic engine from the drug reservoir (Fig. 3). Therefore, the osmotic system

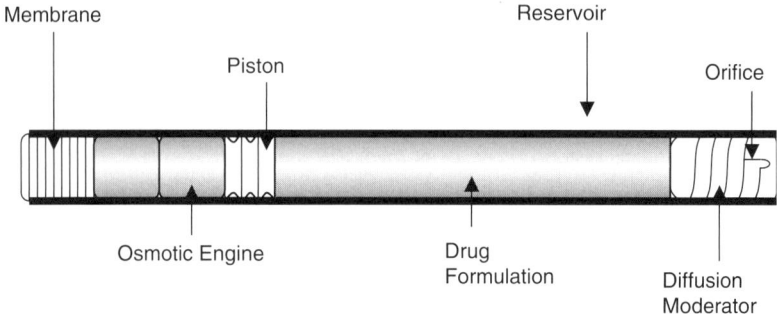

Figure 3   Cross-sectional diagram of the Viadur leuprolide acetate implant.

imbibes water from the body through a semipermeable membrane into the osmotic engine, which swells, resulting in slow and even delivery of drug formulation. For most applications, the osmotic agent in the engine is sodium chloride. This specific salt was chosen after safety considerations and the ubiquitous presence of sodium chloride in the body. In addition, an implant can safely carry excess sodium chloride to maintain a saturated solution throughout the duration of the delivery period. The osmotic engine usually contains gelling polymers [poly(vinylpyrrolidone) or sodium carboxymethylcellulose] in addition to >50 wt% sodium chloride. The elastomeric piston seals the drug formulation from the osmotic engine, is compatible with the drug formulation and osmotic engine, and moves with relatively little resistance.

In operation, water is drawn through the semipermeable membrane in response to an osmotic gradient between the osmotic engine ($\pi = 356$ atm) and moisture in the surrounding interstitial fluid ($\pi \sim 7$ atm) (7,8). The rate of water influx is governed by the permeation characteristics of the semipermeable membrane. As water flows into the implant, the osmotic engine expands as it imbibes water and exerts pressure on the piston. The resulting movement of the piston delivers drug formulation from the orifice at a rate corresponding to the rate of water permeation.

The governing equation for the rate of drug delivery, dm/dt, is given by

$$dm/dt = (A/h)k\Delta\pi C$$

where A is the membrane cross-sectional surface area, h is the membrane thickness, k is the effective permeability of the membrane, $\Delta\pi$ is the osmotic pressure gradient between the engine and the surrounding tissue, and C is the drug concentration in the formulation. The implant is designed for constant drug delivery. If A, h, k, and $\Delta\pi$ are held constant and the drug formulation is stable, so that a constant concentration is maintained, then a constant drug delivery rate is obtained.

DELIVERY

Implantation

The Viadur leuprolide acetate implant is implanted subcutaneously on the inside of the upper portion of the nondominant arm. The implantation is an outpatient procedure requiring a local anesthetic. A 4 to 5 mm incision is made with a scalpel at one end of the anesthetized area. The implant is inserted through the incision, subcutaneously, with the aid of a specifically designed implanter (trocar/cannula). The incision site is closed with a Steristrip and a sterile bandage.

Removal of the implant at the end of the delivery period is also performed in an outpatient procedure. After application of local anesthetic, a 4 to 5 mm incision is made perpendicular to one end of the implant. Finger pressure is applied to the other end of the implant to elevate the removal end. A small slit is made through any surrounding fibrotic tissue to expose the end of the implant. The formation of any fibrotic tissue is not vascularized and does not impact bioavailability. The implant is then expelled with finger pressure on the opposite end, and the incision site is closed with a Steristrip and a sterile bandage.

In Vitro Performance

Leuprolide in vitro release rate studies were performed in test tubes containing phosphate buffered saline (PBS) and 2% sodium azide held at 37°C. The PBS was tested weekly, by RP-HPLC, for leuprolide content (68). Release rate data from the Viadur leuprolide acetate implant demonstrated zero-order delivery for up to one year (Fig. 4). Initially (day 1), a high release of drug was observed due to thermal effects (equilibration to 37°C) during system start up. By day 14, systems deliver at steady state rates, and continue for the remainder of the one-year delivery period.

Implants placed in storage at 25°C for 18 months prior to in vitro release rate testing at 37°C also exhibit the same release rate profile. Similarly, the delivery rates from multiple lots of implants manufactured with varying lots of polyurethane and drug substance were reproducible (68).

Leuprolide release rate was also investigated at more frequent time points. Four implants previously at steady state release, in vitro, for several months were pulled and monitored for a 24-hour period at six minute intervals to investigate the continuity of leuprolide delivery (68). The continuously increasing ultraviolet absorbances demonstrated steady leuprolide delivery from implants. Average

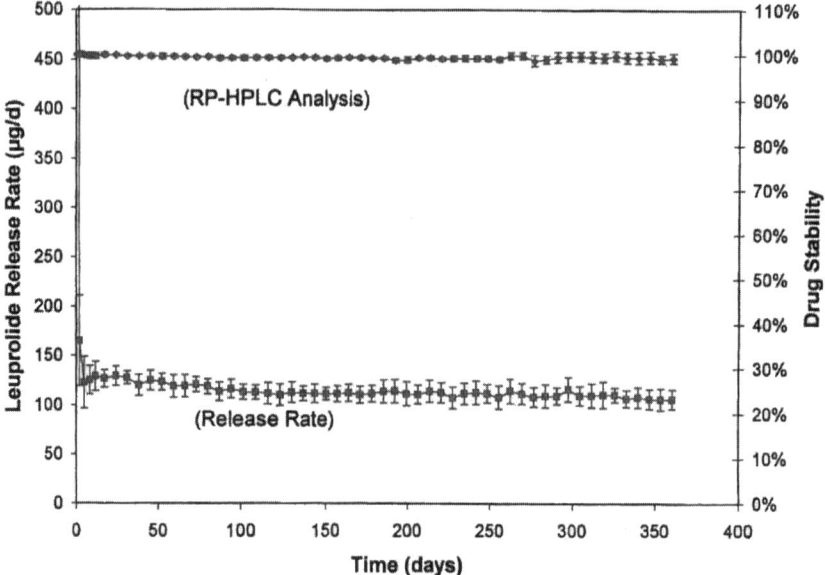

Figure 4   In vitro release rate and formulation stability (37°C) of the Viadur leuprolide acetate implant for one year ($n = 24$).

leuprolide release rates were calculated using the slopes of the absorbance versus time plots determined by linear regression analysis and subtracted against blanks. The average release rate of implants during the 24 hour period was 122.7 ± 2.9 μg/day, which agreed well with the average release rate of 129.5 ± 7.3 μg/day using the weekly RP-HPLC method (68).

In Vivo Performance

The progress of prostate cancer is dependent on circulating androgen levels. Testicular androgen ablation, or decreasing serum testosterone levels to castrate level, has been the standard for primary therapy for advanced prostate cancer for more than 50 years (69). Gonadotropin-releasing hormone (GnRH) analogs have been approved for the treatment of prostate cancer, and have become the preferred choice to orchiectomy or treatment with diethylstilbestrol (70). GnRH analogs, like leuprolide, stimulate the pituitary testicular axis, resulting in a temporary increase in serum testosterone levels. This is followed by downregulation of the pituitary GnRH receptors, resulting in decreased secretion of luteinizing hormone, and suppression of serum testosterone levels to below the castrate level (50 ng/dL) (71–74).

Subsequently, in vivo studies were performed in rats, dogs, and humans. Leuprolide implants were placed into the dorsal subcutaneous space of 120 male Fischer 344 rats. Systems were explanted at 3, 6, 9, and 12 months, and implants were assayed for residual drug content. Linear cumulative delivery was shown to be reproducible in multiple lots of systems implanted into rats (data not shown) (68).

Systems were also implanted into the dorsal subcutaneous space of six sexually mature male beagle dogs (68,75). Serum testosterone and leuprolide levels were monitored, and showed steady release rates for 12 months (Fig. 5). Measurable leuprolide serum levels were observed in the first sample taking, reflecting the rapid onset of leuprolide delivery, consistent with the in vitro delivery profile. Serum testosterone levels exhibited a classical pattern of response to a GnRH analog (68). Prior to implantation, daily testosterone levels were 13 to 630 ng/dL; upon implantation, the average testosterone levels rose and then declined to 50 ng/dL (below castrate level) by day 28 and remained below the castrate level for the duration of treatment.

Similarly, steady serum leuprolide levels were observed in Phase I/II clinical trial patients ($n = 27$) treated with a single implant for one year (Fig. 6) (73,74). Patients were implanted subcutaneously, under local anesthesia, using a specifically designed implanter (trocar). Serum testosterone and leuprolide levels were measured by radioimmunoassay and LC/MS/MS, respectively. Serum testosterone levels exhibited the classical response to continuous administration of a GnRH analog, initially rising above baseline during the first week of therapy and declining to below the castrate level by week 2 to 4, and remained suppressed for the duration of the one-year implant life (73,74).

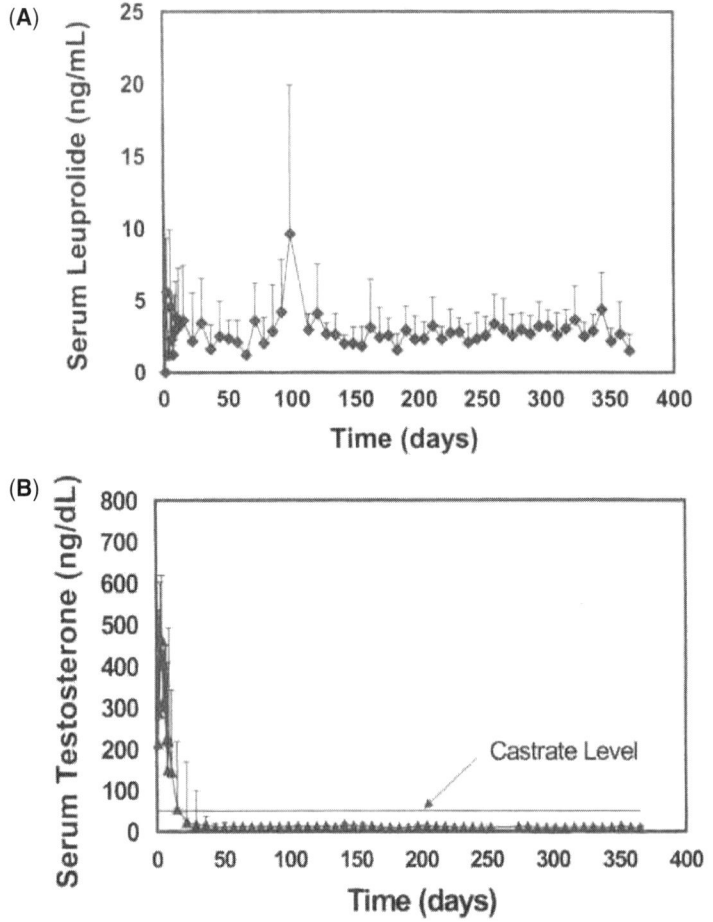

Figure 5   Serum leuprolide and testosterone levels in canines (*n* = 6) with (**A**) serum leu-prolide levels and (**B**) serum testosterone levels (castrate threshold level of 50 ng/dL).

In Vivo/In Vitro Performance Comparison

Finally, a comparison of in vivo and in vitro performance was conducted using two methods: cumulative system leuprolide release rate (amount delivered as calculated from residual drug and initial drug content in systems) and stability of the drug formulation remaining in the explanted system (76). When the in vivo systems were explanted from animals, the corresponding lot of in vitro systems were also terminated and assayed by RP-HPLC.

In rat studies, good in vivo/in vitro correlation was obtained from cumulative drug delivery at explantation times of 3, 6, 9, and 12 months (68). The average ratio of cumulative amount delivered at 12 months in vitro to cumulative

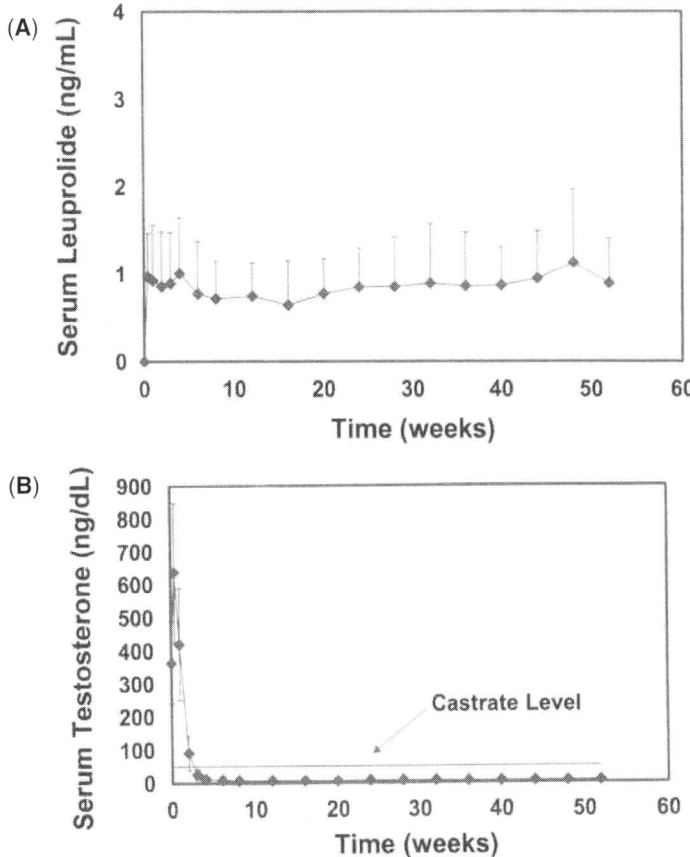

Figure 6 Serum leuprolide and serum testosterone levels in patients receiving on Viadur leuprolide acetate implant in Phase I/II human clinical trials ($n = 27$) with (**A**) serum leuprolide levels and (**B**) serum testosterone levels (castrate threshold level of 50 ng/dL).

amount delivered in vitro was 1.04 ± 0.03. Leuprolide stability remaining in the explanted systems at 3, 6, 9, and 12 months showed no appreciable degradation (68). The observed in vivo/in vitro comparison would not have occurred if the membrane had fouled or was altered in its permeability characteristics.

In dog studies, the in vivo/in vitro comparison results were very similar. Good agreement was observed between the amount of drug delivered in vitro and in vivo: the ratio of cumulative amount delivered in vitro to in vivo was 1.05 ± 0.06. Drug stability in systems explanted at 12 months showed essentially no degradation.

In Phase I/II clinical trials, the cumulative drug release for systems explanted at 12 months showed similar results. The ratio of cumulative amount delivered in vitro to cumulative amount delivered in vivo was 1.03 ± 0.08. Leuprolide stability was unchanged.

OTHER APPLICATIONS

The release rate and duration of the Duros implant depend on the implant size, drug concentration, semipermeable membrane, and osmotic engine design. For example, increasing the membrane permeability can increase the release rate and decrease system duration. Similarly, the implant size and drug reservoir can be increased to accommodate higher therapeutic doses and less potent molecules. For example, by increasing the diameter slightly, but maintaining the same length, a 500 µL drug reservoir system can be obtained.

Continuous zero-order release provides a relatively smooth pharmacokinetic profile, when compared to bolus dosing, and results in the constant delivery of drugs with short half lives. The osmotic technology allows for the delivery of highly concentrated formulations of proteins and peptides stabilized in suspension or nonaqueous conditions. The Duros implant is most often implanted subcutaneously; however, the implant can be adapted to other routes of targeted administration with the attachment of a catheter (8). A variety of pharmaceutically active molecules have been formulated for the implant, and these are discussed below.

Salmon Calcitonin

Salmon calcitonin (sCT) has been used for the treatment of osteoporosis and Paget's disease, with doses ranging from 12 to 25 µg/day (77,78). Assuming a 12-month implant (150 µL implant), a peptide concentration of 30 to 60 mg/mL would be required (8).

Aqueous solution formulations gelled, forming fibrils (79,80). However, when calcitonin was dissolved in DMSO at 50 mg/mL, no gelation was observed, and it resulted in good stability, by RP-HPLC and SEC, after one year at 37°C. Furthermore, in vitro delivery from the Duros implant at 18 µg/day for four months was reported (81).

Glucagon-Like Peptide

Glucagon-like peptide (GLP) has been shown to be useful in the treatment of Type II diabetes at doses of approximately 750 mg/day. Assuming a six-month duration (500 µL implant), a formulation concentration of 270 mg/mL was required. Both solution and suspension formulations were tested. Solution formulations in DMSO gelled above 100 mg/mL; therefore, a suspension formulation was pursued (82). The suspension demonstrated adequate stability for six months at 37°C, as tested by RP-HPLC and SEC (82).

Suspensions present additional challenges for controlled delivery from an implantable pump. Suspension formulations may appear to be stable under static conditions, but can exhibit instabilities under flow conditions. For example, sieving of excipients under extremely low flow rates and preferential delivery of the formulation vehicle must be avoided. Suspension formulation also requires the excipient vehicle to have sufficient viscosity to keep the particles homogenously suspended over the shelf life and implanted delivery period. Since protein suspensions are subject to denaturation, aggregation, and sedimentation, these formulations require characterization of particle charge, size distribution, surface area, wetting, electrostatic double-layer formation, zeta potential, and flocculation profile (27). The GLP suspension formulations effectively resisted settling or sieving over time and resulted in a constant in vitro delivery rate of 480 µg/day for one month (82,83).

α-Interferon

α-Interferon (α-INF) has been utilized for the treatment of hepatitis C, hairy cell leukemia, and Kaposi's sarcoma when dosed at approximately 3 mL U/day (78,84). Assuming a three-month duration (150 µL implant) and 5 to 10 µg/day, a formulation concentration of 5 to 10 mg/mL α-INF would be required. A suspension formulation was pursued since α-INF showed poor solution stability (85,86). Lyophilized α-INF was suspended at 5 mg/mL in perfluorodecalin and showed similar stability to dry powder stored at −80°C after one year at 37°C.

ω-Interferon

ω-interferon (ω-INF)is currently in Phase II clinical trials for the treatment of hepatitis C (87). Patients receive daily injections of ω-INF for up to 48 weeks in order to approximate the blood levels anticipated for a Duros implant. A three-month duration implant (150 µL implant) is targeted.

Factor IX

Factor IX is a serine protease used for the treatment of hemophilia B, and is dosed at 10 to 20 IU/kg/day (78,88,89). Assuming a three-month duration (500 µL implant) and 500 IU/day, a formulation would require a Factor IX concentration of 360 mg/ mL. A suspension formulation was pursued for several reasons: Factor IX solubility was limited, and a high drug loading and stability at 37°C were required. To achieve the required stability and drug loading, Factor IX (pH 6.8) was suspended in perfluorodecalin and perfluorotributylamine for six months at 37°C with little loss of chemical stability (85). The stability profile was similar to lyophilized powder stored at −80°C. Similar suspensions were prepared in methoxyflurane, octanol, and PEG 400 with less promising stability. These results indicated that halogenated systems might give the least interaction between the particulate and the solvent, resulting in chemical and conformation stability similar to lyophilized powders.

Other proteins have also been successfully stabilized by suspending in organic solvents (21). This suspension strategy allows the control of moisture content, increased resistance to thermal denaturation and chemical degradation, and solid state stabilization of the protein structure in the particulate, likely due to minimized protein flexibility (22,90–92). Preservation of this "molecular memory" or "molecular imprint" of the protein can also be used to stabilize structure and activity (93,94). Studies have shown that the presence of the organic solvent has little effect on the protein particulate or its residual bound water (95). However, nonpolar solvents can increase the amount of water bound to the protein, while polar solvents can reduce the amount of bound water, possibly by replacing water with secondary hydration layers (95).

CONCLUSION

Very little work is available in the literature on high-concentration protein formulations; however, this is an emerging area. Classical challenges such as solubility, aggregation, and stability limitations are more familiar. New considerations for high-viscosity delivery, scale-up of novel manufacturing methods, and the cost of goods (fill overage issues) provide additional development challenges. Achieving a successful and scaleable high-concentration formulation requires an integrated approach between formulation, analysis, process, and packaging/device design. The effectiveness of the therapeutic agent can be enhanced by effective drug delivery that activity controls the delivery rate and the site of drug action.

REFERENCES

1. Shire SJ, Shahrokh Z, Liu J. Challenges in the development of high protein concentration formulations. J PharmSci 2004; 93:1390–1402.
2. Stevens RC. Design of high-throughput methods of protein production for structural biology. Structure 2000; 8:R177–R185.
3. Shenoy B, Wang Y, Shan W, Margolin AL. Stability of crystalline proteins. Biotechnol Bioneng 2001; 73:358–369.
4. Cleland JL, Jones AJS. Stable formulations of recombinant human growth hormone and interferon-γ for microencapsulation in biodegradable microspheres. Pharm Res 1996; 13:1464–1475.
5. Costantino HR, Firouzabadian L, Wu C, et al. Proteins spray freeze-drying. 2. Effect of formulation variables on particle size and stability. J PharmSci 2002; 91:388–395.
6. Costantino HR, Andya JD, Nguyen PA, et al. Effect of mannitol crystallization on the stability and aerosol performance of a spray-dried pharmaceutical protein, recombinant humanized anti-ige monoclonal antibody. J PharmSci 1998; 87:1406–1411.
7. Wright JC, Stevenson CL, Stewart GR. Osmotic pumps-DUROS osmotic implants for humans. In: Mathiowitz E, ed. Encyclopedia of Controlled Drug Delivery. New York: John Wiley and Sons, 1999.
8. Stevenson CL, Theeuwes F, Wright JC. Osmotic implantable delivery systems. In: Wise D, ed. Handbook of Pharmaceutical Controlled Release Technology. New York: Marcel Dekker, Inc., 2000.
9. Fara JW, Ray N. Osmotic pumps. In: Praveen T, ed. Drug Delivery Devices. New York: Marcel Dekker, 1988:137–176.

10. Chien YW. Implantable therapeutic systems. Controlled Drug Delivery. New York: Marcel Dekker, 1987:481–522.

11. Zingerman JR, Cardinal JR, Chern RT, et al. The in vitro and in vivo performance of an osmotically controlled delivery system- IVOMEC SR bolus. J Control Release 1997; 47:1–11.

12. Siegel RA, Firestone BA. Mechanochemical approaches to self-regulating insulin pump design. J Control Release 1996; 11:181–192.

13. Baker R. Osmotic and mechanical devices. In: Baker R, ed. Controlled Release of Biologically Active Agents. New York: John Wiley and Sons, 1987:132–155.

14. Kaisary A, Tyrell CJ, Peeling WB. Comparison of LHRH analogue (Zoladex) with orchiectomy in patients with metastatic prostatic carcinoma. Br J Urol 1991; 67:502–508.

15. Huben RP, Murphy GP. A comparison of diethylstilbestrol or orchiectomy with buserelin and with methotrexate plus diethylstibestrol or orchiectomy in newly diagnosed patients with clinical stage D2 cancer of the prostate. Cancer 1988; 62:1881–1887.

16. Soloway MS, Chodak G, Vogelzang NJ. Zoladex versus orchiectomy in treatment of advance prostate cancer: a randomized trial. Urology 1991; 37:46–51.

17. Adjie AL, Hsu L. Leuprolide and other LHRH analogues. In: Wang YF, Pearlman R, eds. Stability and Characterization of Protein and Peptide Drugs. New York: Plenum Press, 1993:159–199.

18. Nikiforovich GV, Marshall GR. Conformation-function relationships in LHRH analogs II. Conformations of LHRH peptide agonists and antoagonists. Int J Pept Protein Res 1993; 42:181–193.

19. Andersen NH, Hammem PK. A conformation-preference/potency correlation for GnRH analogs: NMR evidence. Bioorg Med Chem 1991; 5:263–266.

20. Houen PG. The solubility of proteins in organic solvents. Acta Chem Scand 1996; 50:68–70.

21. Zaks A, Klibanov AM. Enzymatic catalysis in nonaqueous solvents. J Bio Chem 1988; 263:3194–3201.

22. Zaks A, Klibanov AM. The effect of water on enzyme action in organic media. J Biol Chem 1988; 263:8017–8021.

23. Stevenson CL. Characterization of protein and peptide stability in non-aqueous solvents. Curr Pharm Biotechnol 2000; 1:165–182.

24. Chin JT, Wheeler SL, Klibanov AM. Protein solubility in organic solvents. Biotechnol Bioeng 1994; 44:140–145.

25. Frant MS. How to measure pH in mixed and nonaqueous solutions. Today's Chemist at Work January 1995:39–42.

26. Martin A, Sarbrick J, Cammarata A. Physical Pharmacy. Physical Chemical Principles in the Pharmaceutical Sciences. Philadelphia, PA: Lea and Febiger, 1983.

27. Popovych O, Tomkins RPT. Nonaqueous Solution Chemistry. New York: John Wiley and Sons, 1981:221–224.

28. Schulze B, Klibanov AM. Inactivation and stabilization of subtilisins in neat organic solvents. Biotechnol Bioeng 1991; 38:1001–1006.

29. Bromberg LE, Klibanov AM. Transport of proteins dissolved in organic solvents across biomimetic membranes. Proc Natl Acad Sci USA 1995; 92:1262–1266.

30. Stevenson CL, Leonard JJ, Hall SC. Effect of peptide concentration and temperature on leuprolide stability in dimethyl sulfoxide. Int J Pharm 1999; 191:115–129.

31. Tan MM, Corley CA, Stevenson CL. Effect of gelation on the chemical stability and conformation of leuprolide. Pharm Res 1998; 15:1442–1447.

32. Hadden JM, Chapman D, Lee DC. A comparison of infrared spectra of proteins in solution and crystalline forms. Biochim Biophys Acta 1995; 1248:115–122.

33. Haris PI, Chapman D. The conformational analysis of peptide using fourier transform IR spectroscopy. Biopolymers 1995; 37:251–263.

34. Jackson M, Mantsch HH, Spencer JH. Conformation of magainin-2 and related peptide in aqueous solution and membrane environments probed by Fourier transform infrared spectroscopy. Biochemistry 1992; 31:1289–1293.

35. Brauner JW, Mendelsohn R, Prendergast FG. Attenuated total reflectance fourier transform infrared studies on the interaction of melittin, two fragments of melittin and δ-hemolsin with phosphatidylcholines. Biochemistry 1987; 26:8151–8158.

36. Seeback D, Thaler A, Beck AK. Solubilization of peptides on non-polar organic solvents by the additions of inorganic salts: facts and implications. Helvetica Chemica Acta 1989; 72:857–867.

37. Rariy RV, Klibanov AM. Protein refolding in predominantly organic media markedly enhanced by common salts. Biotechnol Bioeng 1999; 62:704–710.

38. Fransson J, Hallen D, Florin-Robertsson E. Solvent effects on the solubility and physical stability of human insulin like growth factor I. Pharm Res 1997; 14:606–612.

39. Kim J, Dordick JS. Unusual salt and solvent dependence of a protease from an extreme halophile. Biotechnol Bioeng 1997; 55:471–479.

40. Collins KD, Washabaugh MW. The hofmeister effect and the behavior of water at interfaces. Q Rev Biophys 1985; 14:323–422.

41. Powell MF, Fleitman J, Sanders LM, Si VC. Peptide liquid crystals: inverse correlation of kinetic formation and thermodynamic stability in aqueous solution. Pharm Res 1994; 11:1352–1354.

42. Powell MF, Sanders LM, Rogerson A, Si V. Parenteral peptide formulations: chemical and physical properties of native luteinizing hormone-releasing hormone (LHRH) and hydrophobic analogues in aqueous solution. Pharm Res 1991; 8:1258–1263.

43. Stevenson CL, Bennett DB, Lechuga-Ballesteros D. Liquid crystals in pharmaceutical systems: the relevance of partially ordered systems. J PharmSci 2005; 94:1861–1880.

44. Samulski ET, DuPre DB. Lyotropic polymeric liquid crystals. J Chim Physique 1983; 80:25–30.

45. Brown GH, Crooker PP. Liquid crystals a colorful state of matter. C&EN January 1983; 24–38.

46. Chandrasekhar S. Liquid Crystals. Cambridge: Cambridge University Press, 1992.

47. Rose S, Nelson JF. A continuous long-term injector. Aust J Exp Biol 1955; 33:415–420.

48. Hall SC, Tan MM, Leonard JJ, Stevenson CL. Characterization and comparison of leuprolide degradation profiles with water and dimethyl sulfoxide. J Pept Res 1999; 53:432–441.

49. Motto MG, Hamburg PF, Graden DA, Shaw CJ, Cotter ML. Characterization of the degradation products of LHRH. J PharmSci 1991; 80:419–423.

50. Oyler AR, Naldi RE, Lloyd JR, Graden DA, Shaw CJ, Cotter ML. Characterization of the solution degradation products of histerelin, a gonatotropin releasing hormone (LH/RH) agonist. J PharmSci 1991; 80:271–275.

51. Okada J, Seo T, Kasahara F, Takeda K, Kondo S. New degradation product of Des-Gly[10]-NH$_2$-LH-RH-Ethylamide (fertirelin) in aqueous solution. J PharmSci 1991; 80:167–170.

52. Adam W, Ahrweiler M, Saiter M, Schmiedeskamp B. Oxidation of indoles by singlet oxygen and dimethyl dioxirane: isolation of indole dioxetanes and epoxides by stabilization through nitrogen acylation. Tetrahedron Lett 1993; 34:5247–5250.

53. Adam W, Ahrweiler M, Peters K, Schmiedeskamp B. Oxidation of N-acylindoles by dimethyldioxirane and singlet oxygen: substituent effects on thermally pesistent indole epoxides and dioxetanes. J Org Chem 1994; 59:2733–2739.

54. Zhang X, Foot CS. 1,2-dioxetane formation in photooxygenation of N-acylation indole derivatives. J Org Chem 1993; 58:47–51.

55. Zhang X, Zhang N, Schuchmann HP, von Sonntag C. Pulse radiolysis of 2-mercaptoethanol in oxygenated aqueous solution. Generation and reactions of the thiylperoxyl radical. J Phys Chem 1994; 98:6541–6547.

56. Itakkura K, Uchida K, Kawakishi S. Selective formation of oxindole and formylkynurenine type products from tryptophan and its peptides treated with a superoxide generating system in the presence of iron (III)-EDTA: a possible involvement with iron-oxygen complex. Chem Res Toxicol 1994; 7:185–190.

57. Li SH, Schoenich C, Borchardt RT. Chemical instability of protein pharmaceuticals: mechanisms of oxidation and strategies for stabilization. Biotechnol Bioeng 1995; 48:490–500.

58. Manning MC, Patel K, Borchardt RT. Stability of protein pharmaceuticals. Pharm Res 1989; 6:903–913.

59. Nashef AS, Osuga DT, Lee HS, Ahmed AI, Whitaker JR, Feeney RE. Effects of alkali on proteins. Disulfides and their products. J Agric Food Chem 1977; 25:245–251.

60. Schrier JA, Kenley RA, Williams R, et al. Huberty. Degradation pathways for recombinant human macrophage colony stimulating factor in aqueous solution. Pharm Res 1993; 10:933–944.

61. Stevenson CL, Leonard JJ, Mabanglo D, et al. Effect of moisture, oxygen and irradiation on leuprolide stability. AAPS PharmSci 1998; 1:S541.

62. Lide DR. Handbook of Chemistry and Physics. 75th ed. Ann Arbor: CRC Press, 1995.

63. Brennan TV, Clarke S. Spontaneous degradation of polypeptides at aspartyl and asparaginyl residues: effects of the solvent dielectric. Protein Sci 1993; 2:331–338.

64. Capasso S, Mazzarella L, Zagari A. Deamidation via cyclic imide of asparaginyl peptide: dependence of salts, buffers and organic solvents. Pept Res 1991; 4:234–238.

65. Fransson J, Florin-Robertsson E, Axelsson K, Nyhlen C. Oxidation of human insulin like growth factor I in formulation studies: kinetics of Methionine oxidation in aqueous solution and in solid state. Pharm Res 1996; 13:1252–1257.

66. Theeuwes F. Elementary osmotic pump. J PharmSci 1975; 64:1987–1991.

67. Theeuwes F, Yum SE. Principles of the design and operation of generic osmotic pumps for the delivery of semisolid or liquid drug formulations. Ann Biomed Eng 1976; 4:343–353.

68. Wright JC, Leonard ST, Stevenson CL, et al. An in vivo/in vitro comparison with a leuprolide osmotic implant for the treatment of prostate cancer. J Control Release 2001; 75:1–10.

69. Crawford E, DeAntonio EP, Labrie F, Schroder FH, Geller J. Endocrine therapy of prostate: optimal form and appropriate timing. J Clin Endocrinol Metab 1995; 80:1062–1078.

70. Goktas S, Crawford ED. Optimal hormonal therapy for advanced prostatic carcinoma. Semin Oncol 1999; 26:162–173.

71. Sharifi R, Knoll LD, Smith J, Kramolowsky E. Leuprolide acetate (30 mg depot every 4 months) in the treatment of advance prostate cancer. Urology 1998; 51:271–276.
72. Sarosdy MF, Schelhammer PP, Soloway MS, et al. Endocrine effects, efficacy and tolerability of a 10.8 mg depot formulation of goserelin acetate administered every 13 weeks to patients with advance prostate cancer. BJU Int 1999; 83:801–806.
73. Fowler JE, Gottesman JE, Bardot SF, et al. Duros leuprolide implantable therapeutic systems in patients with advanced prostate cancer: 114-month results of a phase I/II dose-ranging study. J Urol 1999; 161:S300.
74. Fowler JE, Gottesman JE, Reid CF, Andriole GL, Soloway MS. Safety and efficacy of an implantable leuprolide delivery system in patients with advanced prostate cancer. J Urol 2000; 164:730–734.
75. Cukierski MJ, Johnson PA, Beck JC. Chronic (60 week) toxicity study of DUROS leuprolide implants in dogs. Int J Toxicol 2001; 20:369–381.
76. Wright J, Chen G, Cukierski M, et al. Duros leuporlide implant for continuous one-year treatment of prostate cancer. Proc Int Symp Control Release Bioact Mater 1998; 25:516–517.
77. Lee KC, Lee YJ, Song HM, Chun CJ, DeLuca PP. Degradation of synthetic salmon calcitonin in aqueous solution. Pharm Res 1992; 9:1521–1523.
78. Physicians Desk Reference. Medical Economics. New Jersey: Montvale, 1998.
79. Cudd A, Arvinte T, Gaines Das RE, Chinni C, MacIntyre I. Enhance potency of human calcitonin when fibrillation is avoided. J PharmSci 1995; 84:717–719.
80. Arvinte T, Cudd A, Drake AF. The structure and mechanism of formation of human calcitonin fibrils. J Biol Chem 1993; 268:6415–6422.
81. Stevenson CL, Tan MM. Solution stability of salmon calcitonin at high concentration for delivery in an implantable system. J Pept Res 2000; 55:129–139.
82. Stevenson C, Fereira P, Tan M, Ayer R, Dehnad H, Berry S. Glucagon-like peptide suspension stability and delivery from a DUROS implant. Proceedings of the 27th International Symposium on Controlled Release of Bioactive Materials, 2000, Vol. 27:986–987.
83. Berry SA, Fereira P, Tan M, et al. To investigate zero order delivery profiles of protein/peptide suspension formulations from osmotically driven implantable pumps. AAPS PharmSci 2000; 2:S2248.
84. Johnson HM, Bazer FW, Szente BE, Jarpe MA. How interferons fight disease. Sci Am May 1998; 68–75.
85. Knepp VM, Muchnik A, Oldmark S, Kalashnikova L. Stability of nonaqueous suspension formulations of plasma derived factor IX and recombinant human alpha interferon at elevated temperatures. Pharm Res 1998; 15:1090–1095.
86. Ip AY, Arakawa T, Silvers H, Ransone CM, Niven RW. Stability of recombinant consensus interferon to air-jet and ultrasonic nebulization. J PharmSci 1995; 84:1210–1214.
87. www.intarcia.com
88. Lawn RM. The molecular genetics of hemophilia: blood clotting factors VIII and IX. Cell 1985; 42:405–406.
89. Smith KL. Factor IX concentrates: the new products and their properties. Transfus Med Rev 1992; 6:124–136.
90. Volkin DB, Staubli A, Langer R, Klibanov AM. Enzyme thermoinactivation in anhydrous organic solvents. Biotechnol Bioeng 1991; 37:843–853.
91. Desai UR, Klibanov AM. Assessing the structural integrity of a lyophilized protein in organic solvents. J Am Chem Soc 1995; 117:3940–3945.

92. Griebenow K, Klibanov AM. On protein denaturation in aqueous-organic mixtures but not pure organic solvents. J Am Chem Soc 1996; 118:11698–11700.
93. Santos AM, Vidal M, Pacheco Y, et al. Effect of crown ethers on structure, stability, activity and enantioselectivity of subtilisin carlsberg in organic solvents. Biotechnol Bioeng 2001; 4:295–308.
94. Mishra P, Griebenow K, Klibanov AM. Structural basis for the molecular memory of imprinted proteins in anhydrous media. Biotechnol Bioeng 1996; 52:609–614.
95. Halling PJ. High-affinity binding of water by proteins is similar in air and in organic solvents. Biochim Biophys Acta 1990; 1040:225–228.

8

Freeze-Drying Concepts: The Basics

Larry A. Gatlin
*Parenteral Center of Emphasis, Groton Laboratories, Pfizer Inc.,
Groton, Connecticut, U.S.A.*

Tony Auffret
Sandwich Laboratories, Pfizer Ltd, Kent, U.K.

Evgenyi Y. Shalaev
*Parenteral Center of Emphasis, Groton Laboratories, Pfizer Inc.,
Groton, Connecticut, U.S.A.*

Stanley M. Speaker and Dirk L. Teagarden
*Pharmaceutical Sciences—Global Biologics, Pfizer Inc.,
Chesterfield, Missouri, U.S.A.*

INTRODUCTION

The most important question to ask about freeze-drying is not "what is freeze-drying?" but "why do we freeze-dry?" The answer is simple: "stability." Water is an extremely reactive compound, and many biological and pharmaceutical preparations have only limited stability in aqueous solution. Long-term storage, therefore, needs an alternative, and removing the water does seem to be a reasonable one. If a product is to be dried, then there are two readily available routes: (*i*) evaporation of liquid water by, e.g., vacuum drying (1) or spray drying (2), and (*ii*) sublimation of solid water (ice) following freezing of the solution, i.e., freeze-drying.

Despite the disadvantages of cost and process time over ready-to-use and frozen products, the sublimation route, freeze-drying, has become established in the pharmaceutical industry. There are several very good reasons for this: (*i*) the shelf-life for both chemical and biological entities can be enhanced significantly; (*ii*) it is a proven and trusted aseptic processing operation that meets finished product sterility assurance requirements (as compared with, e.g., spray drying); (*iii*) the product is readily reconstituted at time of use.

The freeze-drying, or lyophilization, process can be broken down into three stages: (*i*) freezing, the crystallization of water; (*ii*) primary drying, the removal of ice by sublimation; and (*iii*) secondary drying, the desorption of residual water from the product. This chapter considers these stages, along with two critical post-lyophilization stages in a freeze-dried product life span, i.e., (*iv*) storage of the freeze-dried product, and (*v*) reconstitution at the point of use, in some detail.

For this process to be successful, particularly at a commercial scale, significant attention must be paid to the practical aspects of engineering a freeze-dried product. This includes selection of excipients, packaging, sterilization, filling, loading of the dryer, process control and operation of the drier, unloading etc., topics that are dealt with in greater detail in other chapters of this book and elsewhere (3–5).

The process itself is characterized by having only three variables, i.e., shelf temperature, chamber pressure, and time. It is one of the few processes where the measurement of a single physical property of a formulation (e.g., collapse temperature) facilitates the design of a process that will operate at all scales of operation. In the next section, we examine each of the five steps to introduce the principles of operation and the process design space that is available for lyophilization development. Scale-up principles are discussed in the latter section of this chapter.

FREEZING

There are two important facts to remember about the freezing operation: (*i*) when a solution freezes it does not completely solidify; (*ii*) when your product is frozen, its fate is usually sealed.[a] From that point, all you can do is remove water. Formulation work, therefore, must take into account the properties of a frozen solution and the behavior of the product at low temperatures.

Although we often talk about the "freezing point" of water being 0°C, in fact, we mean the melting point of water is 0°C. Freezing of pure water at atmospheric pressure may occur anywhere between 0°C and −38°C, with the latter temperature being the homogeneous nucleation temperature of water. If you cool a 1000 vials in a freeze dryer, they freeze randomly, at different times and temperatures.

This is explained by the crystallization process that starts with nucleation followed by crystal growth. The nucleation can be either homogenous or heterogeneous, but in real systems, heterogeneous nucleation predominates. Thus, the random freezing of vials relates to several factors including temperature, number, and nature of foreign particles, imperfections of the contact surface, and vibration. Several of these variables can be difficult to control.

When water freezes, the crystalline ice phase does not include any other molecules within the crystal lattice. Therefore, in a solution, ice forms as a pure

---

[a] Exceptions do exist. For example, crystallization of a solute may occur during warming as discussed in this section and in Chapter 9 of this book in some detail.

water phase and all the solutes remain in any unfrozen liquid. If you lower the temperature, then more ice forms and the solutes become even more concentrated. This behavior is conveniently described by the freezing curve in a phase diagram (see curve AC on diagram, Fig. 1). In a simple solution 100% A = pure water, 100% B = pure solute, the dilution of the initial solution will affect the amount of ice formed, which can be calculated from the phase diagram, but the concentration of the unfrozen solution depends only upon the temperature. Any point along AC has the coordinates "*x,y*" and selecting a temperature (*y*) allows us to read off the *x*-axis value and obtain the composition of the unfrozen solution.

We are all familiar with the concept that, for most solutes at least, solubility increases with solution temperature (this is the curve CB below). As temperature is lowered, however, at some point, the freezing curve and the solubility curve meet. At this point, if you cool the solution further, the solute simply is no longer soluble in the amount of solvent present and it precipitates (crystallizes). This of course leads to a dilution of the unfrozen solution, and since its composition depends only on the temperature, more ice forms, returning us to point C. The point where all this happens is known as the "eutectic" temperature (T_e) and below this temperature, only crystalline solvent (ice) and crystalline solute exist, provided that we are dealing with a system under thermodynamic equilibrium—which is usually not the case in real systems as discussed below. Above the T_e, the ice will melt ("eutectic" is derived from the Greek for "easily melted")—this is an important concept that is further discussed under primary drying below.

In reality, the crystallization of solute is just as fickle as the crystallization of water. It also depends upon a nucleation event, followed by crystal growth, which

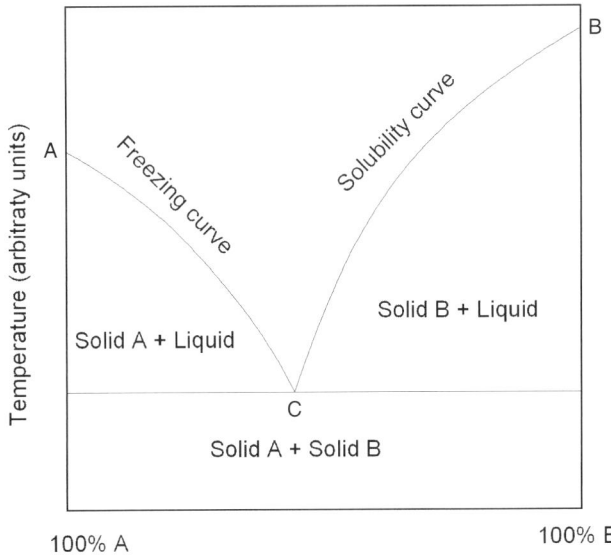

Figure 1 Schematic phase diagram of a binary system with a simple eutectic.

of course at subambient temperatures will be quite slow. Despite the attention that eutectic solidification has attracted in the literature and the considerations given to it in this text, it is not seen with most formulations. In simple solutions (e.g., sodium chloride/water), it does occur but freeze-dried formulations are generally more complex and it is rarely seen. One notable exception is the case of mannitol-containing formulations, where the crystallization behavior of mannitol can be a source of many problems, as described in some detail in Chapter 9. Most solutions simply carry on during cooling below the eutectic temperature producing more ice, increasing the concentration of the unfrozen solution until eventually the residual solution is so viscous that no more ice can form. It has turned into a glass (an amorphous solid solution). The temperature at which any formulation vitrifies (turns to glass) is the glass transition temperature (T_g').

Figure 2 represents a diagram that describes solid–liquid relationships in such a system. Note as this shows both equilibrium, crystal, and nonequilibrium, glass states, it is referred to as a *state diagram* rather than a *phase diagram*. The

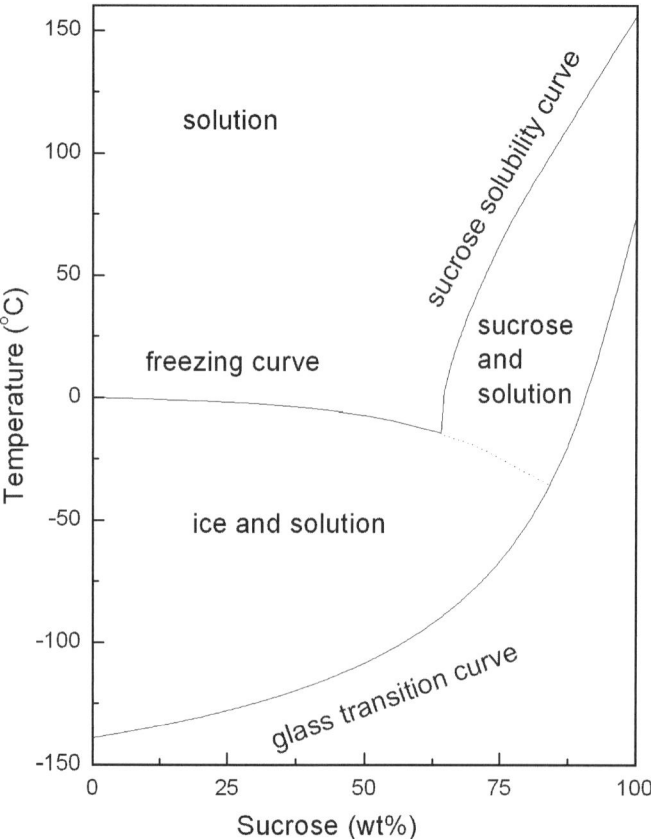

Figure 2 Solid–liquid state diagram of water-sucrose system.

T_g' temperature corresponds to the onset of ice melting (more exactly, ice dissolution into the freeze-concentrated solution), which occurs at either the T_g of the maximally freeze-concentrated solution (6) or several degrees above the glass transition temperature of the maximally freeze-concentrated solution (7). The T_g' can be conveniently measured by differential scanning calorimetry or freeze-drying microscopy. The T_g' corresponds closely to the collapse temperature, the collapse temperature normally being higher by 1°C to 3°C (8).

T_g' values may be found in the literature for many binary water-solute systems [e.g., (3)]. In multicomponent systems, when a frozen solution does vitrify, only one glass, a solid solution of all the solutes together with some residual solvent, is usually formed. A glass has two interesting properties of value for the freeze-drying process; it is mechanically stable and has a superior chemical stability in relation to that in solution.

To summarize the discussion on the freezing behavior, at the end of the freezing process, the product will consist of either (9) (*i*) grains of ice surrounded by a glassy matrix of the residual solution, (*ii*) grains of ice, as well as grains (crystals) of solute, surrounded by a glassy matrix of the residual solution, or (*iii*) grains of ice in an admixture with crystals of solute. By far the most common behavior leads to grains of ice surrounded by a glassy matrix of the residual solution. Mechanical stability means that the glassy network is self-supporting and does not collapse when the ice is removed. This leaves an open network of glassy product that is easily rehydrated. Of course, to the purist, the freezing process, which typically has separated 98% of the original water as ice, may itself be regarded as drying!

PRIMARY DRYING

Primary drying is the removal of ice by the process of sublimation, i.e., transfer directly from the solid to the gas phase. There are two reasons why it is important to avoid the liquid phase: (*i*) The product will lose its mechanical strength and collapse. This will lead to loss of the porous structure of the final product, with subsequent difficulty of rehydration. (*ii*) The product might lose its chemical or physical (e.g., protein aggregation) stability.

The loss of chemical stability can be of particular importance to biologics products. Although we are accustomed to thinking that low temperature results in lower rates of chemical reaction (Arrhenius temperature dependence of kinetics), in a freeze concentrate solution, the rules are different. The kinetics of chemistry in these very concentrated solutions is poorly understood, but it might be controlled by concentration as well as temperature. Literature reports show rate enhancements of up to two orders of magnitude caused by freeze concentration (10,11), and there is evidence that activity loss of enzymes may be increased 3- to 3000-fold (12). Collapse and potential chemical degradation are easily avoided. Whilst ice is present, the product should be kept below the collapse temperature. For an amorphous glass, this conservatively translates to keeping the product temperature below the T_g'. If the product is a eutectic mixture, then it must be kept below T_e. It is as simple as that.

It is important to stress the difference between product temperature, which should be kept below the collapse or eutectic temperature, and shelf temperature. In an idealized process, the shelf temperature would be kept below T_g' whilst ice is present. However, sublimation at such low shelf temperatures would take significantly longer; therefore, this is usually not a practical option. Commercial pressure often encourages us to increase shelf temperature above T_g' in order to increase the heat flux. In such cases, we rely on sublimation to keep the product cold. It would be good practice in such cases to lower the shelf temperature again as the rate of sublimation, or cooling, decreases toward the end of primary drying (see below).

The basic process simply requires that sufficient heat is supplied to the product to exactly balance the heat required to sublime the ice, i.e.,

Heat supplied to the product = Heat used for sublimation

$$K_v(T_s - T_p) = \Delta H(\mathrm{d}m/\mathrm{d}t)$$

where K_v is a heat transfer coefficient, T_s is the shelf temperature, and T_p is the product temperature. ΔH is the heat of sublimation and $\mathrm{d}m/\mathrm{d}t$ is the rate of mass transfer of ice to water vapor.

The driving force for the sublimation process is the temperature difference between the product and the condenser, which results in a water vapor pressure difference. Even without a vacuum, the product would dry, although it would take a longer time and it is customary to use a vacuum to speed the process.

Heat is supplied to the product by three mechanisms, conduction, convection, and radiation. Even though a vacuum is used in the primary drying process, significant heat is supplied by convection, i.e., by gas phase collisions. Reducing the pressure too much, therefore, will slow the rate of heat transfer and of ice sublimation (13). Typically, freeze-drying is carried out at 30% to 50% of the saturation vapor pressure of water over ice at the temperature of the ice–product interface.

Although caution must be applied in setting the shelf temperature, to avoid collapse, the product temperature has a significant effect on the process. The rate of ice sublimation decreases by approximately 13%/°C, or 300%/10°C change in temperature (14). Formulating for a high glass transition temperature can have a significant impact upon the primary drying time.

Ice sublimation is directional, it starts at the top of the vial, and with careful observation, the ice front can be seen to migrate down through the product during the process. As a result, a layer of dried cake builds up above the ice front and it is this cake that provides most of the resistance that limits the rate of sublimation. The rate of sublimation, therefore, gets lower as the process nears completion. At this point, if the cycle uses a shelf temperature greater than T_g', and relies upon the rate of sublimation to cool the product, there is a consequential increase in the risk of the product temperature exceeding T_g'. If this happens before sublimation is complete, ice will melt back in accordance with the freezing curve, and the product will collapse. Such cycles, which rely on a balanced heat flux to cool the product, may need further development on scale-up or transfer to different equipment. Therefore, it is important to understand resistance of the dry layer; such measurements can be performed using a modification of a pressure rise method (15).

An efficient way to optimize the shelf temperature during primary drying is through the use of a so called "Smart Freeze Dryer." This system is intended to develop a freeze-drying cycle with a single laboratory scale run (15,16). The basis of the Smart Freeze-Drying technology is a manometric temperature measurement, which measures the product temperature on the ice/vapor interface during primary drying by quickly isolating the freeze-drying chamber from the condenser and analyzing the pressure rise during this period.

SECONDARY DRYING

As can be seen in the state diagram for sucrose (Fig. 2), the freeze-concentrated glass contains a significant amount of residual unfrozen water (it is incorrect to call this bound water; see Ref. 17). Typically, a frozen glass may contain as much as 20% to 50% by weight of water (18). If we warmed this composition above T_g, it would lose its mechanical strength and collapse. The secondary drying phase of the process is concerned with removal of this residual water. (Note: We have used the term T_g here to refer to the general case of heating a glassy product. All points on the glass curve may be designated T_g. The term T_g' refers exclusively to that point where the glass curve is intersected by the freezing curve, i.e., it is the glass temperature of the maximally freeze-concentrated solution. Warming a solution to T_g' in the presence of ice leads to melt back. In the absence of ice, warming above T_g' results in softening and collapse.)

If the product has formed a eutectic mixture, then when all the ice is sublimed, there is no residual water and the product may be quickly warmed before unloading. It should be borne in mind, however, that eutectic solids may precipitate as hydrates (NaCl and mannitol are familiar examples). Such forms may not be stable at ambient temperatures and care should be exercised in dehydrating the crystals.

Most products, however, will be amorphous and the residual water may be removed by desorption. It is customary to carry out this process at a low pressure, although what evidence there is in the literature suggests that pressure is not a significant factor in controlling the rate of desorption (19).

If we remove water from the glass, its composition changes and as the state diagram shows, as the water content is lowered, the glass temperature, T_g, of the product increases in value. Even though, in almost all cases, the position and shape of the glass curve will not be known, in practice, at the end of primary drying, the product T_g will be slightly higher than T_g'. This leads to a practical method for conducting secondary drying as we may now increase the product temperature. This, in turn, leads to further desorption, and a higher T_g, which in turn allows for another increase in temperature and further desorption.

If we gradually ramp the temperature and balance the rate of drying (desorption) with the increase in product temperature, we can desorb the residual water without significantly exceeding T_g at any time. In practice, the rate of collapse in the immediate region of T_g is quite low (20) and as long as the rate of temperature ramp is not excessive (<5°C/hr), problems are rarely encountered in secondary drying.

The kinetics of water desorption are controlled by diffusion of water within the glass and at any given temperature, the loss of water will slow with time (plateau). In many cases, the plateau is achieved after 6 to 12 hours at common secondary drying temperatures (25–40°C). The temperature dependence of the extent of desorption also suggest, with a properly dried product, if the temperature is decreased for unloading, there is nothing to be gained by an extended hold at the final temperature.

STORAGE

The product must be stored below T_g to avoid mechanical collapse, to meet the appearance and reconstitution specifications. Also, despite well publicized exceptions [e.g., indomethacin (21)], the crystallization of components, with subsequent changes to composition and stability of the remaining amorphous product, can generally be avoided by storage below T_g. It is not possible to predict the final glass temperature nor the water content, of the product, though both are easily measured. Water acts as a plasticizer and will decrease the T_g of the dried product.

Factors that control the chemical stability of the product are poorly understood. It is clear that despite being essentially isoviscous, all glasses are not equivalent in their ability to prevent chemical degradation. The glass transition temperature is an important reference point for stability of amorphous materials, and determination of the T_g in formulations would be a necessary step in formulation development. Formulations with low T_g (e.g., formulations having a T_g lower than 40°C) are expected to have low chemical and physical stability. However, the T_g alone cannot predict chemical stability of freeze-dried formulations—indeed, higher T_g does not necessary mean better stability. There are cases when formulations with lower T_g had higher chemical stability (22,23). Other properties sometime provide better correlations with chemical stability, such as enthalpy relaxation, which is another measure of molecular mobility (24), and Hammett acidity function, which reflects acid–base relationships in lyophilized state (22,25,26). The current state of the art does not, therefore, allow the *ab initio* formulation of a stable product.

It is, therefore, generally necessary to conduct stability trials. Although it is reasonable to apply Arrhenius temperature kinetics for storage below T_g, care must be shown when interpreting data obtained at $T > T_g$. On reversion to a mobile and highly concentrated solution, chemical kinetics do deviate from the Arrhenius model of temperature dependence. Unfortunately, the temperature dependence of chemical reactions within T_g is poorly understood and there is no generally accepted model.

RECONSTITUTION

Reconstitution is often considered to be a routine item that does not deserve separate consideration in discussion of such a sophisticated process as freeze-drying. As a result, reconstitution is often omitted from freeze-drying texts. However, there are several important points to consider at this last stage in the life of an injectable protein formulation, which are as follows:

- Reconstitution time should be reasonable from a user's perspectives, i.e., usually less than one minute.
- Although "off-the-shelf" diluents (e.g., saline) are usually preferred, use of a specially designed diluent may be warranted in certain cases. For example, protein recovery can be significantly improved when reconstituted with a diluent containing the surfactant Tween 20 (27). Also, certain types of excipients such as antimicrobial preservatives or tonicity modifiers are commonly included in a diluent rather than in a lyophile cake.
- In-use stability of the reconstituted solution should be evaluated as a part of formulation development. It is advised to perform in-use stability testing on aged formulations, to avoid unforeseen reconstitution and in-use stability (e.g., particulate formation) problems.
- The reconstituted solution should be suitable for the intended use, for example, solution viscosity should be appropriate to allow injection of the intended dose with the needle gauge selected. Guidance on relationships between solution viscosity and syringeability for any particular needle size can be found in (28).

Overall, lyophilization process development on laboratory scale is becoming a predictable and routine exercise if one uses state-of-the art knowledge and equipment. Scale-up of a freeze-drying process from the laboratory to production scale, on the other hand, is less predictable. In addition, the laboratory-to-production scale transfer may require more extensive documentation as such projects are usually closer to regulatory filing. In the next section, we will consider scale-up of freeze-drying cycles in more detail.

SCALE-UP AND TECHNOLOGY TRANSFER OF LYOPHILIZED DRUG PRODUCTS

Scale-up and technology transfer involve all activities dealing with modification of a laboratory manufacturing process to accommodate the larger scales of pilot/ clinical and International Conference on Harmonisation stability manufacturing and eventually commercial production. These activities include experimental studies and trial engineering or scale-up batches that must be well defined and carefully executed. Although delivery of the process is the ultimate goal of technology transfer, complete and accurate documentation of the scale-up activities and their transfer to the commercial manufacturing site is extremely important. This will not only ensure a robust process but also will aid in the regulatory review process since documentation is an intense focus of regulatory scrutiny from government agencies. Effective communication between research and production groups is the cornerstone of efficient and successful process scale-up and technology transfer. Information exchange and gap analysis exercises should be executed early and reviewed often within the research/production team to ensure complete and thorough identification and evaluation of all critical issues.

Basic Understanding of Freeze-Dry Equipment

It is important to have a fundamental knowledge of the design and operation of the freeze-dry equipment that is being utilized for development and production of freeze-dried products. Nail and Gatlin (29) provide an extensive discussion on the design and operating components of most pharmaceutical freeze-dryers. These systems consist of a chamber containing shelves through which a heat transfer fluid can be circulated; a system for pumping, heating, and cooling the fluid; a vacuum pumping system; a condenser for trapping water vapor; and a refrigeration system for cooling the condenser. Additionally, most new freeze-dryers contain a system for sterilization of the chamber and condenser.

The vacuum system employed in most freeze-dryers is typically a rotary oil pump. These pumps are capable of attaining vacuum as low as 1 μm of Hg. However, for these pumps to operate properly, the condenser must remove water vapor to prevent oil contamination of the pump oil by moisture and hence reduce pumping efficiency. An additional redundant pump such as a Roots-type pump is frequently utilized in many commercial freeze-dryers in order to increase the speed of the pumping system and to achieve the lowest attainable vacuum.

A refrigeration system is needed to cool the shelves during freezing and the condenser during drying. The condenser is cooled by direct expansion of a refrigerant such as a fluorocarbon in the condenser coils. Cooling of the heat transfer fluid (typically silicone oil) in the shelves is achieved by heat exchangers. Typically, the refrigeration can be switched from the shelves during freezing to the condensers during drying.

The condensers are either the internal type (i.e., inside the drying chamber along the walls) or external type with a separate chamber. Most production scale dryers have an external condenser design, whereas some small-scale freeze-dryers (e.g., Virtis bench freeze-dryer) utilize internal condensers. The use of internal versus external condensers can have a significant impact on the warm versus cool zones typically present during drying and hence influence the drying rates throughout the dryer. It should be noted that use of an external condenser design is required for use of pressure rise tests should this method be used to monitor the drying process.

The main drying chamber and external condensers are constructed of type 304 or 316 stainless steel. The smaller freeze-dryers that have Plexiglas-type doors may experience different drying rates near this door due to additional radiant heat. The internal surfaces are polished to facilitate cleaning. Most freeze-dryers have the capability to move their shelves within the dryer to enable internal stoppering.

Characterization of a freeze-dryer should include determination of the condenser capacity and the maximal sustainable sublimation rate. Such tests are commonly performed using pure water. In addition, shelf mapping, for both product temperature and sublimation rate, is essential in order to be able to develop efficient and robust lyophilization cycles and to ensure high quality product. It should be mentioned that temperature mapping is often performed with empty freeze-dryers. Although this is an important step in the equipment characterization, one should not stop there. It is recommended that the temperature mapping should also be performed with a loaded freeze-dryer using, e.g., water.

Scaling Up and Technology Transfer of the Freeze-Dry Cycle

Usually, some modification of a laboratory-scale freeze-dry cycle is needed to properly run the cycle on a production size freeze-dryer. It is critical to recognize the differences in equipment design for the various size dryers used as the scale-up process evolves. Potential differences include the following: variable dryer size/architecture, control of pressure (e.g., absolute vs. comparative or nitrogen bleed vs. valve, etc.), use of trays or trayless systems, tray material if utilized (e.g., stainless steel or aluminum), variable package types (e.g., standard glass vials, two-chamber vials/syringes, trays, etc.), internal stoppering capability, and process monitoring capability (thermocouples, Pirani gauge, dew point sensor, etc.). These differences influence the resulting product temperatures that are attained for each size freeze-dryer as a function of corresponding shelf temperatures and chamber pressure set points. It must be emphasized that the goal of scale-up is that the product should experience the same conditions (e.g., product temperature) as process scale increases. This does not mean that the cycle parameters (shelf temperature, chamber pressure, or drying time) are the same between scales. Cycle parameters are a means to an end to achieve the desired product temperature profile and it is highly likely that these parameters can and will change as scale increases. For example, Figure 3 shows average product temperature profiles obtained on the laboratory and pilot scale at the same shelf temperatures and chamber pressure.

It should be noted that in each freeze-dryer there will be heterogeneity of product temperatures during the drying process. Typically the perimeter samples (i.e., near the walls and door) are warmer due to radiant heat than the internal insulated samples. As a result, the drying times between different vials of the same batch can vary significantly. For example, Figure 4 illustrates the higher product temperatures (i.e., higher drying rates) for the perimeter samples as compared to

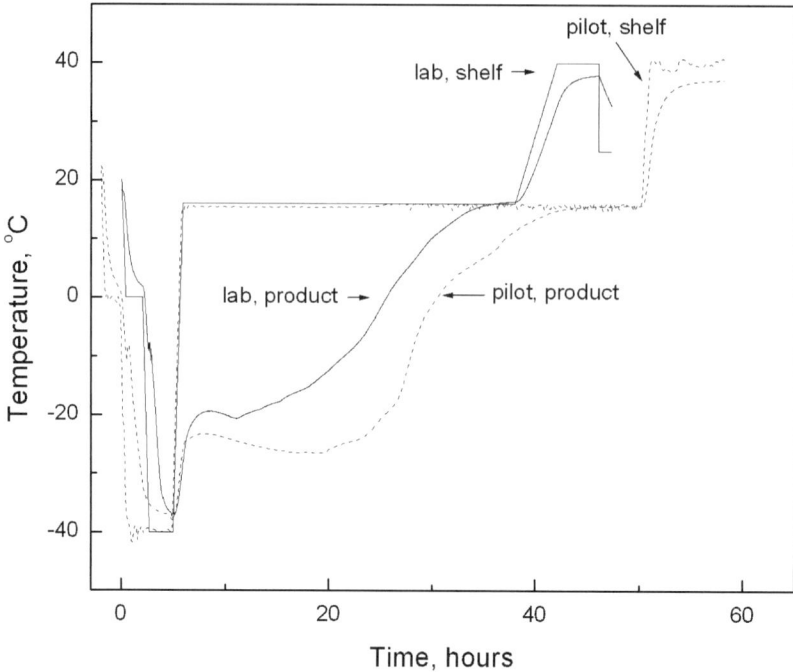

Figure 3   Average product temperature and the shelf temperature from two freeze-drying cycles for the same product on the laboratory and pilot scales.

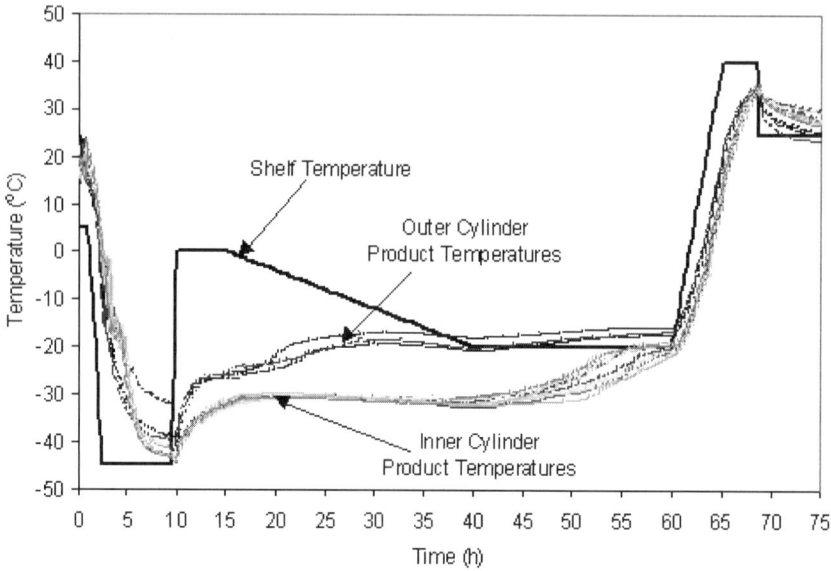

Figure 4   Product temperature data comparison of edge versus center of shelf load.

the lower product temperatures (i.e., lower drying rates) for the insulated inside samples. This heterogeneity in product temperature can differ significantly with dryer design and scale. Properly designed development experiments are the key for guidance during this scale-up process. Historical data including temperature/ pressure set points and profiles for actual process feedback should be analyzed as a part of transferring the process from lab scale to pilot scale to production scale. Additionally, sublimation and secondary drying endpoint determinants should be implemented to detect critical quality attributes (CQAs) of any particular freeze-dried product, e.g., moisture content, appearance (both cake and reconstituted product), reconstitution time, particles upon reconstitution, potency, and purity.

Efforts to maintain analogous processing dynamics (e.g., shelf temperature-time profile) between similarly sized lyophilizers should be performed to the extent they are experimentally practical or included in other aspects of the development work. Acceptable range/limit (and/or edges of failure) for the product temperature profile should be established in development.

Basic requirements for a robust freeze-dried product depend upon the chemical and physical characteristics of the lyophilized drug product. Key in the determination of an appropriate lyophilization cycle is "at-scale" parameters that affect these characteristics. Successful pilot lot planning and manufacture evaluates the relationship between critical cycle parameters and product characteristics. Available development data for lyophilized products from the lab scale should be inclusive of analyses of critical process parameters (CPP) that affect CQA. CPP may include shelf-loading temperature, shelf-cooling rate, freeze-hold and annealing time/temperature, shelf temperature, and chamber pressure during primary and secondary drying, primary and secondary drying endpoints, and stoppering pressure at the end of the cycle. Targets and ranges for process parameters should be justified with development data, and their relationship to product quality at release and upon stability should be established for key process parameters. Parameters that are important in this context should be preliminarily identifiable from research data. It is recommended to perform a risk assessment to define which parameters are critical and which parameters should be explored during development studies. Identification of CQA and CPP should be an explicit process with defined outcomes.

Correlation between scales should be performed for products being scaled from lab to pilot and final manufacturing scale. A direct relationship should not be assumed. According to Kuu et al. (30), when developing a strategy of correlation, it is important to consider the characteristics of the following three areas: (*i*) the dryer, (*ii*) the container, and (*iii*) the formulation. The characteristics of the dryer include shelf-temperature control, the chamber vacuum level, and the efficiency of the condenser. The dryer capability can be quantitatively determined by the shelf temperature heat transfer coefficient, radiation emissivity, shelf temperature and heat transfer mapping, and condenser efficiency. The characteristics of the container include the heat transfer coefficients of the container itself. The characteristics of the formulation include the critical product temperature (e.g., collapse temperature for amorphous formulations) during primary drying and the dry layer mass transfer resistance.

Tsinontides et al. (31) concluded that scale-up of a freeze-drying process can be achieved in a cost effective and efficient manner provided one employs smart use of experimental tools to monitor the drying process of product in limited experiments at manufacturing conditions. Use of appropriate modeling can enhance the success rate of scale-up by evaluation of the robustness of freeze-dry cycle around target set points. Several modeling techniques have been developed that may help to optimize cycle parameters and to validate parameters for robustness and edge of failure. One example is the Passage® FreezeDrying software developed by Technalysis (32). This computer program uses finite-element analysis for the simulation of freeze-drying processes in vials and pans. It is intended for modeling of containers with asymmetric boundary conditions, as in the case of vials placed in corners or near the walls of freeze-drying ovens, with both primary and secondary drying simulation capabilities. Other examples of freeze-drying models are described in (33,34). Despite the recent success in freeze-drying modeling, real lyophilizer behavior at scale under full load should be determined to refine the final dynamics of the process. Lyophilizer type, size, age, and other factors can all affect performance of the equipment and thereby impact the final product quality. Therefore, it is important to support the modeling results by performing experimental tests using real manufacturing equipment.

Regardless of development efforts to alleviate issues of scale, one should be prepared for scale-up issues to be encountered. One should note that if it is necessary to use a surrogate formulation to bulk out the dryer during scale-up (due to lack of available raw materials or due to high cost of the active ingredient), careful selection of the surrogate is critical because it may significantly influence the product temperature and drying rates of active ingredient–containing vials in close proximity. It is also important to remember that lyophilization is a process inherently dependent on and influenced by the unit in which the process is run and the scale of that unit. The total shelf area, refrigeration/compressor capacity, condenser area, chamber wall thickness, etc., have a tremendous influence on the product temperatures encountered. As a lyophilization process is transferred to a larger scale unit (whether from a development lyophilizer to a pilot or clinical dryer, or from a lab or pilot dryer to production), one should expect to see differences and plan engineering and qualification runs in advance to address these concerns. The issue that will most commonly be encountered is that product temperature as a function of shelf temperature and chamber pressure may be different (usually lower) in a production unit. Therefore, as a result of the lower-than-expected product temperatures, the total time required for primary drying might be longer than expected. The implication of this is that, if a scale-up study is planned, implementing a cycle designed and optimized in a small scale lyophilizer into a commercial scale unit may be a problem since it is possible that secondary drying may be initiated before the end of sublimation and thus resulting in significant melt back of ice that was not removed from the product.

Although differences in product temperature between manufacturing and laboratory scale may be predicted with a reasonable level of confidence by measuring the heat transfer coefficients in different freeze-dryers and using an appropriate freeze-drying model, direct experiments on a real manufacturing freeze-dryer are still needed in the majority of cases. Engineering run(s) should be performed in order to evaluate the scale issues between the units before process qualification is initiated. Generally, although engineering runs are recommended and useful, commercial engineering runs can be minimized and/or avoided by the effective use of pilot scale experiments.

Scale-up issues are less significant during secondary drying, primarily due to the fact that secondary drying temperatures are almost universally higher than ambient temperatures. As a result, the factors that introduce differences between units such as radiation effects, convective heat from the chamber atmosphere, etc. are less critical factors due to the higher operating temperatures. One important issue to bear in mind is the final stoppering conditions in cases when internal stoppering cannot be employed. This is often the case with new delivery systems, such as dual-chambered vials and syringes that may require external stoppering. In such cases, the additional moisture uptake between lyophilizer unloading and stoppering must be considered when determining the total secondary drying time (i.e., target moisture content at end of secondary drying).

Number, size, and conditions of lyophilization scale-up lots will be determined by the type and complexity of the lyophilized product. A gap analysis between desired and available development data should be performed. Studies needed should cover gaps and address issues related to the product and process as necessary. A risk assessment approach should aid in the screening of key issues and allow for only the necessary studies to be conducted. Bracketing can be used as an approach to perform best and worst case scenarios for the process and determine the edge of failure. Additionally, it is necessary to be aware of the regulatory filing implications of the additional data that is generated and ensure that studies are done in a manner that could facilitate answers to questions/issues that may arise during regulatory review or inspection.

Process Qualification and Validation

Process qualification is not only a requirement by all regulatory agencies but is also an important part in the understanding of the manufacturing process and its limitations. Qualification and validation of a process is used to set acceptable processing limits for the product. Deviations from these limits require an investigation and review by the technical support and quality assurance departments.

Because of the complexity of the processing steps in freeze-drying, validation can be difficult. In the freeze-drying process, there are many factors that are critical to an acceptable product. Examples of critical factors include shelf temperature, chamber pressure, processing time (freezing, primary and secondary drying), stoppering pressure (both chamber pressure and shelf ram pressure), and condenser temperature. Deviations from any one of these factors could result in product collapse, melt back, high moisture content, and potentially lead to an adverse effect on stability of the product. The validation of any transfers should be based on a process dependent parameter such as product temperature (35). The final step in development of the freeze-dry cycle is the confirmation of an acceptable freeze-drying process via validation. Several authors adequately describe in detail how to properly perform validation of the freeze-dry process (36,37).

Terminal Sterilization of Freeze-Dried Products

Traditionally, freeze-dried products are processed using established sterilizing filtration and aseptic filling methods. Recently, however, European regulatory guidance requires the evaluation of terminal sterilization for freeze-dried products (38). Although it is not clear at this point if this requirement is extended to protein products, it would be appropriate to discuss this in some detail in this chapter. Undoubtedly, terminal sterilization is a reasonable request from a microbiological safety standpoint. However, there are major technical and scientific challenges associated with the application of terminal sterilization to freeze-dried products. For example, dry powder heat sterilization, which is the first technique required for evaluation, is not expected to work for freeze-dried protein products because of the high temperatures involved (usually 160°C). Sterilization by ionizing irradiation may be more feasible from a technical standpoint, although there are major challenges associated with this route as well. Proteins are known to be sensitive to ionizing radiation (39,40) although the extent of degradation might be lower than that associated with the high temperature treatment during dry-heat sterilization. Additionally, lyophilization stoppers are often incompatible with irradiation whereas radiation-resistant stoppers may not be compatible with lyophilized products that require special rubber formulations with low water retention (41). At this point, we are not aware of any marketed freeze-dried products (proteins or small molecules) that are terminally sterilized.

CONCLUSIONS

The current level of understanding of physical, chemical, and engineering aspects of freeze-drying allows a formulator to design an acceptable laboratory-scale freeze-drying process with minimal experimental work. Scaling-up a freeze-drying process from laboratory to manufacturing scale, however, is still a challenging, risky, and expensive process that may require extensive experimentation at production scale. Recent advancements in modeling of freeze-drying processes may reduce the need to do large-scale experiments but not eliminate them. Additionally, there is a significant challenge remaining which is associated with the considerable heterogeneity in product drying rates and temperatures during the cycle across the freeze-dryer. This heterogeneity, as well as unit-to-unit differences between lyophilizers is especially important with lyophilization of less traditional ("novel") delivery systems, e.g., lyophilization in syringes. Such variations create significant uncertainty in determination of sublimation endpoint and secondary drying endpoint. One way to deal with this problem is detailed mapping of any particular freeze-dryer with respect to both product temperature and drying rates. Also, 100% nondestructive tests of finished product, e.g., water content by near infrared, may help to confirm if vial-to-vial variations represent a potential issue for any particular product and lyophilization cycle.

ACKNOWLEDGMENT

The manuscript is based in part on information provided by approximately 30 Pfizer colleagues.

REFERENCES

1. Franks F. Protein hydration. In: Franks F, ed. Protein Biotechnology Isolation, Characterization, and Stabilization. Totowa, NJ: Humana, 1993:437–465.
2. Lee G. Spray-drying of proteins. Pharm Biotechnol 2002; 13:135–158.
3. Pikal MJ. Freeze-drying. Encyclopedia of Pharmaceutical Technology. Marcel Dekker, 2002:1299–1326.
4. Gatlin LA, Nail SL. Protein purification process engineering. Freeze-drying: a practical overview. Bioprocess Technol 1994; 18:317–367.
5. Carpenter JF, Chang BS, Garzon-Rodriguez W, Randolph TW. Rational design of stable lyophilized protein formulations: theory and practice. Pharm Biotechnol 2002; 13:109–133.
6. Chang L, Milton N, Rigsbee D, et al. Using modulated DSC to investigate the origin of multiple thermal transitions in frozen 10% sucrose solutions. Thermochim Acta 2006; 444(2):141–147.
7. Shalaev EY, Franks F. Structural glass transition and thermophysical processes in amorphous carbohydrates and their supersaturated solutions. J Chem Soc Faraday Trans 1995; 91:1511–1517.
8. Pikal MJ, Shah S. The collapse temperature in freeze-drying: dependence on measurement methodology and rate of water removal from the glassy phase. Intern J Pharm 1990; 62:165–186.
9. Shalaev E, Franks F. Solid-liquid state diagrams in pharmaceutical lyophilization: crystallization of solutes. In: Levine H, ed. Progress in Amorphous Food and Pharmaceutical Systems. Cambirdge: The Royal Society of Chemistry, 2002:200–215.
10. Fennema O. Reaction kinetics in partially frozen aqueous systems. In: Duckworth RB, ed. Water Relationships of Food. London: Academic Press, 1985:539–556.
11. Vajda T. Cryochemistry. Large acceleration of the oxidation of hydroxylamiine by iodate in frozen solution. Int J Chem Kinetics 1993; 25:1015–1018.
12. Hately RHM, Franks F, Mathias S. The stabilization of labile biochemicals by undercooling. Process Biochem 1987; 169–172.
13. Xiang J, Hey JM, Liedtke V, Wang DQ. Investigation of freeze-drying sublimation rates using a freeze-drying microbalance technique. Int J Pharm 2004; 279(1–2):95–105.

14. Franks F. Effective Freeze-drying: a combination of physics, chemistry, engineering and economics. Proc Inst Refrigeration1994; 3-1–3-6.
15. Gieseler H, Lee H, Mulherkar B, Pikal MJ. Applicability of manometric temperature measurement (MTM) and Smart™ freeze dryer technology to development of an optimized freeze-drying cycle: Preliminary investigation of two amorphous systems. 1st European Congress on Life Science Process Technology, Nuremberg, Germany, Oct 11–13, 2005.
16. Tang XC, Pikal MJ. Design of freeze-drying processes for pharmaceuticals: practical advice. Pharm Res 2004; 21:191–200.
17. Franks F. Improved freeze-drying: an analysis of the basic scientific principles. Process Biochem 1989; 24:iii–vii.
18. Levine H, Slade L. Thermomechanical properties of small-carbohydrate-water glasses and "rubbers." Kinetically metastable systems at sub-zero temperatures. J Chem Soc Faraday Trans 1 1988; 84:2619–2633.
19. Pikal MJ, Shah S, Roy ML, Putman R. The secondary drying stage of freeze-drying: drying kinetics as a function of temperature and chamber pressure. Int J Pharmaceutics 1990; 60(3):203–217.
20. Levi G, Karel M. Volumetric shrinkage (collapse) in freeze-dried carbohydrates above their glass transition temperature. Food Res Int 1995; 28(2):145–151.
21. Andronis V, Zografi G. Crystal nucleation and growth of indomethacin polymorphs from the amorphous state. J Non-Cryst Solids 2000; 271(3):236–248.
22. Chatterjee K, Shalaev EY, Suryanarayanan R, Govindarajan R. Correlation between chemical reactivity and the Hammett acidity function in amorphous solids using inversion of sucrose as a model reaction, J Pharm Sci 2007 in press.
23. Allison SD, Chang B, Randolph TW, Carpenter JF. Hydrogen bonding between sugar and protein is responsible for inhibition of dehydration-induced protein unfolding. Arch Biochem Biophys 1999; 365:289–298.
24. Pikal MJ. Mechanisms of protein stabilization during freeze-drying and storage: the relative importance of thermodynamic stabilization and glassy state relaxation dynamics. In Drugs and the Pharmaceutical Sciences 137. Freeze-Drying/Lyophilization of Pharmaceutical and Biological Products. Rey L and May JC. Dekker, 2004:63–107.
25. Li J, Chatterjee K, Medek A, Shalaev E, Zografi G. Acid-base characteristics of bromophenol blue-citrate buffer systems in the amorphous state. J Pharm Sci 2004; 93:697–712.
26. Govindarajan R, Chatterjee K, Gatlin L, Suryanarayanan R, Shalaev EY. Impact of freeze-drying on ionization of sulfonephthalein probe molecules in trehalose-citrate systems. J Pharm Sci 2006; 95(7):1498–1510.
27. Jones LS, Randolph TW, Kohnert U, et al. The effects of Tween 20 and sucrose on the stability of anti-L-selectin during lyophilization and reconstitution. J Pharm Sci 2001; 90(10):1466–1477.
28. Chien YW, Przybyszwski P, Shami EG. Syringeability of nonaqueous parenteral formulations-development and evaluation of a testing apparatus. J Parent Sci Tech 1981; 35(6):281–284.
29. Nail S, Gatlin L. Freeze-drying: principles and practice. In: Avism KE, Lieberman HA, Lachman L, eds. Pharmaceutical Dosage Forms: Parenteral Medications. Vol. 2. New York: Marcel Dekker, 1992:163–233.
30. Kuu W, Hardwick L, Akers AJ. Correlation of laboratory and production freeze-drying cycles. Int J Pharm 2005; 302:56–67.

31. Tsinontides SC, Rajniak R, Pham D, Hunke WA, Placek J, Reynolds SD. Freeze drying—principles and practices for successful scale-up to manufacturing. Int J Pharm 2004; 280:1–16.

32. Pikal MJ, Cardon S, Bhugra C, et al. The nonsteady state modeling of freeze-drying: in-process product temperature and moisture content mapping and pharmaceutical product quality applications. Pharm Dev Technol 2005; 10(1):17–32.

33. Hottot A, Peczalski R, Vessot S, Andrieu J. Freeze-drying of pharmaceutical proteins in vials: modeling of freezing and sublimation steps, Drying Technol 2006; 24:561–570.

34. Tang MM, Liapis IA, Marchello JM. A mulit-dimensional model describing the lyophilisation of a pharmaceutical product in a vial. In: Mujumdar AS, ed. Advances in Drying. Hemisphere Publish Corp, 1986:57–64.

35. Jennings TA. Transferring the lyophilization process from one freeze-dryer to another. Amer Pharm Rev 2002 (spring); 1:34–39.

36. Bindschaedler C. Lyophilization process validation. Freeze-Drying/Lyophilization of Pharmaceutical and Biological Products. 2nd ed. New York: Marcel Dekker, 2004:535–573.

37. Trappler EH. Validation of lyophilized products. In: Berry IR, Nash RA, eds. Pharmaceutical Process Validation. 2nd ed. New York: Marcel Dekker, 1993:445–477.

38. The European Agency for Evaluation of Medicinal Products. Committee for Propriety of Medicinal Products (CPMP). Decision Trees for the Selection of Sterilization Methods (CPMP/QWP/054/98). Annex to note for guidance on development pharmaceutics (CPMP/QWP/155/96). London, Apr 5, 2000.

39. Yamamoto O. Effect of radiation on protein stability. In: Ahern TJ, Manning MC, eds. Stability of Protein Pharmaceuticals, Part A. Chemical and Physical Pathways of Protein Degradation. New York and London: Plenum Press, 1992:361–421.

40. Shalaev E, Reddy R, Kimball RN, Weinschenk MF, Guinn M, Margulis L. Protection of a protein against irradiation-induced degradation by additives in the solid state. Rad Phys Chem 2003; 66:237–245.

41. Kiang P, Ambrosio T, Buchanan R, et al. Technical Report No. 16: effects of gamma irradiation on elastomeric closures. J Parent Sci Tech 1992; 46(S2):S1–S13.

9

Rational Choice of Excipients for Use in Lyophilized Formulations

Evgenyi Y. Shalaev

*Parenteral Center of Emphasis, Groton Laboratories, Pfizer Inc.,
Groton, Connecticut, U.S.A.*

Wei Wang

*Pharmaceutical Sciences—Global Biologics, Pfizer Inc.,
Chesterfield, Missouri, U.S.A.*

Larry A. Gatlin

*Parenteral Center of Emphasis, Groton Laboratories, Pfizer Inc.,
Groton, Connecticut, U.S.A.*

INTRODUCTION

The majority of protein drugs are delivered by the injection route, although there is an increasing interest in alternative delivery routes, e.g., pulmonary. Ready-to-use liquid formulations are preferred injectable dosage forms because they are considered easier to manufacture and administer. However, the majority of proteins are not sufficiently stable in aqueous media to provide adequate commercial shelf-life and this limits the development of protein pharmaceuticals as ready-to-use injectables. Freeze-drying is an established process to increase long-term stability of proteins and achieve an acceptable shelf-life (1). In some cases, as with proteins intended for administration by inhalation, spray-drying is used (2). It is also possible to simply dry protein solutions slowly at ambient temperatures under vacuum (3). This chapter deals with freeze-dried protein formulations as they are the most common commercial dosage forms. However, general principles can be applied to other dehydration processes such as spray-drying and vacuum-drying.

Essentially all protein formulations contain one or more inactive ingredients (excipients). Excipients are used to facilitate the formulation manufacturing process, ensure stability of the active ingredient during processing, storage, and administration, minimize adverse effects upon administration (e.g., minimize pain upon injection), and ensure desirable bioavailability and (for sustained release dosage forms) release profiles. Each excipient in the formulation requires justification for its use and an appropriate rationale for the level selected. Only excipients that are essential for performance and/or stability of a dosage form and suitable for injectable products are allowed to be included.

The majority of lyophile protein dosage forms contain buffer and a bulking agent, the latter often playing a dual role for both pharmaceutical elegancy and cryo- and lyoprotection, to achieve stability during processing and the shelf-life. In addition, many protein formulations contain additional stabilizers, e.g., a surfactant, and occasionally an antioxidant or a chelating agent. In some cases, a tonicity modifier, a solubilizer, a processing aid, or an antimicrobial agent may be used. It is notable that the active ingredient level in the formulation can range from as high as close to 100% to as low as a few parts per million. Therefore, it is also possible to have a large range of excipient levels in the final formulation. It should be mentioned also that excipients, which are important for the reconstituted solution, e.g., antimicrobial agents or tonicity modifiers, can be added with the diluent rather than being incorporated into the lyophile cake. Generally, selection of a proper excipient should take into account (*i*) the type of product, (*ii*) the delivery route, dose, and administration frequency, (*iii*) the chemical and physical properties of the excipient, (*iv*) potential interactions with other product components, and (*v*) the container/closure system.

It is typically advantageous to choose formulation excipients that will not only enable the product to meet its critical quality parameters but also facilitate the freeze-drying process because of the high cost/long processing times for this unit operation. This is especially critical when developing formulations for unique package systems such as dual chamber syringes, because of the difficulties encountered in uniform drying in these packages. Therefore, selection of excipients that can potentially increase the collapse or eutectic temperatures of the frozen solution can greatly facilitate the drying process, thereby reducing cost and processing times. It is also important to select excipients whose vapor pressure is sufficiently low so as not to permit its removal during the lyophilization process.

There are a number of reviews available on different aspects of protein freeze-drying (1,4–10). In particular, it has been recognized that understanding phase behavior is a key for lyophile formulation and process development. Therefore, we start with a discussion of phase behavior of excipients during manufacturing and storage. Phase transitions have a major impact on stability and performance of protein dosage forms. For example, crystallization of a lyoprotector may result in protein destabilization during freeze-drying and storage. Description of excipients based on their functional role in protein formulations is given in the section titled Role and Properties of Excipients, followed by practical advice on rationale excipient choice (based on both functional and physical chemical properties of excipients).

PHASE BEHAVIOR OF EXCIPIENTS DURING LYOPHILIZATION AND STORAGE: GENERAL CONSIDERATIONS

Phase Transitions During Lyophilization and Storage

During initial cooling of protein formulation solutions, water is normally the first component to crystallize. At this stage, a biphasic system is formed, consisting of ice and residual freeze-concentrated solution (FCS), which contains protein drug, excipients, and remaining water. The composition of the FCS after initial (also known as primary) water crystallization depends on the ratio of solutes and the temperature, but is independent of total solid content (11). As cooling (and water crystallization) proceeds further, the FCS may either remain in the amorphous state or partially crystallize, depending on the composition of the system and the cooling rate (11) as described below:

1. The FCS may form a kinetically stable (but thermodynamically unstable) amorphous phase. A typical example of such behavior is sucrose-rich formulations. In this case, solutes do not usually crystallize during freezing, drying, and storage, provided that the freeze-dried cake is protected from water uptake and the storage temperature is well below the glass transition temperature.

2. The FCS may form a "doubly unstable" (i.e., both thermodynamically and kinetically unstable) state. Mannitol- and glycine-based formulations are typical examples of such behavior. In these systems, secondary excipient + ice crystallization[a] would occur either during cooling (if the cooling rate is slower than the critical cooling rate) or subsequent heating/annealing of the frozen solution (if the cooling rate was higher than the critical cooling rate). The critical cooling rate depends on the composition of the solution (e.g., glycine/sucrose ratio) and increases with an increase in the fraction of a crystallizable component (12). It should be noted that crystallization of an excipient is often incomplete, i.e., the maximal FCS contains usually all the components (water and all the solutes including protein and the excipients), although the relative fraction of the partially crystallized excipient remaining in the FCS is significantly reduced.

In addition to the common cases described above, it was proposed that a liquid–liquid (amorphous–amorphous) phase separation might take place, resulting in two amorphous phases of different chemical composition (3,9,13). For example,

---

[a] Note also that the secondary solute+water crystallization is referred in the pharmaceutical literature as "eutectic" crystallization, although this is not a strictly correct term to apply to a multicomponent system.

protein–excipient amorphous–amorphous phase separation would result in protein-rich and excipient-rich amorphous phases. Such phase separation is expected to compromise the stabilizing activity of excipients. It should be stressed, however, that there is a lack of experimental reports on such demixing behavior between protein and amorphous excipient during lyophilization, and it is not clear if this is a common behavior for protein formulations.

Although the phase state of excipients is usually "fixed" during freezing and annealing, further phase transformations may take place during primary and secondary drying as well as during shelf storage, depending on the properties and the concentration of excipients as well as the storage temperature. For example, if an excipient crystallizes as a crystallohydrate (i.e., a crystal with water in the crystal lattice) during freezing, the water of hydration might be removed during either primary or secondary drying. Such removal of water of hydration can result in either amorphous [e.g., sodium phosphate (14)], or crystalline anhydrous (e.g., mannitol) excipient. In addition, an amorphous excipient [e.g., inositol (15)] may crystallize during the shelf-life, especially if the water content in the lyocake increased because of water transfer either from the stopper or (if the stopper was not properly sealed) from the environment.

Significance of Excipient Crystallization

Phase transitions of excipients during manufacturing and storage have a major impact on both stability and performance of protein dosage forms. In particular, crystallization of either a buffer or lyo- and a cryoprotector is usually undesirable because of the negative impact on protein stability. Indeed, crystallization of buffer components is often accompanied by significant changes in the pH of the FCS (16), which often causes destabilization of a protein. Also, crystallization of a lyoprotector can compromise its protective function. For example, it was shown that inositol stabilized a protein when it existed as an amorphous form, whereas loss of protein activity was observed when inositol crystallized during storage (15). Another example is crystallization of a lyoprotector, raffinose, during freezing, causing destabilization of lactate dehydrogenase (LDH) (17). The negative impact of a lyoprotector (e.g., sugar) crystallization on protein stability can be attributed to two different factors: (*i*) Crystallization results in a physical separation of sugar molecules from protein molecules, i.e., an increase in intermolecular sugar/protein distance, from several angstroms [which is a typical hydrogen-bond length (18)] in molecular mixtures to micrometers in physical mixtures of crystalline sugar and amorphous protein; such separation would be expected to eliminate any protection imparted by sugars irrespective of the exact mechanism (e.g., water substitution vs. the glass transition hypothesis, or thermodynamics vs. kinetics mechanism). (*ii*) Crystallization of a sugar in an anhydrous form would result in a redistribution of water and in a significant increase in the local water content of the remaining amorphous protein-containing phase, which could be detrimental to long-term stability. It should be noted, however, that a sugar crystallohydrate (e.g., a pentahydrate as with raffinose) might serve as a water "scavenger" during crystallization, thus preventing an increase in local water content (19,20).

From the manufacturing process point of view, however, crystallization of an excipient may be beneficial because it allows primary drying at a higher temperature without visible collapse, which results in a shorter and more robust freeze-drying cycle. In addition, partially crystalline formulations have higher drying rates (i.e., shorter cycle) than amorphous formulations of a similar composition, possibly because a higher fraction of water is isolated as ice, with ice easier to remove than nonfrozen water associated with the amorphous phase (11).

A compromise between a desire to improve freeze-drying cycle efficiency and robustness (which is achieved by using a crystalline bulking agent) and sustaining protein protection (which needs an amorphous lyoprotector) can be achieved by using partially crystalline–partially amorphous formulations (21). It has been proposed that the crystalline portion provides a physical support even at relatively high product temperatures (i.e., higher than the collapse temperature of the amorphous phase) whereas the amorphous portion provides lyoprotection for protein, allowing aggressive primary drying conditions (21). Feasibility of such crystalline–amorphous formulations was demonstrated using glycine–sugar formulations with a lyophilization-sensitive enzyme, LDH (22). In this system, freeze-drying at a product temperature more than 10°C above the T_g' resulted in a freeze-dried cake without any evidence of macroscopic collapse and with a retained enzymatic activity, when the crystalline/amorphous ratio was higher than 1.2/1 (raffinose) or 1.6/1 (trehalose) (22). One should be aware, however, that timing of crystallization of a crystalline bulking agent, i.e., whether the crystallization takes place during cooling or annealing, may influence protein stability. An example of the significance of crystallization conditions was given in Ref. (11), where the stability of a freeze-dried conjugate of immunoglobulin G and horseradish peroxidase in a partially crystalline glycine/sucrose matrix was reported. In this case, the activity recovered was significantly higher in the material that crystallized during annealing as compared with material in which crystallization occurred during cooling. Therefore, although crystalline–amorphous formulations may be beneficial, the phase behavior and protein stability need to be investigated in each particular case in order to ensure a stable and robust freeze-dried product.

Amorphous–amorphous (liquid–liquid) phase separation of excipients also may cause protein destabilization, probably because of creation of an interface (1). In addition, it is also possible that a similar amorphous–amorphous demixing may occur between protein and lyoprotectant, with expected loss of protection (4,9), although there is a lack of experimental data on such protein–excipient amorphous–amorphous phase separation. Note that excipient–excipient demixing in FCS can be studied by differential scanning calorimetry DSC (23), whereas no reliable methods to detect such transitions exist for protein–lyoprotectant amorphous–amorphous demixing.

Phase State of Excipients: Analytical Aspects

Excipients can undergo phase transitions during different stages of freeze-drying and over the shelf-life of the freeze-dried product. To develop a robust and stable formulation, it is essential to understand and monitor such changes. The most common methods to detect crystalline–amorphous changes are DSC and X-ray diffraction (XRD), whereas other methods such as polarized light microscopy (PLM) and different spectroscopic techniques can also be used, depending on the formulation properties and stage in the formulation "life" when such transitions may occur.

Excipient crystallization may usually be expected to occur during the freezing and annealing stages of freeze-drying. Such events are commonly and conveniently studied by DSC. On a DSC curve, crystallization can be detected as a second exotherm during cooling (with a first exotherm being water crystallization), an exotherm during heating, and/or an additional endotherm during heating preceding the main ice melting peak. If either of these events is observed on DSC cooling–heating curves, one may conclude that a component(s) of the formulation would likely crystallize during freeze-drying. The reverse statement, however, is not always correct, i.e., lack of a crystallization event in a DSC experiment does not necessarily mean that crystallization would not occur in vials during freeze-drying. Indeed, relatively high scanning rates and small sample volume in a DSC study would provide less favorable crystallization conditions as compared with a larger sample volume and slower temperature ramping during a real freeze-drying run. Low-temperature XRD is another main method used to study crystallization in aqueous solutions (12,24), and is especially well suited both to distinguish between crystalline and amorphous structures and to identify the nature of any crystalline phase(s) present.

X-ray powder diffraction (XRPD) is probably the most common and convenient method to detect crystalline structures in a freeze-dried cake. PLM can also be used to confirm the amorphous nature of a freeze-dried cake. If birefringence is observed, it usually means that the cake is at least partially crystalline. However, PLM does not usually allow for the identification of the specific crystalline phase. Note that a sample is often exposed to an ambient atmosphere during both XRPD and PLM experiments, which may result in water uptake followed by crystallization. Therefore, precautions should be taken to minimize sample exposure to ambient relative humidity (RH) during measurements, to prevent erroneous conclusions. As indicated previously, DSC is another common tool to distinguish between crystalline and amorphous formulations. Spectroscopic methods such as Fourier transform infrared (FTIR) and Raman, and solid state nuclear magnetic resonance (NMR) may also be used for structure characterization of lyophile cakes.

Confirmation of the amorphous nature of a formulation is essential during the formulation development process, e.g., when a formulator needs to choose a buffer and/or lyoprotector. For both buffer and lyoprotector, retention of the excipient in an amorphous state is desirable, and can serve as a key criterion for the selection of an excipient. Usually, a combination of evaluation of the pre-lyo solution using DSC and evaluation of the finished cake using XRPD is sufficient to conclude if any crystallization would occur during freeze-drying. For example, if no crystallization is observed by DSC and a freeze-dried cake is amorphous by X-ray, one could conclude that no crystallization of either buffer or lyoprotector occurred during lyophilization. However, although the solution–DSC/freeze-dried-cake–XRPD combination is usually sufficient to make a reliable conclusion about excipient(s) crystallization, it is not always the case. In particular, an erroneous conclusion might be made in cases when the following conditions are met: (*i*) an excipient crystallizes as a hydrate; (*ii*) water of hydration is removed during drying (e.g., secondary drying); (*iii*) loss of water of hydration causes crystal-to-amorphous conversion. Although it might appear that such a combination of events is unlikely, it was shown by in situ freeze-drying XRD that this scenario can take place in real systems, such as phosphate buffer (14) and the lyoprotectant raffinose (17). As a result of such findings, use of the in situ XRPD method is attracting increased attention in the field. In addition, it is important to note a recent improvement in XRPD through the use of a high-intensity synchrotron radiation source, which provides an excellent signal-to-noise ratio and superior sensitivity as compared to the traditional XRPD method. Use of synchrotron XRPD (sXRPD) is especially important when one needs to detect crystallization of a low-concentration excipient, e.g., buffer, where sensitivity is a major issue. Application of sXRPD in the analysis of both freeze-dried cakes and phase transitions in frozen solutions and during freeze-drying can be found elsewhere (25,26).

ROLE AND PROPERTIES OF EXCIPIENTS

Buffers

Control of pH is often needed to ensure optimal solubility and stability of a product during manufacturing, storage, and upon reconstitution. In most cases, an appropriate amount of buffer is needed to provide adequate buffering capacity. Buffer type and concentration, as well as solution pH before lyophilization, are important formulation variables. Buffering capacity and the possibility of buffer catalysis are the major buffer properties to be considered in the development of liquid pharmaceutical formulations (27). There are additional requirements for buffers for freeze-drying, i.e., they should be nonvolatile, have a high collapse temperature (T_c or T_g') in the FCS, remain amorphous during freeze-drying, and have a high glass transition temperature in the solid state (28). Several buffers that are common for parenteral formulations have unfavorable freeze-drying properties. For example, acetate—a common buffer—is not a preferred buffer for lyophilization because it is volatile and can be partially lost during freeze-drying, resulting in a significant pH change. Hydrochloric acid is another example of a pH modifier that is volatile and should be used with caution. In addition, several common buffers have a high tendency to crystallize during freezing. Buffer crystallization is usually undesirable because it can lead to substantial pH shifts during freezing and therefore could destabilize the protein. In particular, sodium phosphate buffer demonstrated significant pH changes (several pH units) during freezing, as a result of freeze concentration and crystallization of the buffer components (16). Similarly, tartrate and succinate buffers crystallize readily whereas citrate, glycolate, and malate are more resistant to crystallization (29). It should be noted that both crystallization and collapse behavior depend on solution pH. For example, collapse temperatures of several common buffers are presented in Figure 1 as a function of solution pH. The figure illustrates two interesting features of the collapse behavior of buffers, i.e., significant changes in the collapse temperature with solution pH, and an influence of a counter ion (e.g., sodium citrate vs. potassium citrate).

Overall, buffers with a higher collapse temperature (T_c or T_g') and a lower crystallization potential are preferred for lyophilized formulations. However, it should be stressed that buffers with a relatively low collapse temperature and relatively high crystallization potential (e.g., phosphate buffer) can still be used in lyophilized formulations. Both collapse temperature and crystallization ability can be modified using other excipients. For example, if a formulation contains a

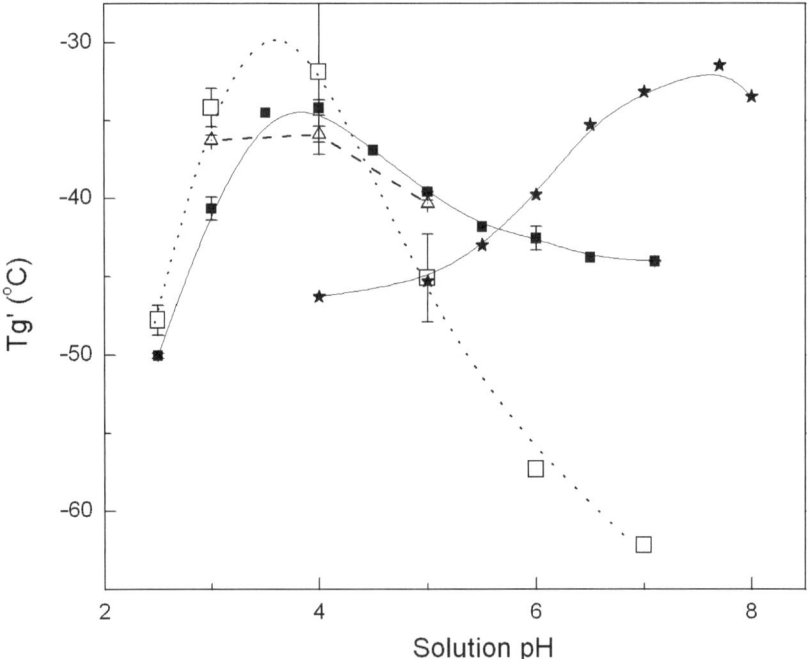

Figure 1   Collapse temperature (T_g') as a function of pH. ■: sodium citrate; □: potassium citrate; Δ: sodium tartrate; ★: L-histidine. Error bars represent standard deviation. Lines are given as a visual help. *Source*: From Refs. 29, 30.

significant amount of a solute with a high T_c (e.g., a protein), the collapse temperature of the formulation would increase and buffer crystallization may be suppressed. However, the amount of an amorphous component required to suppress crystallization and/or to provide an acceptable T_c would be lower in the case of a buffer with a higher T_c or with a lower crystallization potential. Citrate buffer appears to be a good choice for lyophilized formulations that are prepared at acidic or near neutral pH. Frozen solutions of citrate buffer have a low crystallization potential, relatively high collapse temperature, and minimal pH changes during freezing. In addition, citrate has a reasonably high T_g' especially when prepared from solutions at higher pH values. Glycine may be a reasonable choice for the alkaline pH region based on a relatively low crystallization potential of sodium glycinate. Other details on freeze-drying properties of different buffers can be found in Ref. (28).

Bulking Agents/Lyoprotectors

In many cases, the dose of a drug is quite small, and a bulking agent (or filler) is needed to provide a matrix to carry the active ingredient. Common bulking agents include mannitol, lactose, sucrose, dextran, trehalose, and glycine. In protein formulations, bulking agents often play a dual role as both bulking agent and lyoprotector. These bulking agents range both in their ability to crystallize (mannitol and glycine) or remain amorphous (e.g., sucrose) and in their impact on the formulation collapse or eutectic temperature. An appropriate choice of a bulking agent results in optimal product quality (e.g., physical and chemical stability, reconstitution time, moisture levels) and facilitates freeze-drying and scale-up to production size. The level of bulking agent utilized will vary depending on the rationale for use, e.g., as a matrix-forming agent, as the collapse temperature modifier, or stabilizer. As a "rule of thumb," use of a bulking agent can be considered if the active concentration in the fill solution is less than 2 wt%.

Mannitol is the most common bulking agent used in protein freeze-dried formulations. Usually, mannitol crystallizes during cooling or annealing of frozen solutions, which, combined with a high mannitol–ice eutectic temperate of −1.5°C (31), allows one to freeze-dry such formulations at a relatively high primary drying temperature, without a macroscopic collapse. Therefore, mannitol-based formulations are known to be easy to lyophilize, with a shorter and robust lyophilization cycle. There are two potential complications associated with mannitol-based formulations. Mannitol forms a crystallohydrate during freeze-drying (32,33) that hinders removal of water of hydration and requires elevated secondary drying temperature. Vial breakage is another potential problem a formulator might encounter while working with mannitol-rich formulations (34,35). It has been shown that vial breakage is likely associated with volume expansion that occurs during warming of a frozen solution at approximately −25°C to −20°C (36) probably because of secondary mannitol + water crystallization. Vial breakage is affected by mannitol concentration, cooling rate, and filling volume (34), as well as presence of amorphous solutes and vial configuration (35). Glycine is another common crystalline bulking agent. Because of ionizable groups of glycine, it may also serve as a buffer. At neutral pH, glycine crystallizes as anhydrous beta polymorph, and its eutectic with ice has a relatively high eutectic temperature of $-3.6°C$ (37). Both solution pH and counter ions have a significant impact on glycine crystallization behavior (38), by influencing both ionization of glycine in solution and a precipitating solid form. Usually, the presence of other solutes (such as a lyoprotector or the protein itself) hinders crystallization of both glycine and mannitol. For example, sucrose inhibited the crystallization of mannitol at a sucrose/mannitol ratio of 2:1 (39). Inhibition of glycine crystallization by sugars (sucrose, trehalose, and raffinose) was reported in Refs. (12,40,41). For example, a critical sucrose/(glycine + sucrose) ratio above which glycine does not crystallize during either cooling or annealing was reported to be 0.8 (12).

Sucrose is probably the most popular lyoprotector used in lyophile protein formulations. Because of its low collapse temperature, however, lyophilizing sucrose-rich formulations can be a challenge. In addition, the presence of other components with low collapse temperatures (e.g., buffers) may lower the T_c even further. Possible ways to overcome this challenge are reducing the concentration of components with low T_c, addition of an excipient with a high collapse temperature (e.g., dextran), or addition of a crystalline bulking agent. A critical factor for the selection of a lyoprotector is its impact on physical and chemical stability of proteins, both during freeze-drying and the product shelf-life. Sucrose is known to substantially increase the stability of proteins both during freeze-drying and during subsequent storage. Lactose is another popular bulking agent although there is a potential for chemical interaction with amino groups of a protein forming a Schiff base, known as the Maillard reaction, or nonenzymatic browning. Recently, another disaccharide lyoprotector was introduced with claims of superior stabilization properties, i.e., trehalose (42). The most obvious advantage of trehalose is its higher glass transition temperature as compared to sucrose (120°C vs. 74°C). However, side-by-side comparison of stabilization of proteins by sucrose and trehalose revealed that sucrose might provide a comparative or, in some cases, better protection than trehalose. For example, sucrose appeared to be more effective in stabilizing the native structure of lysozyme during spray-drying (43), and a comparable level of protection to lysozyme and catalase during freeze-drying (44). An example of better stabilization of a protein by sucrose relative to trehalose during freeze-drying was reported in Ref. (45). To explain the difference in the protective action between different sugars, it was suggested that trehalose might have a greater tendency for protein–sugar phase separation as compared to sucrose (9). Amorphous/amorphous phase separation between a sugar and a protein (i.e., when a sugar forms a separate sugar-dominant phase thus leaving protein unprotected in a protein-rich phase) is considered to be undesirable because it would result in a physical separation of protein molecules from a lyoprotectant. Such a separation would have a detrimental impact on both reconstitution recovery and the shelf-life.

Polymeric bulking agents such as dextran have good manufacturing properties (i.e., high T_g'), and can protect proteins. For example, dextran, carboxy methylcellulose (CMC), DEAE-dextran, and polyethyleneglycol (PEG) have been shown to reduce aggregation of lyophilized bovine serum albumin (BSA) significantly during storage at 37°C (46). Hydroxypropyl-ß-cyclodextrin was found to minimize inactivation of LDH during freeze-drying (47), stabilize a mouse monoclonal antibody (MN12) during storage at 56°C (48), inhibit moisture-induced aggregation of solid insulin (49), stabilize interleukin (IL)-2 against aggregation during storage at 5°C (50), and inhibit the dimerization of tumor necrosis factor (TNF) during storage at 37°C (51). Dextran 40 increased the activity of lyophilized elastase (10% dextran in solution of 10 mM sodium acetate at pH 5.0 at protein concentration of 20 mg/mL) from 33% to 82% during storage for two weeks at 40°C and 79% RH (52). Dextran (162 kDa) at 3.5% and 5% (w/v) improved the storage stability of lyophilized rFXIII and Humicola lanuginosa lipase, respectively, at 40°C and 60°C (53,54). Both poly(vinylpyrrolidone) (PVP) and maltodextrin stabilized lyophilized invertase during incubation at 90°C (55). Polyethyleneimine was shown to increase the storage stability of lyophilized LDH in a concentration-dependent manner at 36°C (56). However, polymers may not provide the same level of protection as disaccharides and, in certain cases, they may have an adverse effect. For example, dextran had a lower extent of protection than either trehalose or sucrose to lysozyme and catalase during freeze-drying (44). Inclusion of dextran 40 in a lyophilized IL-6 formulation containing sucrose significantly increased protein aggregation during storage at 40°C for nine months (57). The destabilization can be attributed to a failure of inflexible dextran molecules to interact with the protein effectively by hydrogen bonding. Apparently for the same reason, the activity of lyophilized bilirubin oxidase in a dextran formulation decreased faster than that in a polyvinyl alcohol formulation during storage at 70°C (58). Finally, while PVP, dextran, and other sugar polymers form an amorphous state, PEG readily crystallizes during freezing, although one may speculate that the extent of PEG crystallization could depend on the molecular weight and presence of other solutes (59).

Other Excipients

Surfactants

Surfactants are often used in liquid protein formulations to inhibit protein aggregation during shipping and storage. For lyophilized protein formulations, shipping is not considered a significant stress factor to cause protein denaturation and/or aggregation. However, both the lyophilization process and long-term storage can still cause protein denaturation and/or aggregation. Between the two factors, lyophilization seems to be the major destabilizing stress for lyophilized protein formulations. The lyophilization process consists of two major steps—freezing and drying. Both steps may cause protein denaturation and/or aggregation. The formation of ice–water interfaces during freezing may cause surface protein adsorption, leading to denaturation and/or aggregation of proteins. Surfactants may decrease the driving force of protein adsorption, reducing denaturation and/ or aggregation of proteins at these interfaces. Tween 80 is one of the commonly used surfactants for protein stabilization during freezing. For example, Tween 80 protected several proteins from freezing denaturation including LDH and gluta- mate dehydrogenase (60), TNF-binding protein, interleukin-1 receptor antagonist (IL-1ra), bFGF, malate dehydrogenase, aldolase, and phosphofructokinase (61). Other nonionic and ionic surfactants have also been reported in cryoprotection of proteins, such as Brij-35, Brij-30 (polyoxyethylene lauryl ether), Lubrol-px, Triton X-10, Pluoronic F127 (polyoxyethylene-polyoxypropylene copolymer), and sodium dodecyl sulfate (60,62).

The sole effect of drying stress on protein denaturation and/or aggrega- tion during lyophilization is scarcely reported, partly because the drying step is coupled to the freezing step. Nevertheless, proteins in an aqueous solution are fully hydrated and removal of the hydration shell may disrupt the native state of a protein and cause denaturation and/or aggregation. Nonionic surfactants have been shown to bind weakly to proteins, and the binding can potentially block the aggregation-prone hydrophobic sites on the protein surface (63). It is reason- able to expect that replacement of water molecules by surfactant molecules on the protein surface during drying could potentially inhibit protein denaturation and/or aggregation. Since the common nonionic surfactants have low glass transition temperatures [e.g., below −60°C for Tween 80 (64)] and make the lyophilization process difficult, it is generally not recommended to use a large amount of surfac- tants in the protein formulation.

Although surfactants can effectively protect protein denaturation and/or aggregation during lyophilization, they do not seem to stabilize proteins effec- tively during long-term storage based on a limited number of studies. Bush et al. demonstrated that Tween 80 could not provide significant protection for the lyophilized factor IX during storage (64). Inclusion of 0.002% Tween 20 in a lyophilized rFXIII formulation did not improve its storage stability at 40°C and 60°C (53). Tween 80 at 0.05% or 0.1% actually destabilized the spray-dried LDH during storage at 25°C, 40°C, and 60°C (65). Therefore, the role of surfactants in the formulation of lyophilized products is not straightforward. Long-term stabil- ity studies are usually needed to decide whether surfactants should be included in the product.

Antimicrobial Preservatives

Antimicrobial preservatives are often needed for multidose presentations, to ensure microbiological safety of reconstituted solution during in-use storage. They can be added with a diluent rather than being incorporated into the lyophile cake. If a preservative has to be included into the lyophile cake, a formulator should be aware of potential compatibility issues between proteins and antimicro- bial preservatives (1). For example, an increase in the amount of IL-1R aggregates correlates roughly with a gradual decrease in the protein unfolding temperature in the presence of one of the three preservatives, phenol, *m*-cresol, and benzyl alcohol (66). In addition, proteins may inhibit antimicrobial activity of preservatives. Therefore, antimicrobial effectiveness of preservatives needs to be tested in a market-image formulation.

Salts

Physical stability of proteins in solution may require presence of salt to achieve an optimal ionic strength. In addition, salts have been shown to stabilize proteins in a lyophilized cake. Liu et al. (46) found that NaCl or sodium phosphate could significantly inhibit aggregation of lyophilized BSA on incubation at 37°C. rHA co-lyophilized with NaCl at a NaCl:protein weight ratio of 1:6 did not aggregate upon incubation at 37°C and 96% RH for four days, while the protein without NaCl lost over 80% solubility in just one day under the same conditions (67). Sodium chloride at an excipient:protein weight ratio of 1:5 was also able to reduce aggregation of lyophilized proteins during storage at 37°C and 86% RH for six days (68). Calcium ions have been shown to protect solid recombinant human deoxyribonuclease significantly against aggregation during storage at 40°C (69).

Miscellaneous Compounds

In some cases, stability of proteins may be improved by using other, less common, types of excipients. A combination of 2.0% arginine and 2.3% carnitine significantly decreased aggregation of lyophilized IL-2 during storage at 37°C for four weeks (70). The combined use of phenylalanine, arginine, and a mineral acid inhibited aggregation of vacuum-dried rhG-CSF or LDH during storage at 40°C (71). Several excipients, such as D-glucaric acid and D-gluconic acid, have been shown to inhibit aggregation of lyophilized albumin during storage (67).

Antioxidants are added to a product to minimize or slow down the oxidative processes that may occur with some drugs or other excipients. The oxidative process may be catalyzed by light, temperature, metal ions, or peroxides and the formulator should consider that metal ions or peroxides might be contained in the active pharmaceutical ingredient or excipients. Antioxidants should be used at the lowest effective concentration and are typically used at low concentrations from 0.001% to 0.2%. In many cases, an antioxidant is used in concert with a chelating agent to sequester metal ions. The selection of an antioxidant system is generally a trial-and-error process; however, commonly used antioxidants and chelating agents (72) may provide a starting point. One should keep in mind that certain antioxidants that are used to stabilize proteins in the liquid state or during lyophilization might destabilize proteins in the solid state. For example, ascorbic acid is a frequently used antioxidant, but at 5 mM, it reduced the storage stability of lyophilized elastase (20 mg/mL in 10 mM sodium acetate, pH 5.0) significantly at 40°C and 79% RH (52).

An interesting approach to stabilization of proteins by a plasticizer was suggested recently. Cicerone et al. (73,74) and Pikal's group (75) demonstrated that a small amount of a plasticizer, e.g., glycerol and sorbitol, could increase the storage stability of protein formulations. These findings contradict a "common wisdom" that plasticizers are expected to compromise stability by enhancing molecular mobility, i.e., large-scale cooperative α-like motions associated with the glass transition. However, a more detailed analysis of complex molecular mobility processes suggest that plasticizers, when present in a relatively low amount, might inhibit local motions (also known as the high-frequency relaxations and β-like motions), and therefore certain degradation processes.

PRACTICAL ADVICE

"The simpler, the better." An ideal pharmaceutical injectable formulation would contain only two ingredients, i.e., the active ingredient and water. However, in almost all cases, excipients are needed in order to achieve appropriate stability, manufacturability, injectability, efficacy, and safety. For freeze-dried proteins, a formulator's task starts usually with determining if a bulking agent and a buffer are needed. As a "rule of thumb," a bulking agent may be needed if the total solute content in prelyophilization solution is below 2%w/w. In this case, a formulator has to choose between crystalline bulking agents such as mannitol and glycine, amorphous low-molecular-weight bulking agents such as sucrose and trehalose, and polymers such as dextran and PVP. While either a crystalline or a polymeric amorphous bulking agent is beneficial from the processing standpoint because of a high eutectic or collapse temperature, they might not provide sufficient protein protection against lyophilization stresses and/or during the long-term storage. An amorphous low-molecular-weight bulking agent (e.g., disaccharide), on the other hand, is usually expected to improve the stability of proteins; however freeze-drying such formulations may be more difficult because of relatively low collapse temperatures. A compromise between ease of lyophilization and protein stabilization may be achieved by using amorphous/crystalline formulations as described earlier.

In parallel with the bulking agent decision, a target solution pH should be determined, and the need for a buffer should be evaluated. Selection of a solution pH should be based on both solution and lyophile cake stability requirements. Optimal pH for lyophile stability can be similar to the optimal solution pH, although in some cases, a significant difference (one or more pH units) is observed. For example, significant differences in the "pH-stability" profiles between solution and lyophiles was reported for two model hexapeptides (76). Therefore, the impact of solution pH on stability of the lyophile cake should be evaluated over a relatively wide pH range. In some cases, the protein itself may provide sufficient buffering capacity, and inclusion of a buffer may not be needed. However, in the majority of formulations, this is not the case, and a buffer is required.

When there is a need for a bulking agent and/or a buffer, a formulator needs to choose which bulking agent and buffer to use and at what concentration. Selection of specific excipients is commonly based on stability considerations, i.e., by identifying a formulation providing better stability during both manufacturing (compounding and freeze-drying), shelf-life as a lyo cake, and upon reconstitution. While solution and freeze-drying stability can be explored under real time and temperature conditions, lyophile cake stability evaluation may require accelerated aging to shorten the drug development process. Often, a more stable formulation can be identified by accelerated stability evaluation of different formulations, sometimes with the help of an experimental design (77,78), even though temperature dependency of protein stability may not follow Arrhenius behavior. Recent success in understanding mechanisms governing chemical processes in protein formulations [e.g. Refs. (6,9,10)] may provide additional tools to predict relative stability of different formulations (i.e., stability rank-order) based on a series of relatively simple tests performed on freeze-dried formulations. Such tests include the following:

1. *Crystallinity of excipients, e.g., by either XRPD or PLM.* Crystallization of a cryoprotector and a buffer is usually associated with protein destabilization; therefore formulations in which either lyoprotector or buffer crystallize can be expected to be less stable compared with formulations with amorphous lyoprotector and buffer. Note, however, that this consideration should not be applied to crystalline–amorphous formulations (e.g., mannitol–sucrose), where an amorphous lyoprotector (sucrose) may provide adequate protection despite the use of a bulking agent (e.g., mannitol) that crystallizes.

2. *Protein secondary structure by FTIR.* Formulations with a significant disturbance of a protein secondary structure are expected to have compromised long-term stability (79). For example, it was reported that lyophilized antibodies with more "native-like" structure had a better storage stability in respect to both aggregation and chemical degradation (75).

3. *Residual water content.* The statement "the drier, the better" generally does not apply to protein formulations, although high moisture content is often associated with poor protein stability. It is essential to know the effect of different water contents in formulations on protein stability. In some cases, there is a "threshold" water content, which is associated with a significant change in stability (10). Water content should be measured through a stability study to check if there is any change with time because of potential water transfer from or to a stopper. Such stopper-related change in water content during storage would complicate interpretation of long-term stability data, and may lead to significant vial-to-vial variations.

4. *Glass transition temperature (e.g., by DSC).* Although formulations with higher T_g do not necessarily have better stability, development of a formulation with a high T_g is more desirable. Knowing the T_g is essential to choosing the appropriate accelerated storage conditions,

in particular, to ensure that the storage temperature is well below (at least 10°C) the T_g. Note that DSC does not have a sufficient sensitivity to detect the T_g in pure proteins and protein-rich formulations (80). In such cases, alternative tools could be used such as the thermally stimulated current method (81).

5. *Advanced measurements of molecular mobility.* For example, enthalpy relaxation (82) and NMR spin-lattice relaxation times correlate with rates of chemical degradation in some systems (83).
6. *Solid state acidity, e.g., expressed as the Hammett acidity function.* This is a relatively new concept in the freeze-drying community, although the Hammett acidity function is widely used in other fields, e.g., physical organic chemistry and heterophase catalysis. The Hammett acidity function is determined through the ionization of a probe molecule that is colyophilized with a formulation (84). Correlations between Hammett acidity function and chemical stability of lyophiles were studied using a model acid-catalyzed reaction of sucrose inversion (85), although such relationships with protein physical or chemical stability are yet to be established.

CONCLUSIONS

Use of excipients allows a formulator to design a protein dosage form with desirable properties such as protection of biological activity, manufacturability, stability, safety, and injection site toleration. The current knowledge allows a formulator to "engineer" a formulation with processing-favorable properties based primarily on understanding the phase behavior of excipients and formulations during the lyophilization process. The choice of excipients to achieve optimal solid-state chemical and physical stability over the shelf-life of the product, on the other hand, is a semiempirical enterprise. Although there has been significant progress in understanding the main factors that govern solid-state reactivity, it is still difficult to predict the stability of different formulations based on their properties; therefore, accelerated stability studies are essential to evaluate and choose more stable formulations. Since the stability of proteins often exhibit non-Arrhenius-like behavior, real-time stability studies are needed to select the ultimate commercializable formulation. One parameter that is particularly difficult to predict is the impact of excipient-related impurities on product stability. Because of this, it is beneficial to use excipients of the same grade and from the same vendor in formulation development, clinical manufacture, and International Conference on harmonization (ICH) stability studies, to ensure consistency.

ACKNOWLEDGMENT

The manuscript is based in part on information that was provided by approximately 30 Pfizer colleagues.

REFERENCES

1. Wang W. Lyophilization and development of solid protein pharmaceuticals. Int J Pharm 2000; 203:1–60.
2. Lee G. Spray-drying of proteins. Pharm Biotechnol 2002; 13:135–158.
3. Franks F. Protein hydration. In: Franks F, ed. Protein Biotechnology: Isolation, Characterization, and Stabilization. Totowa, NJ: Humana, 1993:437–465.
4. Pikal MJ. Freeze-drying of proteins: process, formulation, and stability. In: Cleland JL, Langer R, eds. Formulation and Delivery of Proteins and Peptides. Washington: ACS Symposium Series, 567, 1994:120–133.
5. Gatlin LA, Nail SL. Protein purification process engineering. Freeze-drying: a practical overview. Bioprocess Technol 1994; 18:317–367.
6. Carpenter JF, Chang BS, Garzon-Rodriguez W, Randolph TW. Rational design of stable lyophilized protein formulations: theory and practice. Pharm Biotechnol 2002; 13:109–133.
7. Costantino HR. Excipients for use in lyophilized pharmaceutical peptide, protein, and other bioproducts. Biotechnology: Pharmaceutical Aspects, 2(Lyophilization of Biopharmaceuticals). AAPS Press, 2004:139–228.
8. Franks F, ed. Protein Biotechnology: Isolation, Characterization, and Stabilization. Totowa, NJ: Humana, 1993:592.
9. Hill JJ, Shalaev EY, Zografi G. Thermodynamic and dynamic factors involved in the stability of native protein structure in amorphous solids in relation to levels of hydration. J Pharm Sci 2005; 94:1636–1667.
10. Pikal MJ. Freeze-drying. Encyclopedia of Pharmaceutical Technology. Marcel Dekker, 2002:1299–1326.
11. Shalaev E, Franks F. Solid-Liquid state diagrams in pharmaceutical lyophilisation: crystallisation of solutes. In: Levine H, ed. Progress in Amorphous Food and Pharmaceutical Systems. Cambridge: The Royal Society of Chemistry, 2002:200–215.
12. Shalaev EY, Kanev AN. Solid-liquid state diagram of the water-glycine-sucrose system. Cryobiology 1994; 31:374–382.
13. Heller MC, Carpenter JF, Randolph TW. Application of a thermodynamic model to the prediction of phase separations in freeze-concentrated formulations for protein lyophilization. Arch Biochem Biophys 1999; 363:191–201.
14. Pyne A, Chatterjee K, Suryanarayanan R. Crystalline to amorphous transition of disodium hydrogen phosphate during primary drying. Pharm Res 2003; 20:802–803.
15. Izutsu K, Yoshioka S, Kojima S. Physical stability and protein stability of freeze-dried cakes during storage at elevated temperatures. Pharm Res 1994; 11:995–999.
16. Gomez G, Pikal MJ, Rodriguez-Hornedo N. Effect of initial buffer composition on pH changes during far-from-equilibrium freezing of sodium phosphate buffer solutions. Pharm Res 2001; 18:90–97.
17. Chatterjee K, Shalaev EY, Suryanarayanan R. Raffinose crystallization during freeze-drying and its impact on recovery of protein activity. Pharm Res 2005; 22:303–309.
18. Jeffrey GA. An Introduction to Hydrogen Bonding. New York, Oxford: Oxford University Press, 1997:303.
19. Saleki-Gerhardt A, Stowell JG, Byrn SR, Zografi G. Hydration and dehydration of crystalline and amorphous forms of raffinose. J Pharm Sci 1995; 84:318–323.
20. Franks F, Auffret AD, Aldous BJ. Compositions In Glassy Phase, Stabilized By a Sugar. In: The Patent Cooperation Treaty, ed. International: WO 96/33744; 1996:40.

21. Shalaev EY, Franks F. Changes in the physical state of model mixtures during freezing and drying: impact on product quality. Cryobiology 1996; 33:14–26.
22. Chatterjee K, Shalaev EY, Suryanarayanan R. Partially crystalline systems in lyophilization: II. Withstanding collapse at high primary drying temperatures and impact on protein activity recovery. J Pharm Sci 2005; 94:809–820.
23. Izutsu K-I, Kojima S. Freeze-concentration separates proteins and polymer excipients into different amorphous phases. Pharm Res 2000; 17:1316–1322.
24. Cavatur RK, Suryanarayanan R. Characterization of frozen aqueous solutions by low temperature X-ray powder diffractometry. Pharm Res 1998; 15:194–199.
25. Nunes C, Mahendrasingam A, Suryanarayanan R. Quantification of crystallinity in substantially amorphous materials by synchrotron x-ray powder diffractometry. Pharm Res 2005; 22:1942–1953.
26. Varshney DB, Kumar S, Shalaev EY, Kang S-W, Gatlin LA, Suryanarayanan R. Solute crystallization in frozen systems–use of synchrotron radiation to improve sensitivity. Pharm Res 2006; 23(10):2368–2374.
27. Flynn G. Buffers-pH control within pharmaceutical systems. J Parenteral Drug Assoc 1980; 34:139–162.
28. Shalaev EY. The impact of buffer on processing and stability of freeze-dried dosage forms, Part 1. Solution freezing behavior. Am Pharm Rev 2005; 8:80–87.
29. Shalaev E, Johnson-Elton T, Chang L, Pikal MJ. Thermophysical properties of pharmaceutically compatible buffers at sub-zero temperatures: implications for freeze-drying. Pharm Res 2002; 19:195–201.
30. Osterberg T, Wadsten T. Physical state of L-histidine after freeze-drying and long-term storage. Eur J Pharm Sci 1999; 8:301–308.
31. Gatlin LA, Deluca P. A study of the phase transitions in frozen antibiotic solutions by differential scanning calorimetry. J Parenteral Drug Assoc 1980; 34:398–408.
32. Cavatur RK, Suryanarayanan R. Characterization of phase transitions during freeze-drying by in situ X-ray powder diffractometry. Pharm Dev Technol 1998; 3:579–586.
33. Yu L, Milton N, Groleau EG, Mishra DS, Vanisickle RE. Existence of a mannitol hydrate during freeze-drying and practical implications. J Pharm Sci 1999; 88:196–198.
34. Williams NA, Lee Y, Polli GP, Jennings TA. The effects of cooling rate on solid phase transitions and associated vial breakage occurring in frozen mannitol solutions. J. Parenteral Sci Tech 1986; 40:135–141.
35. Williams NA, Dean T. Vial breakage by frozen mannitol solutions: correlation with thermal characteristics and effect of stereoisomerism, additives, and vial configuration. J Parenteral Sci Tech 1991; 45:94–100.
36. Williams NA, Guglielmo J. Thermal mechanical analysis of frozen solutions of mannitol and some related stereoisomers: evidence of expansion during warming and correlation with vial breakage during lyophilization. J Parenteral Sci Tech 1993; 47:119–123.
37. Shalaev EY, Malakhov DV, Kanev AN, et al. Study of the phase diagram water fraction of the system water-glycine-sucrose by DTA and X-ray diffraction methods. Thermochim Acta 1992; 196:213–220.
38. Akers MJ, Milton N, Byrn SR, Nail SL. Glycine crystallization during freezing: the effects of salt form, pH, and ionic strength. Pharm Res 1995; 12:1457–1461.
39. Lueckel B, Bodmer D, Helk B, Leuenberger H. Formulations of sugars with amino acids or mannitol-influence of concentration ratio on the properties of the freeze-concentrate and the lyophilizate. Pharm Dev Technol 1998; 3:325–336.

40. Suzuki T, Franks F. Solid-liquid phase transitions and amorphous states in ternary sucrose-glycine-water systems. J Chem Soc Faraday Trans 1993; 89:3283–3299.
41. Chatterjee K, Shalaev EY, Suryanarayanan R. Partially crystalline systems in lyophilization: I. Use of ternary state diagrams to determine extent of crystallization of bulking agent. J Pharm Sci 2005; 94:798–808.
42. Hatley RHM, Blair JA. Stabilisation and delivery of labile materials by amorphous carbohydrates and their derivatives. J Mol Catalysis B: Enzymatic 1999; 7:11–19.
43. Liao YH, Brown MB, Nazir T, Quader A, Martin GP. Effects of sucrose and trehalose on the preservation of the native structure of spray-dried lysozyme. Pharm Res 2002; 19:1847–1853.
44. Liao YH, Brown MB, Quader A, Martin GP. Protective mechanism of stabilizing excipients against dehydration in the freeze-drying of proteins. Pharm Res 2002; 19:1854–1861.
45. Allison SD, Chang B, Randolph TW, Carpenter JF. Hydrogen bonding between sugar and protein is responsible for inhibition of dehydration-induced protein unfolding. Arch Biochem Biophys 1999; 365:289–298.
46. Liu WR, Langer R, Klibanov AM. Moisture-induced aggregation of lyophilized proteins in the solid state. Biotechnol Bioeng 1990; 37:177–184.
47. Izutsu K, Yoshioka S, Kojima S. Increased stabilizing effects of amphiphilic excipients on freeze-drying of lactate dehydrogenase (LDH) by dispersion into sugar matrices. Pharm Res 1995; 12:838–843.
48. Ressing ME, Jiskoot W, Talsma H, van Ingen CW, Beuvery EC, Crommelin DJ. The influence of sucrose, dextran, and hydroxypropyl-beta-cyclodextrin as lyoprotectants for a freeze-dried mouse IgG2a monoclonal antibody (MN12). Pharm Res 1992; 9:266–270.
49. Katakam M, Banga AK. Aggregation of insulin and its prevention by carbohydrate excipients. PDA J Pharm Sci Technol 1995; 49:160–165.
50. Hora MS, Rana RK, Wilcox CL, Katre NV, Hirtzer P, Wolfe SN, Thomson JW. Development of a lyophilized formulation of interleukin-2. Dev Biol Stand 1992; 74:295–303.
51. Hora MS, Rana RK, Smith FW. Lyophilized formulations of recombinant tumor necrosis factor. Pharm Res 1992; 9:33–36.
52. Chang BS, Randall CS, Lee YS. Stabilization of lyophilized porcine pancreatic elastase. Pharm Res 1993; 10:1478–1483.
53. Kreilgaard L, Frokjaer S, Flink JM, Randolph TW, Carpenter JF. Effects of additives on the stability of recombinant human factor XIII during freeze-drying and storage in the dried solid. Arch Biochem Biophys 1998; 360:121–134.
54. Kreilgaard L, Frokjaer S, Flink JM, Randolph TW, Carpenter JF. Effects of additives on the stability of Humicola lanuginosa lipase during freeze-drying and storage in the dried solid. J Pharm Sci 1999; 88:281–290.
55. Cardona S, Schebor C, Buera MP, Karel M, Chirife J. Thermal stability of invertase in reduced-moisture amorphous matrices in relation to glassy state and trehalose crystallization. J Food Sci 1997; 62:105–112.
56. Andersson MM, Hatti-Kaul R. Protein stabilizing effect of polyethyleneimine J Biotechnol 1999; 72:21–31.
57. Lueckel B, Helk B, Bodmer D, Leuenberger H. Effects of formulation and process variables on the aggregation of freeze-dried interleukin-6 (IL-6) after lyophilization and on storage. Pharm Dev Technol 1998; 3:337–346.

58. Nakai Y, Yoshioka S, Aso Y, Kojima S. Solid-state rehydration-induced recovery of bilirubin oxidase activity in lyophilized formulations reduced during freeze-drying. Chem Pharm Bull (Tokyo) 1998; 46:1031–1033.

59. Izutsu K-i, Yoshioka S, Kojima S. Phase separation and crystallization of components in frozen solutions: effect of molecular compatibility between solutes. Therapeutic Protein and Peptide Formulation and Delivery. ACS, 675, 1997:109–118.

60. Chang BS, Kendrick BS, Carpenter JF. Surface-induced denaturation of proteins during freezing and its inhibition by surfactants. J Pharm Sci 1996; 85:1325–1330.

61. Kendrick B, Chang BS, Carpenter JF. Detergent stabilization of proteins against surface and freezing denaturation. Pharm Res 1995; 12(suppl):S-85.

62. Nema S, Avis KE. Freeze-thaw studies of a model protein, lactate dehydrogenase, in the presence of cryoprotectants. J Parenter Sci Technol 1992; 47:76–83.

63. Bam NB, Randolph TW, Cleland JL. Stability of protein formulations: investigation of surfactant effects by a novel EPR spectroscopic technique. Pharm Res 1995; 12:2–11.

64. Chu C-H, Berner B. Thermal analysis of Poly(Acrylic Acid)/Poly(Oxyethylene) Blends. J Appl Polymer Sci 1993; 47:1083–1087.

65. Adler M, Lee C. Stability and surface activity of lactate dehydrogenase in spray-dried trehalose. J Pharm Sci 1999; 88:199–208.

66. Bush L, Webb C, Bartlett L, Burnett B. The formulation of recombinant factor IX: stability, robustness, and convenience. Semin Hematol 1998; 35(2 suppl 2):18–21.

67. Costantino HR, Langer R, Klibanov AM. Aggregation of a lyophilized pharmaceutical protein, recombinant human albumin: effect of moisture and stabilization by excipients. Biotechnology (N Y) 1995; 13:493–496.

68. Schwendeman SP, Costantino HR, Gupta RK, Siber GR, Klibanov AM, Langer R. Stabilization of tetanus and diphtheria toxoids against moisture-induced aggregation. Proc Natl Acad Sci USA 1995; 92:11234–11238.

69. Chen B, Costantino HR, Liu J, Hsu CC, Shire SJ. Influence of calcium ions on the structure and stability of recombinant human deoxyribonuclease I in the aqueous and lyophilized states. J Pharm Sci 1999; 88:477–482.

70. Hora MA, Rana RK, Wilcox CL, et al. Development of a lyophilized formulation of interleukin-2. Dev Biol Stand 1992; 74:295–303.

71. Mattern M, Winter G, Kohnert U, Lee G. Formulation of proteins in vacuum-dried glasses. II. Process and storage stability in sugar-free amino acid systems. Pharm Dev Technol 1999; 4:199–208.

72. Waterman KC, Adami RC, Alsante KM, et al. Stabilization of pharmaceuticals to oxidative degradation. Pharm Dev Technol 2002; 7:1–32.

73. Cicerone MT, Soles CL. Fast dynamics and stabilization of proteins: binary glasses of trehalose and glycerol. Biophys J 2004; 86:3836–3845.

74. Cicerone MT, Tellington A, Trost L, Sokolov A. Substantially Improved Stability of Biological Agents in Dried Form. BioProcess Int 2003; 1:2–9.

75. Chang L, Shepherd D, Sun J, Tang X, Pikal MJ. Effect of sorbitol and residual moisture on the stability of lyophilized antibodies: implications for the mechanism of protein stabilization in the solid state. J Pharm Sci 2005; 94:1445–1455.

76. Li B, Gorman EM, Moore KD, et al. Effects of acidic N+1 residues on asparagine deamidation rates in solution and in the solid state. J Pharm Sci 2005; 94:666–675.

77. Lee YC, Nelson J, Sueda K, Seibert D, Hsieh WY, Braxton B. The protective effect of lactose on lyophilization of CNK-20402. AAPS-PharmSciTech (epub) 2005; 6:E42–E48.

78. Chen B, Bautista R, Yu K, Zapata GA, Mulkerrin MG, Chamow SM. Influence of histidine on the stability and physical properties of a fully human antibody in aqueous and solid forms. Pharm Res 2003; 20:1952–1960.

79. Carpenter JF, Prestrelski SJ, Dong A. Application of infrared spectroscopy to development of stable lyophilized protein formulations Eur J Pharm Biopharm 1998; 45:231–238.

80. Chang L, Shepherd D, Sun J, et al. Mechanism of Protein Stabilization by Sugars during freeze-drying and storage: native structure preservation, specific interaction and/or immobilization in a glassy matrix? J Pharm Sci 2005; 94:1427–1444.

81. Shalaev EY, Chang L, Reddy R, Collins G, Luthra S, Pikal MJ. Molecular mobility and the glass transition in globular lyophilized proteins. AAPS PharmSci 2003; 5(4): Abstract T2300.

82. Liu J, Rigsbee DR, Stotz C, Pikal MJ. Dynamics of pharmaceutical amorphous solids: the study of enthalpy relaxation by isothermal microcalorimetry. J Pharm Sci 2002; 91:1852–1862.

83. Yoshioka S. Molecular mobility of freeze-dried formulations as determined by NMR relaxation, and its effect on storage stability. In Drugs and the Pharmaceutical Sciences 137 (Freeze-Drying/Lyophilization of Pharmaceutical and Biological Products), 2004:187–212.

84. Govindarajan R, Chatterjee K, Gatlin L, Suryanarayanan R, Shalaev EY. Impact of freeze-drying on ionization of sulfonephthalein probe molecules in trehalose-citrate systems. J Pharm Sci 2006; 95:1498–1510.

85. Chatterjee K, Shalaev EY, Suryanarayanan R, Govindarajan R. Correlation between chemical reactivity and the Hammett acidity function in amorphous solids using inversion of sucrose as a model reaction, J Pharm Sci 2007. In press.

10

Formulation of Proteins for Pulmonary Delivery

Andrew R. Clark and Cynthia L. Stevenson
Nektar Therapeutics, San Carlos, California, U.S.A.

Steven J. Shire
Genentech, Inc., San Francisco, California, U.S.A.

INTRODUCTION

Formulation of proteins and peptides often is more challenging than formulation of small molecules, because of the important role of protein conformation as well as the potential for numerous chemical degradation pathways (1,2). This fact, coupled with the necessity of using a device to generate an aerosol, augments the challenge considerably. A developed formulation must provide one to two years of stability on storage (preferably room temperature), and it also must meet additional requirements that are unique to its delivery as an aerosol. First, the formulation must not cause adverse pulmonary reactions such as cough or bronchoconstriction; it must be safe for delivery to the lungs; and any excipients used should preferably be generally-regarded-as-safe (GRAS). Second, the formulation may also have to be designed to minimize interactions with the inhaler component materials, where the drug-contacting materials should be of medical grade with acceptable leachable profiles. Finally, the formulation must stabilize the protein sufficiently to ensure that the protein survives the rigors of the aerosol-generation process. In addition to these challenges, the development and ultimate approval of an aerosol formulation often goes hand in hand with the development and/or use of a particular device, often requiring approval as, and product release as, a drug/device combination.

EVALUATION OF DEVICES AND FORMULATION COMPATIBILITY

Aerosol Parameters and Device Performance That Impact Drug Delivery

The delivery efficiency of a device/formulation system relies essentially on two key performance parameters: the efficiency of delivering the dose to the patient and the generation of an aerosol containing suitable particles or droplets for deposition in the airways. The efficiency of a delivery system is usually expressed in terms of the emitted dose (ED). This is defined as the quantity of drug delivered from the mouthpiece of the device, and it is a measure of how much of the drug product would actually reach the patient. ED can be expressed as a dose (mg or µg) or as an efficiency E in Equation 1:

$$E = \frac{ED}{\text{Nominal (or loaded) dose}} \tag{1}$$

While there is no expected minimum value for E, particular values being a function of formulation and device design, higher efficiencies obviously mean lower nominal drug per dose is needed with the concomitant decrease in cost of goods. However, for metered dose and dry powder inhalers (DPIs), the Food and Drug Administration (FDA), in its 1998 guidance (3), set expectations for variability in ED in terms of dose content uniformity (DCU).

The deposition of aerosol particles, or droplets, in the airways is a function of airway geometry, breathing pattern, and aerosol characteristics. As aerosol particles penetrate into the airways, they encounter branching passageways of smaller and smaller diameters. An aerosol particle's ability to traverse these structures is dependent upon its inertial characteristics. The probability of inertial impaction is dependent upon aerodynamic drag (a function of size and shape) and the particle's inertia (mass). Thus, the relevant parameter governing aerosol deposition is not physical size but rather aerodynamic diameter d_{aer}, which is related to the particle diameter d and particle density ρ by Equation 2 (4):

$$d_{aer} = d\rho^{1/2}. \tag{2}$$

Aerodynamic diameter is defined as the diameter of a unit-density spherical particle having the same settling velocity in air as the particle under consideration. It is generally accepted that particles with aerodynamic diameters larger than 5 to 6 µm deposit in the oropharyngeal region during normal breathing, whereas particles smaller than 0.5 to 1.0 µm tend to penetrate into the lower airways but are exhaled without significant deposition (5,6). Although it is somewhat arbitrary, the portion of the size spectrum between 1 and 5 µm is generally considered to contain the bulk of the particles with the potential to penetrate and deposit in the lungs, and this is referred to as the fine particle fraction

$(FPF)_{<5 \mu m}.^a$ (7) This is an obvious oversimplification, since particles within this size range will be deposited with varying efficiencies dependent upon their particular aerodynamic characteristics. If the target is the alveolar region, it would be more appropriate to generate a size distribution on the lower end of the 1 to 5 μm region, say fine particle dose (FPD)<3 μm. An FPD can be defined from FPF and ED, and this FPD can be used as an indicator of the potential dose that will be deposited in the lung (Eq. 3):

$$FPD = FPF \times ED \tag{3}$$

However, it should be noted that FPD is usually proportional to, but not necessarily equal to, the amount of drug actually deposited in the lungs.

Clearly, as noted above, the actual deposition of aerosol will depend on additional factors in individual patients (breathing pattern, lung anatomy, pulmonary obstruction, etc.). However, in vitro comparisons of devices/formulations in terms of ED, FPF, and FPD can guide the development of a formulation and the choice and/or design of a device.

Characterization of Aerosols

There are then clearly four key questions that need to be answered in regard to delivery performance:

1. ED: How much of the drug is converted to an aerosol that ultimately exits from the mouthpiece?
2. FPF: What is the size distribution of particles or droplets in the aerosol?
3. DCU: What is the reproducibility of the aerosol-generation process? (Note. DCU is usually a measure of ED reproducibility not FPF or FPD).
4. Stability and integrity of the ED: What effect does the device and/or aerosolization process have on the protein drug quality?

The first three questions require methods to analyze the ED and the size distribution of the aerosol. Although a variety of methods have been developed to characterize aerosol size distributions, including laser diffraction (8), holography (9), static (10) and dynamic light scattering (11), and time-of-flight aerosol beam spectrometry (12), with the exception of the use of laser diffraction for the assessment of the size distribution of nebulizer clouds (13), cascade impaction (14) and impinger methods (15) are considered to be the most reliable methods for the assessment of aerosol particle size distribution and device performance (16).

---

[a] The fraction of the aerosol with this size range has been referred to in the past as the "respirable fraction"; however, during in recent years, the general convention has been to refer to it as the "fine particle fraction" (7). Often the range of sizes chosen for FPF or FPD is operational in nature and will typically be chosen so as to match the cutoff values for the stages of the impactor or liquid impinger used by a particular investigator.

Indeed, a number of apparatus are described in the European and U.S. pharmacopeias (17,18). The advantages of the alternate methods include their speed and ease of analysis in comparison to impactor and impinger technologies. Generally, they can be used as part of a research and design program, provided their limitations are understood. However, the world's regulatory agencies expect to see drug-specific impactor data as part of an aerosol product's filing documentation and as part of any product-release specifications.

A great advantage of the impinger and impactor technology is that it readily allows for a determination of aerosol mass balance. This is particularly critical when one is addressing device efficiency and size distribution. Clearly, if a majority of the aerosol is made up of large particles that are not collected because of impaction onto surfaces prior to entry into the measuring device, then the distribution will be skewed to lower sizes. Similarly, small particles that are not collected or recorded by the measuring device will lead to a distribution skewed to larger sizes. The Pharmaceutical Quality Research Institute and the FDA are currently working on defining criteria for mass balance when impaction methods are being used as part of quality control release (19). As a representation of the human airways, cascade impactors have their limitations and, since the target is the human airways, it is usual to add an artificial throat or bend at the impactor inlet to collect the larger-size particles that would normally be deposited in the oropharyngeal region. In recent years, these bends have developed from simple right angles as detailed in the pharmacopeias (18) to complex oropharyngeal models obtained by magnetic resonance imaging of human volunteers (20). Similarly, a filter is placed after the final stage of the size detector to trap the smallest particles/droplets.

The determination of the reproducibility of the ED delivered can be a challenging exercise, especially in the case of devices that deliver small amounts of drug. The limitations of such an analysis will depend on the sensitivity of the assays used to detect the protein as well as the ability to recover the aerosol reproducibly. Again the pharmacopeias recommend an apparatus that can be used for ED collection (17,18). However, care should be taken with these filter-type collection devices when nebulized proteins are being tested, as drying on the filter may cause aggregation, denaturation, and potentially low recoveries.

The chemical characterization of protein that has been aerosolized also requires the collection of most of the protein exiting the aerosol-generation device. Often, this is done by impaction, but small particles/droplets are difficult to collect by impaction. A successful characterization of the protein drug will, of course, require an orthogonal set of stability-indicating assays, but the full recovery of protein for analysis is critical to ensure that protein contained in the small particles/droplets (<1 μm diameter) has not undergone any chemical or physical degradation. A device and technique for increasing the size of nebulized droplets for impaction has been developed and was successful in collecting ~96% of the aerosol (21).

PROTEIN FORMULATIONS FOR AEROSOL DELIVERY

Basic formulations of proteins for aerosol delivery can be developed as either liquid or solids, as in the case of formulations for parenteral administration. However, subsequent inclusion of the formulation into a delivery system such as a nebulizer, DPI, or propellant-driven metered dose inhalers (MDIs) has its own unique set of challenges. The following sections discuss the development issues for both liquid and solid dosage forms for aerosol administration.

PROTEIN LIQUID FORMULATIONS FOR AEROSOL DELIVERY

Choice of Device

As already mentioned, the approval of a pulmonary formulation often is linked to the device used for generation of the aerosol. In the case of nebulizer products, the solution formulation is linked to performance via the clinical trails conducted with specific nebulizers and subsequent documentation in the package insert. In the case of DPIs or MDIs, the clinical data are linked to the specific device/drug by the combination product-release criteria. From a formulation perspective, the device, and hence method used to generate the aerosol, will dictate what components are required to ensure protein stability.

A common device for generation of aerosols is the so-called jet nebulizer (Fig. 1A). A portable air compressor generates a high-velocity air stream through a jet nozzle, and liquid is drawn up from the reservoir as a result of the partial pressure drop at the orifice. A droplet spray is generated upon contact with the air stream. The larger droplets impact on an appropriately placed baffle and are returned to the liquid reservoir, whereas sufficiently small droplets remain in the air stream and exit from the inhalation port of the nebulizer. The result is that during the course of nebulization, more than 99% of the solution is essentially refluxed and undergoes repeated stress and exposure to air–water interfaces, an experience that may promote protein denaturation (22,23).

Another device for nebulization is the ultrasonic generator. Ultrasonic nebulizers essentially come in two forms (Fig. 1B and C). Those that transmit high-frequency sound waves through a reservoir of solution to generate aerosol droplets and those that use vibrating screens. In the former type, the transfer of high energy, as well as the potential build-up of heat in the solution, may lead to degradation of a protein drug via thermal rather than surface exposure (24). However, attempts have been made to directly generate small droplets by ultrasonic nebulization in single inhalations in devices such as the original Respimat®,[b] which generated respirable aerosols from a small volume of solution, typically 25 to 50 μL in one atomization cycle lasting about one second (25). This circumvents the potential degradation that may occur upon recirculation of the solution within

---

[b] The original Respimat® was discontinued, and a mechanical device formerly known as the BINEB® is now referred to as Respimat.

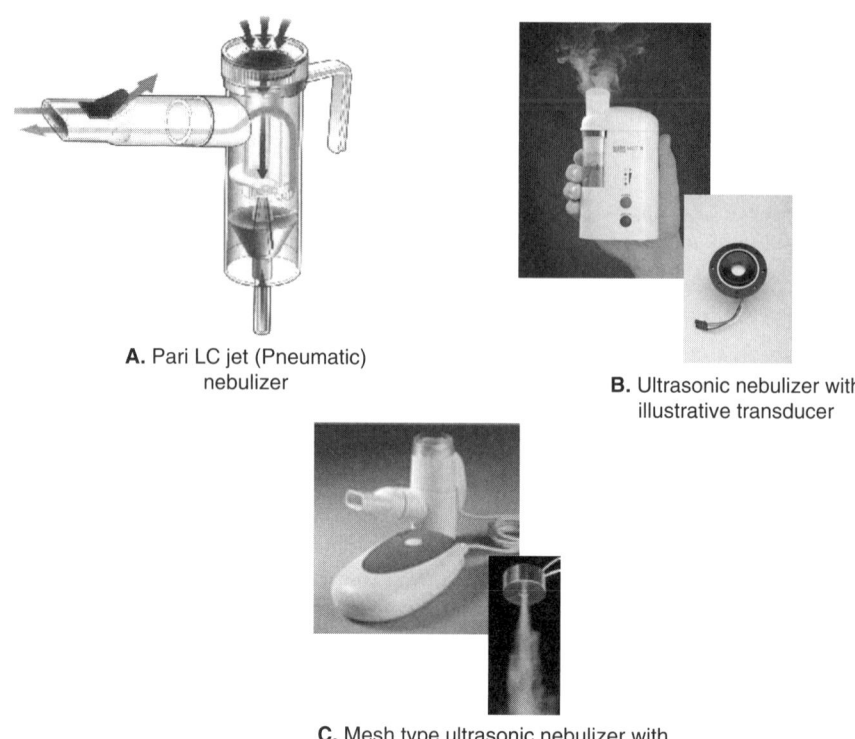

A. Pari LC jet (Pneumatic)
nebulizer

B. Ultrasonic nebulizer with
illustrative transducer

C. Mesh type ultrasonic nebulizer with
illustration of mesh atomizer

Figure 1   Nebulizer   designs: (**A**)   conventional   pneumatic   (jet);   (**B**)   ultrasonic;
(**C**) ultrasonic mesh.

a nebulizer or heating, but it would necessitate the development of formulations
that support a high concentration of protein. As an example, assume that 2.5 mL
of Pulmozyme® at 1 mg/mL is delivered by jet nebulizer with an estimated delivery efficiency of 10% (26). Pulmozyme is a recombinant human deoxyribonuclease (rhDNase) used in the treatment of cystic fibrosis. To deliver a similar amount
of drug to the lung by means of a small-volume ultrasonic nebulizer would require
a Pulmozyme concentration of approximately 50 mg/mL.

In the latter type of system, a solution is atomized as it is forced to flow
through tiny holes in a vibrating mesh or plate. While typically droplet sizes
are approximately twice the "hole" diameter, recent developments have shown
that it is possible to produce droplets that are approximately half the diameter
of the mesh hole size (27). This method is currently used in a number of commercial nebulizer devices, for example the Nektar Aeroneb, the Pari E-flow, and
the Omron NEU05, but it is yet to be incorporated into a "single breath" inhaler.
However, a number of prototype devices, such as the Aerodose (28), have been
tested in humans.

One alternate strategy for greater delivery efficiency while avoiding the rigors of a continuous nebulization or potential degradation due to high-energy input by ultrasonics is to generate the aerosol mechanically. Such a strategy is used with the AER_X^{TM} (Fig. 2) (29) and the new Respimat (30). In the AER_X device, a blister pack containing 25 to 100 μL of drug is rapidly compressed with a mechanical piston, forcing the solution through a series of laser-drilled holes (1–3 μm in diameter). Novo-Nordisk currently have this device in Phase III clinical testing with inhaled insulin. An advantage of such a system is the unit dose packaging. Other systems, such as the Respimat, that use a reservoir to supply small volumes to an aerosol generator require development of a multidose formulation. Multidose formulations of proteins, in turn, call for the use of preservatives that may interact with the protein, either reducing the effectiveness of the preservative or affecting the proteins activity. The Respimat also uses a mechanical piston to generate the aerosols but is actually designed as an MDI (31). The drug is stored in a collapsible plastic bag, and metered quantities are delivered to the atomization chamber. Nonreturn valves prevent back-flow and control the flow of liquid through the device. However, as with most devices of the multidose liquid reservoir type, it is probably essential that the formulations be preserved. To date, this type of "single inhalation" liquid system has not been successfully applied to protein delivery and has been restricted to the delivery of small molecules.

Choice of Excipients

The osmolality and the pH of solutions are critical variables that affect tolerability of inhaled aqueous solution and may contribute to adverse reactions during pulmonary delivery of drugs, such as bronchoconstriction (22–37). It has been recommended that, whenever possible, solutions for pulmonary delivery, especially

Figure 2 The AER_X^{TM} iDMS insulin inhaler.

those delivered in large volumes by nebulizer, be formulated as isotonic solutions at pH values exceeding 5.0 (34). Recent studies have shown that if the formulation is not isotonic, then the droplet size distribution of an aerosol may be altered during delivery as the result of a loss of or uptake of water vapor from the airways, leading to changes in deposition patterns (38,39). It is also not uncommon to find that buffer components cause adverse reactions such as cough (40,41), and, therefore, many inhalation products have been formulated without buffer components. The control of the pH in an unbuffered formulation is a major concern, especially since many protein degradation pathways are highly pH dependent (2). However, at high-enough protein concentrations, the titratable amino acid residues of the protein may provide sufficient buffering capacity to stabilize the pH of the formulation (42).

One of the biggest challenges in developing a liquid formulation for aerosol delivery is the exposure of the protein to an air–water interface. As described earlier, the generation of respirable droplets greatly increases the protein exposure to this interface. Solvent–protein interactions are critical for maintaining the native conformation of a protein, and removal of the aqueous phase can have profound effects on protein structure. In particular, the unique properties of water, such as its ability to form an extensive hydrogen bond network, are believed to be essential for the entropically driven hydrophobic forces that play a major role in folding of proteins (43). Hydrophobic amino acid residues tend to organize water structure by forming cavities in the bulk solvent, leading to a large decrease in the entropy of the system. Removal of the hydrophobic residues from the solvent phase and coalescence into an interior hydrophobic phase result in a decrease of the protein surface area, as well as an increase in the entropy of the solvent phase large enough to exceed the configurational entropic decrease due to folding of the protein into a more compact form.

A protein exposed to an air–water interface during nebulization may become denatured, forming both soluble and insoluble aggregates (44). Often, the inclusion of an acceptable surfactant such as polysorbate 20 can minimize this degradation (Fig. 3). The grade of polysorbate may be critical in this application. In particular, polysorbates have been shown to contain trace quantities of peroxide (45). The protein is stressed by the aerosol-generation process, especially in the case of jet and large-volume ultrasonic nebulization, and exposure to large quantities of oxygen may lead to oxidation of susceptible amino acid residues such as cysteine, methionine, tryptophan, and histidine (2,23). Contamination with peroxides may increase the rate of oxidation, which may destabilize the protein and lead to an increase in aggregation during aerosolization. The increased potential for oxidation is not necessarily a problem. Whereas many proteins have altered conformation or altered activity, or undergo aggregation, there are also many examples of oxidation having no perceptible effect on the protein (2).

If oxidation does pose a problem, excipients such as antioxidants or amino acids such as methionine can be added as oxygen scavengers. However, the

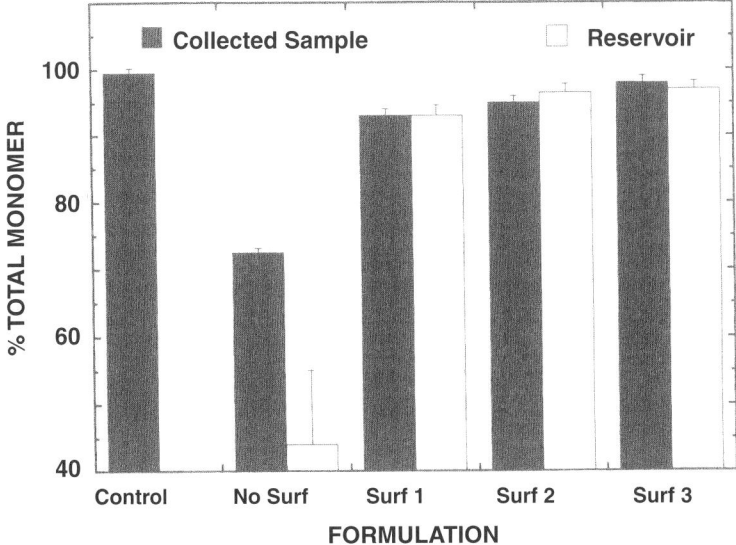

Figure 3 Effect of surfactant on hGH aggregation following 18 minute nebulization of 25 mg/mL hGH. Unaerosolized control (*left*), collected sample by impaction into test tube (■), sample remaining in reservoir (□). *Abbreviation*: hGH, human growth hormone. *Source*: From Ref. 44.

addition of these excipients may cause adverse events in clinical situations, depending on the sensitivity of the patient population. This, in fact, is generally the major limitation on formulation development of proteins for pulmonary delivery. The problem is particularly difficult if the targeted population involves pulmonary diseases such as cystic fibrosis or asthma. The potential sensitivity of asthmatics to excipients makes the development of a multiuse formulation even more problematic than in the development of parenteral formulations.

The potential use of an aerosol delivery system, such as the Respimat described above which uses an "open system", will require the addition of a preservative to the formulation. Interaction of the preservatives with the protein is a major concern. If the interactions are substantial, protein denaturation and possibly soluble aggregate formation of precipitation may result (46). The interactions of protein with preservatives may also result in a decrease in the effectiveness of the preservative. This presents significant challenges, especially if the protein drug is being developed for world markets, since the preservative challenge requirements in Europe are significantly more stringent than in the United States. A variety of preservatives delivered as aerosols have been tested in people. Among these preservatives are benzalkonium chloride, chlorbutol, benzyl alcohol, sodium metabisulphite, chlorocresol, and phenol, and some of them have been implicated in adverse effects such as bronchoconstriction (47–51).

Manufacturing Issues

As already discussed, the development of a formulation for aerosol administration of a protein drug depends highly on the type of device that will be used. The concern regarding adverse events from excipients or preservatives in patients with compromised lung function may be minimized by the use of devices that administer small volumes. The trade-off, of course, is that the protein may need to be formulated at higher concentrations. The ease of accomplishing this will be related to the solubility and tendency of the protein to aggregate. In addition, although the target concentrations, perhaps as high as 50 mg/mL, may be attainable, the long-term stability of the product may be compromised at the higher concentrations. In particular, the bulk product will need to be stored prior to loading (into cartridges, blister packs, etc.), for use with the device. If long-term storage is required, it may be necessary to freeze formulated bulk. If, however, the developed liquid formulation lacks the necessary stability to undergo repeated freeze/thaw cycles, the formulation may have to be stored at controlled temperatures of 2°C to 8°C. Generally, bulk formulations in large quantities are stored in stainless steel tanks, and this practice poses several additional challenges. In particular, a useful pharmaceutical tonicifier such as sodium chloride may cause problems in long-term storage because the well-known interactions of stainless steel with halides can result in metal catalyzed oxidation of proteins (52). In such a case, alternate tonicifiers may have to be explored. Sugars such as mannitol are generally acceptable and have been used in commercially approved products (53). There have been no published toxicology data that suggest that there is a long-term safety concern on the use of mannitol or other sugars in respirable products. Indeed, lactose has been used for many years in pulmonary products with no reported problems. Moreover, any concerns may be less problematic for formulations for small-volume aerosol delivery devices, since the total amount of excipients delivered is quite low in the small volumes (<50 µL) used.

Additional complications may occur if the formulation is designed for multiple uses, and hence it requires preservatives. This has been demonstrated in studies of the effect of metals and the preservative phenol on human growth hormone degradation (54). Although a single-use formulation does not require preservatives, compliance with recent FDA proposals, nevertheless, requires sterility for liquid formulations intended for inhalation (55).

If the pharmaceutical company possesses the technology to make and fill the devices or drug closures, the required storage time from formulation to final packaging may be minimized. Unfortunately, if the pharmaceutical company developing a product for inhalation therapy often does not possess this capability, it is compelled to form an alliance with the company that has the technology. Such an arrangement usually necessitates the development of appropriate manufacturing steps for long-term storage and shipment of the product to the filling site for loading into drug reservoirs, cartridges, blister packs, or other delivery-ready closures. The shipment of formulated bulk places additional stresses on the product and may require the allocation of resources and improvements to allow for the additional manipulations of the protein drug. Any alterations in formulation will still need to be compatible with the airways and should not interfere with the aerosol-generation process.

Examples

Novo Nordisk, using a technology originally developed by the Aradigm Corporation (29), are developing the only inhaled liquid insulin formulation (AERx iDMS); it is in later-stage Phase III clinical trials. The system utilizes a sterile solution packaged in disposable strips that provide dose adjustability. Each disposable strip contains insulin in a cold-formed blister that is heat-sealed to a lid-nozzle laminate structure (56).

At a much earlier stage of development, Phase I clinical trials, Coremed, Inc. is developing a liquid polymer/bioadhesive aqueous insulin formulation with GRAS excipients (57). The product uses a generic handheld device that purports to generate an aerosol with a mass median aerodynamic diameter of approximately 2 µm.

PROTEIN POWDER FORMULATIONS FOR AEROSOL DELIVERY

Choice of Device

As described above for liquid systems, the registration and approval of a protein aerosol product is intimately linked to the aerosol-generation system chosen for its delivery. Hence the choice of a delivery system is crucial to the success of the product.

For the purposes of protein applications, DPIs may be placed into two major categories: multidose devices, where drug powder is stored in bulk and metered inside the device before inhalation, and unit dose devices, where the drug powder is stored as a premetered dose in an individual storage unit. These two categories can be further subdivided into patient-driven devices, where a patient's inspiratory effort provides the energy to disperse the powder, and powered systems, where an external source provides the energy.

The first category, multidose devices, presents some major difficulties with protein powders. Multidose powder reservoirs with their associated metering valves are notoriously difficult to seal effectively during use. Thus moisture ingress can be a major problem. As described below, one of the general techniques for stabilizing proteins in the solid state is the use of amorphous glasses. The Achilles' heel of amorphous glasses, however, is moisture. The excipients used as solid-state stabilizers, when stored in the amorphous state, have the potential to crystallize; and as the moisture content of the powder increases and plasticizes the solid, the probability of crystallization, and hence the probability of destabilization of the formulation, increases. Although it is possible to develop formulations that are reasonably stable at ambient humidities, the general physical instability of amorphous solids severely limits the application of multidose-type DPIs. In general, the choice for protein powders is therefore the unit dose approach.

Figure 4 presents a schematic illustration of the various approaches to DPI design, indicating some proprietary devices and some manufacturers. The unit dose approach in terms of device design can come in many forms. These range from the original DPI designs such as the Spinhaler and Rotahaler, where the drug is stored in hard gelatin capsules, to the more complex foil blister devices, where the drug is stored as individual doses either singly or on multidose disks or tapes. Essentially, the latter devices offer unit dose drug storage with multidose convenience for the patient. In essence, because of the humidity instabilities described above, the foil systems become the packaging of choice for protein inhalation powders. Even if the primary packaging is not foil, the products should be overwrapped with foil to maintain the moisture content necessary to keep the powder stable.

Again, it is possible to develop formulations with ambient humidity stability that probably would allow the use of systems based on conventional gelatin capsules or other simple capsule technologies such as hydroxy methyl propyl cellulose (HPMC), but the extra security guaranteed by the foil blister technology, or a foil overwrap, will usually be necessary. It should be noted that for satisfactory operation,

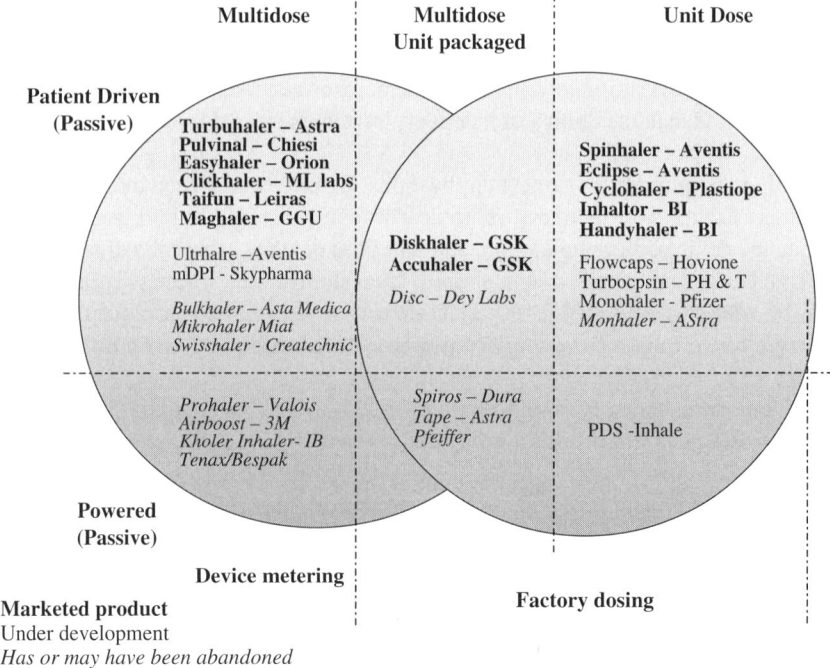

Figure 4   Current and future dry powder inhalers grouped by dosing method, packaging, and dispersion power source.

capsule technologies, gelatin or HPMC, require the capsules to be maintained within a relatively narrow range of water content in order to retain the mechanical properties necessary for piercing; if a capsule becomes too dry, it may shatter instead of opening, and if it becomes too wet, it may not be possible to open at all.

The other packaging issue, which is not directly related to powder stability, is microbial contamination and growth. However, to date, there have been no published reports to suggest that protein powders are any more or less susceptible to microbial growth than are the small-molecule powders currently used in commercial inhalation formulations. Both the U.S. and European pharmacopoeias impose microbial limits for inhalation powders but do not require sterility (55). Table 1 presents the current proposed microbial limits for powder inhalation products. In general, powders maintained below 50% to 55% relative humidity will not sustain the growth of organisms; organisms already present may survive in stasis, however.

The remaining choice is then between a patient-driven system (passive) and a powered device (active). While the majority of the products on the market are patient-driven, recently Pfizer Inc. obtained approval for Exubera®, which

Table 1 Assignment of Microbial Limit Tests for Nonsterile Finished Dosage Forms According to Route of Administration

Route of administration	Total aerobic microbial count (cfu/g or mL)	Combined yeasts and molds count (cfu/g or mL)	Examples of objectionable microorganisms[a]
Inhalation	≤10[b]	≤10	*Escherichia coli* *Pseudomonas aeruginosa* Salmonella species
Vaginal	≤100	≤10	*E. coli* Staphylococcus aureus *P. aeruginosa* *Candida albicans*
Nasal/optic/ Rectal/topical	≤100	≤10	*E. coli* *S. aureus* *P. aeruginosa*
Oral liquids	≤100	≤10	*E. coli* Salmonella species
Oral solid	≤1000	≤100	E. coli Salmonella species

[a]It is virtually impossible to list every microorganism that may be objectionable for a specific product class. The microorganisms listed are merely examples of those microorganisms usually found to be objectionable in the respective product class.
[b]Except for nonpressurized powders for oral inhalation for which the total aerobic mircrobial count does not exceed 100 colony-forming units per gram.
Source: Pharmacopoeial Forum, 1996.

uses a pneumatically driven powder inhaler developed by Nektar Therapeutics of San Carlos (Fig. 5) (58). The choice between patient-driven and powered systems for protein delivery depends less on the physicochemical stability of the formulation and more on its dispersibility and the delivery requirements (large vs. small doses, reproducibility of deposited dose, etc.). With conventional type formulations, patient-driven DPIs have traditionally exhibited lung delivery that is both flow-rate dependent and less efficient than that obtainable from a powered system. For example, for typical asthma systems, lung delivery may vary by a factor of three or more depending on the flow rate achieved by the patient during any particular inhalation, and lung deposition is only of the order of 10% to 20% of the nominal dose. Whereas this set of conditions may not be problematic with some protein therapeutics, where raw material cost and dose consistency are not of concern, it could be disastrous for an expensive protein with a narrow therapeutic window.

With proteins, efficiency, *E,* and FPD are important because of their impact on the cost of goods. Bioavailability of proteins from the gastrointestinal tract is very poor (59,60). Therefore, while avoidance of oropharyngeal deposition is obviously desirable, it is not as important as it is for some small molecules, where oral deposition might cause unwanted side effects, for example, candidiasis, or thrush, and/or serious increases in total unwanted systemic load due to systemic availability from the gut. Avoiding oropharyngeal deposi-

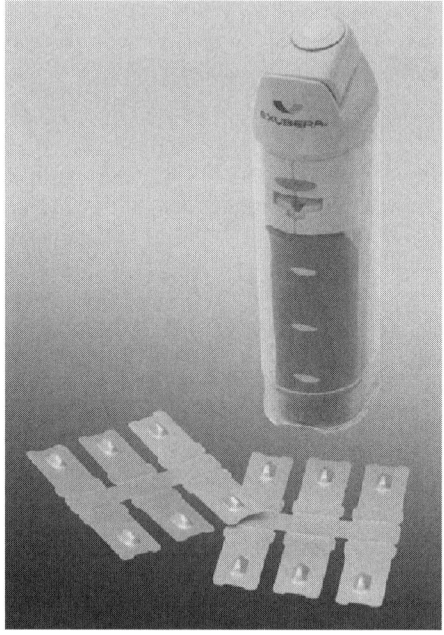

Figure 5   Exubera® inhaler.

tion, by generating a high FPF, with proteins results in an overall increase in delivery efficiency to the lung and reduces the nominal dose required for an effective lung dose.

In contrast to the patient-driven passive systems, powered active systems decouple aerosol generation from the patient's inspiratory effort. Hence, they are in principle more robust and they reduce the flow-rate dependence of the dose delivered to the lung. However, it should be realized that such dependence is not completely eliminated. Since the capture efficiency of the mouth and oropharynx is dependent on the inhaled flow rate, even if aerosol generation is independent of inspiratory flow, the deposition pattern in the body, and hence the lung dose, is not. However, this depositional dependency is far less dramatic than the variations that can be brought about by the variations in particle size and flow rate that can be produced by patient-driven DPIs. A number of authors have used the following function (Eq. 4), reported by Rudolph et al. (61), to estimate the oropharyngeal filtering term η_{oral}, for mono-disperse aerosols:

$$\eta_{\text{oral}} = 1 - \left[1.1 \times 10^{-4} \left(V_t^{-0.2} d_{\text{aer}}^2 Q^{0.6} \right)^{1.4} + 1 \right]^{-1} \tag{4}$$

where V_t is tidal volume, d_{aer} is the aerodynamic diameter of the inhaled particle, and Q is the inspiratory flow rate. Examination of Equation 4 shows that as either d_m or Q decreases, the probability of penetration beyond the oropharynx and deposition in the lung increases. It should be noted that at a sufficiently small d_{aer}, Q becomes unimportant and η_{oral}, tends to zero.

From a protein delivery perspective then, the most desirable design attributes of stability, delivery efficiency, and reproducibility lead in general to unit dose packaging with a device featuring either powered aerosol generation or a patient-driven device incorporating particles engineered to a have high levels of dispersibility and flow-rate independence. However, in an imperfect world, it may be that all these attributes will not be present in one particular technological combination and a number of compromises between delivery performance and stability may be necessary. It may also be that for certain molecules, where expense and therapeutic window are not important, simpler, less-efficient devices may be acceptable. Indeed, from a patient compliance perspective, they may be even more desirable.

Choice of Excipients

As described above, aerosol particles intended to penetrate and deposit in the lungs must be sufficiently fine to pass through the oropharynx and upper airways. In general, this requirement leads to the need for a high FPF in the delivered aerosol. This, in turn, leads to the requirement for fine powders of the active drug moiety. In general, fine powders are cohesive and do not flow or disperse easily. The major objective and challenge of powder inhalation formulation is therefore to manufacture formulations that both flow, for ease of filling and dispensing, and disperse, to form aerosols fine enough to penetrate and deposit in the lungs. (The term "fine" generally refers to particles having aerodynamic diameters in the size range 1–5 or 1–3 μm.)

Early solutions to this problem resulted in two basic approaches to meet these almost mutually exclusive requirements. The first, developed in the late 1950s, by the then Fisons Pharmaceuticals, is to blend the active drug with a carrier consisting of large particles. This approach produces powders that flow and dispense well and, when a suitable shear force is applied, disperse to produce respirable aerosols of reasonable quality. The coarse carrier used in these systems, typically 30 to 100 μm in diameter, is usually such that a large fraction of it deposits in the mouth and oropharynx and only a small amount reaches the lung. The second approach, developed in the early 1960s, is to palletize the drug particle to make loose agglomerates, which flow for filling and dispensing, but which break up when an aerodynamic shear force is applied to produce a respirable aerosol. These two approaches currently account for most of the inhalation products on the market.

Another approach developed more recently is to engineer particles so that they intrinsically possess the characteristics necessary for inhalation products. While this has been accomplished in a variety of ways, from volume exclusion precipitation (62) to coprecipitation with novel excipients (63), spray drying is by far the most popular technique. For example, spray drying has been used to produce particles of large physical size and low absolute density (64). This approach results in particles that behave as if they have large physical diameter as far as powder flow and dispersion are concerned, but because of their low absolute density, possess a suitably small aerodynamic diameter when dispersed [Eq. 2]. This type of formulation, in combination with a simple patient-driven inhaler, is currently being used by Eli Lilly in Phase III clinical trials with inhaled insulin (64). Small porous particles have also been investigated (65). In both of these cases, the formulations are based on the use of lipids, and besides their aerodynamic characteristics, the lipids produce particles with hydrophobic surfaces, which appear to enhance dispersibility. Work has also been reported on the use of hydrophobic amino acids in spray dried particles. Although these molecules do not necessarily produce a porous structure, they do enhance surface hydrophobicity and hence produce highly dispersible powders (66).

The ability of a protein powder to produce a respirable aerosol, whether blended with a coarse carrier or in its "raw" engineered state, is controlled by numerous influences. As described above, the diameter of the primary powder particles and their absolute density is of obvious importance. However, the cohesive forces that hold the powder particles together and prevent dispersal of the powder, and the numerous environmental influences that affect these cohesive forces, also play a major role. As an example of how formulation and particle size can influence aerosol performance, Figure 6 shows data from a study report by Chan et al. (67) detailing the effects of sodium chloride content and powder particle size [expressed as mass median diameter (MMD) of the raw powder] on the aerosol properties of rhDnase powder blends delivered via a Rotahaler. It can

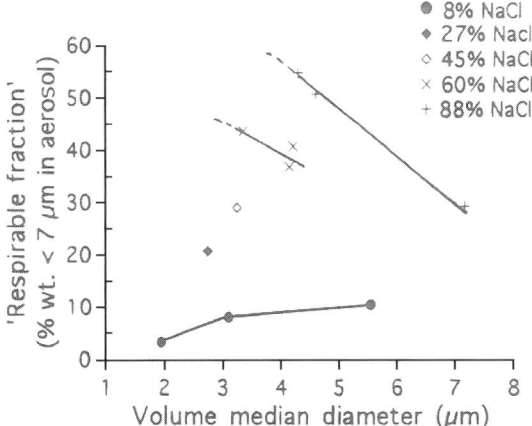

Figure 6 Summary of the dispersion properties (as FPF) versus particle size for powders containing co-spray-dried NaCl and rhDNase (67). *Abbreviations*: FPF, fine particle fraction; rhDNase, recombinant human deoxyribonuclease.

be seen that as the sodium chloride content is increased, the FPF increases. That is, the powder disperses more easily and generates a finer aerosol cloud. It can also be seen that the FPF is a function of powder particle size: for the powders that disperse easily, the FPF appears to decrease as MMD increases, whereas for poorly dispersing powders, FPF appears to go through a maximum. The effect of particle size is related to cohesion, which is a measure of the force needed to separate a unit mass of particles from their stable agglomerated state. Although the contributions of the various forces to the cohesion of the powder in this study are uncertain, as they are in most studies of this type, if it is generally assumed that van der Waals forces are dominant, then the cohesive force would be expected to be proportional to the square of the particle diameter (68). Thus the powders with smaller MMD are more difficult to disperse, and as the MMD increases, the powder disperses more easily. However, as the MMD increases the amount of fine particles avialable in the power decreases. Hence, these two competing phenomena result in an optimum powder size for any given formulation, the particular optimum size being dependent on the relative cohesiveness of the powder. Figure 7 presents data for FPF with a constant powder MMD and varying degrees of crystallinity. For this particular formulation–device combination, an increase in crystallinity results in an increase in the fine particle mass generated from the powder blend. This increase presumably occurs because the interparticulate forces have been modified by the change in physical state of the solid and the external morphology of the powder particles.

As an example of reducing the interparticulate forces by the addition of an excipient, data on the effects of the addition of trileucine to cromolyn sodium and albuterol powders are presented in Fig. 8 (69). It can be seen that the addition of trileucine at levels of around 15% by weight increases ED by 33% and FPF by

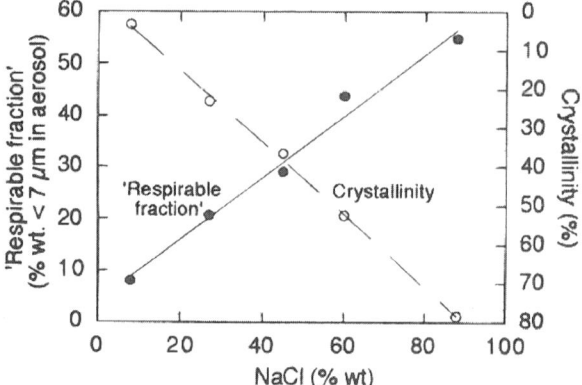

Figure 7   Relationship between NaCl content, the corresponding crystallinity (pure rhD-Nase powder is amorphous), and the dispersing properties (as FPF) of rhDNase powders. All powders have similar primary size distributions before aerosolization, with median diameters of 2.7 to 3.3 μm (span 1.04–1.63 mm) (67). *Abbreviations*: FPF, fine particle fraction; rhDNase, recombinant human deoxyribonuclease.

45%. The mechanism for this effect is postulated to be preferential coating of the surface of the particles with the hydrophobic amino acid: 15% by weight trileucine achieves 60% surface coverage.

In addition to the overall requirements for an inhalation powder formulation, proteins require stabilization in the solid state. The task of manufacturing powders containing proteins that have both the required protein-stabilizing properties and the physical characteristics necessary for generation of fine aerosols can be particularly daunting. The formulation approaches involve the technical challenges of first manufacturing "engineered" fine powder containing the protein and then either working with the powder and overcoming any filling and

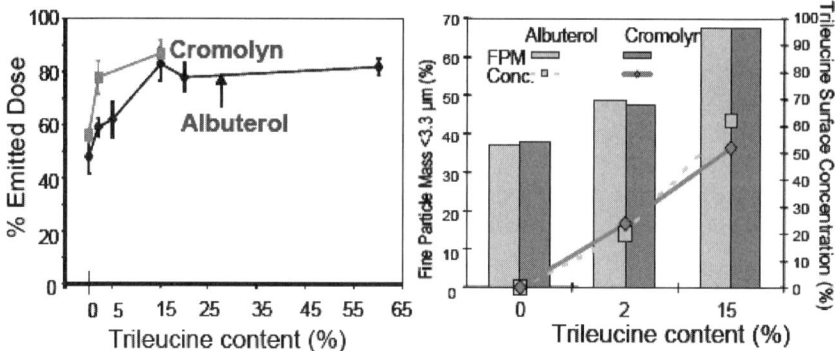

Figure 8   Relationship between the addition of trileucine, surface coverage, and powder performance for cromolyn sodium and albuterol powders. *Source*: From Ref. 69.

dispensing problems via mechanical means or, as described above, blending the protein-containing powder with a coarse carrier. The former approach is the one that has been adopted for most protein powder products, resulting in particles engineered to have the required combination of stability and dispersibility for particle device platforms. From a stability perspective, solid state stabilizers used thus far in the development of powder inhalation formulations are those used in stabilization during the lyophilization process. The use of mannitol, lactose, trehalose, and various other excipients such as citrate have been reported (67,69,70). However, the exact mechanism by which protein stabilization is achieved with these excipients is still debatable. Theories range from water replacement (73), where the sugar substitutes for the water hydrogen bonds and thus keeps the protein in its native conformation, to mere physical hindrance (74), where the protein molecules are simply immobilized in the glassy matrix and are prevented from unfolding and being spatially separated, to prevent aggregation. In these solids, it is important to achieve both chemical and physical stability.

As described above, amorphous powders, which are desirable from a protein stability point of view, possess the thermodynamic desire to crystallize. Crystallization can lead to the physical degradation of the powder, in terms of flow and dispersibility, as well as to aggregation of the protein (67). Thus, in general, if the physical explanation of stability is to be accepted, the approach is to produce powders with the highest glass transition temperature (T_g) possible (73). An alternative approach to glass stabilization is to produce powders that contain only small quantities of stabilizing material and are predominantly protein. High protein content appears to prevent crystallization (74). However, reducing the excipient content with some molecules can increase physical stability, but it may lead to increasing damage to the protein, with aggregation as a result. Another approach is to stabilize the protein as a thermotropic liquid crystal (75,76). Liquid crystals are characterized by having order in two dimensions, as compared with a three-dimensional crystal structure. A delicate balance between physical stability, aerosol performance, and protein stability may have to be struck (70).

Table 2 presents the effect of lactose on the biochemical stability of spray dried powders of rhDNase over a 40-week storage period. It can clearly be seen that as lactose content is increased, aggregation decreases. However, Figure 9 presents FPF as a function of lactose content and humidity over four weeks: clearly, as lactose content is increased, the powder performance decreases. It was demonstrated that this decrease in performance is due to crystallization of the lactose (70). From a stability perspective, then, designing a protein powder for inhalation may entail the art of choosing a particular compromise between physical and biochemical stability. In general, the use of lactose as an excipient for protein powders can be contraindicated because of the propensity of this sugar to react with lysine residues, producing lactosylated protein molecules (77).

Table 2 Biochemical Stability Data for Spray-Dried Powders Containing rhDNase and Lactose After Storage for 40 Weeks at 38% Relative Humidity

Formulation (rhDNase: lactose)	Temperature (°C)	Monomer (%)[a]	Δ Deamidation (%)[b]	Relative activity[c]
100:00	4	97.1	0.3	0.91
	25	93.1	0.6	0.89
84:16	4	99.4	0.5	0.99
	25	97.4	2.0	0.85
66:34	4	99.6	0.1	1.00
	25	98.6	1.5	0.94
50:50	4	99.7	0.1	1.01
	25	99.2	1.6	0.96

[a]Determined by size exclusion chromatography using TSK2000SWXL column. Immediately after spray drying, samples contained 100% monomer. All values are mean of duplicate determination.
[b]Determined by tentacle cation exchange chromatography using a LiChronsphere 1000 SO_5 column. All values are mean of duplicate determinations.
[c]Determined by rhDNase methyl green assay. Immediately after spray drying, samples were fully active (i.e., relative activity of 1).
Abbreviation: rhDNase, recombinant human deoxyribonuclease.
Source: From Ref. 70.

A further complication is added when a combination of excipients that stabilize the protein and enhance dispersion is considered. Here the aim would be to stabilize protein within the bulk of the particle and then "surface coat" the particle with a hydrophobic material that enhances dispersion. This approach relies on a complex interplay between solution droplet drying rates, molecule diffusivity, and surface activity. Although not yet used in a product, this has been exemplified by Lechuga et al. (69).

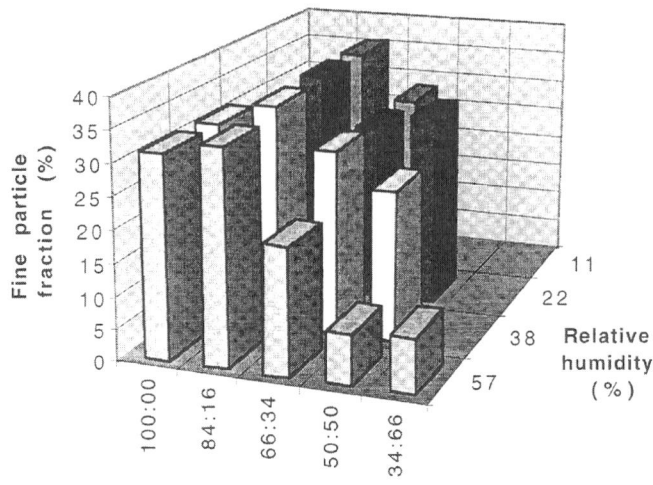

Figure 9 Fine particle fraction as a function of protein/lactose content and storage relative humidity after one month at 25°C. *Source*: From Ref. 70.

Careful assessment of the physical and toxicological acceptability of excipients must also be made. Only a few excipients have GRAS status from the FDA for use in the lung, and toxicology studies may be required to demonstrate the acceptability of a chosen excipient. Fortunately pulmonary delivery of proteins in general appears to be reasonably safe (78). Also, as with the liquid solution formulations described above, the effects of pH on reconstitution and induced tonicity changes on dissolution of the powder on the airway walls must all be considered. For example, Anderson et al. (79) recently showed that delivering large quantities of sodium chloride, lactose, or mannitol to the airways can induce a bronchoconstrictor response in asthmatic patients, presumably due to a hypertonic challenge. Although even in severe asthmatics, lung doses on the order of 5 to 10 mg had to be delivered before a significant clinical response was seen, the ever-wary formulator should keep these issues in mind when developing protein inhalation formulations.

Manufacturing Issues

A number of approaches have been taken in the manufacture of fine protein powders. These involve lyophilization followed by micronization (80), and coprecipitation followed by lyophilization (71) or spray drying. Although in general, lyophilization of proteins is a well-established process (82), micronization is not, and thus the first approach can be problematic. Conventional jet mills produce local heating due to the particle–particle collisions necessary for comminution and can introduce metallic contamination, which can affect product stability. Although not excluded, this process has had limited success.

The second approach is the basis for the Technospheres® formulation that is being developed by MannKind (63). Insulin is coprecipitated with diketopiperazine to produce spherical particles that are then extracted from suspension by filtration and lyophilization.

The third approach, spray drying aqueous solutions, has been very successful and has been used for many years to prepare dry protein powders (67,69, 82–87). Similar to liquid formulation stabilization issues, where proteins exposed to an air–water interface during nebulization may denature and aggregate, the atomization step of the protein solution during spray drying can also cause interfacial denaturation (82,83,85). Trileucine has been used as a surface active agent to protect proteins from interfacial denaturation (69). Spray drying from aqueous/organic solvents, such as ethanol, has also been demonstrated for poorly soluble peptides (75,76). As described above, the spray drying process is the basis for a number of different formulation approaches, ranging from porous particle to "M&M" type particles containing protein stabilizers inside and surface energy reducing agents as a coating.

Examples

Eli Lilly and Alkermes AIR® are in Phase III trials with inhaled insulin (64). The Alkermes AIR technology consists of large, porous, insulin particles of low density, containing dipalmitoyl phosphatidylcholine (DPPC) (86). The insulin formulation is spray dried from an ethanol solution. The particles (insulin, DPPC, lactose, albumin) have a mass median aerodynamic diameter (MMAD) of 1 to 5 μm, with a mass density of 0.05 to 0.3 g/cm^3 and an ED of 45% to 96% (86). The high ED values are related to low particle mass densities, suggesting that the large, porous particles are useful for inhalation applications.

MannKind Pharmaceuticals (Technospheres) is currently in Phase III trials with inhaled insulin. Technospheres uses a diketopiperazine derivative (3,6-bis(*N*-funaryl-*N*-(n-butyl) amino-2,5-diketopiperazine)) formulation that self-assembles into an ordered lattice at low pH and encapsulates the insulin (89–91). Following lyophilization, the particles have an MMAD of 2 to 4 μm. The powder is delivered with a Medtone DPI and dissolves at physiologic pH. Technospheres have also been applied for pulmonary delivery of parathyroid hormone as well as insulin (91).

A large number of protein powders have been investigated. Researchers have also spray dried formulations of cyclosporine, a highly water-insoluble molecule from an ethanol solution (77). The feasibility of delivering inhalation powders of granulocyte colony stimulating factor produced by spray drying and combined with coarse carrier lactose has been demonstrated (92). Spray dried powders containing sodium chloride and rhDNase have been described by Chan et al. (67) . Spray dried liposomes of leuprolide have also been utilized in a DPI (94). Human growth hormone has also been spray dried from a DPPC formulation (94). Growth hormone, tissue plasminogen activator, and interferons have all been characterized for surface-induced aggregation during spray drying (82–86). Furthermore, the crystalline content, water content, secondary structure, and aggregation rates of salmon calcitonin spray dried with ratios of mannitol were characterized (87). The effect of mannitol on recombinant humanized anti-IgE monoclonal antibodies has also been investigated (95), and finally, both humanized anti-IgE monoclonal antibodies and rhDNase powders have been prepared by spray drying and spray freeze-drying techniques (95,96).

PROTEIN MDI FORMULATIONS

Choice of Device

In general, the MDI has not been the delivery system of choice for proteins and peptides. This is presumably because of poor efficiency and reproducibility of delivery, coupled with the relatively harsh propellant environment. However, issues concerning the phase-out of chlorofluorocarbon (CFC) propellants and their replacement with hydrofluoroalkanes (HFAs) may be involved as well. The pharmaceutical industry has experienced difficulties in reformulating existing small molecules, and one can only suspect that the formulation of proteins and peptides in HFAs could be even more difficult.

Choice of Excipients

Despite these difficulties, a number of early feasibility studies have been reported, and recently KOS Pharmaceuticals have presented clinical data on an insulin MDI

(97). Indeed, a technically successful, although not commercially viable, formulation has been developed for leuprolide, a potent agonist of luteinizing hormone–releasing hormone (LHRH) (leuprolide is discussed as a case study in the section titled Formulation Development and Characterization of Inhaled Insulin Exubera). Brown et al. (98,99) have reported a number of MDI preparations involving an antigenic protein, an enzyme, and an antibody. In their first series of experiments, they reported two MDI formulations based on dimethyl ether (DME). The first contained either alkaline–phosphatase or an IgG_1 κ murine monoclonal antibody. In both cases, the proteins were lyophilized with either Laureth-9, Tween 40, or Tween 80 before being added to the DME. In the case of alkaline–phosphatase, the lyophilization process alone reduced enzymatic activity fivefold. In case of the antibody, the use of the surfactant appeared to offer protection against denaturation. The MDIs were shown to produce FPFs of the order of 10% to 25%. Alkaline–phosphatase appeared to be stable for up to 10 days regardless of the surfactant used, whereas the IgG denatured quite rapidly with Tween-40 or Laureth-9. Their second report (99) describes formulations involving bovine γ-globulin suspended in CFC 11/12 (trichlorofluoromethane/dichlorodifluoromethane), DME or propane. A variety of surfactants were investigated, and FPFs from the MDIs were again around 25%. However, in all cases except Tween 80 and DME, denaturation and a loss of antigenicity occurred rapidly. These early studies were aimed at utilizing surfactant micelle formation to produce solution aerosols of the protein.

The KOS product utilizes insulin suspended in HFA propellant. To deal with stability issues, the formulation's water content is controlled, and to deal with delivery efficiency, a breath actuation and spacer mechanism is employed (97).

Examples

To date, there have been no successful products delivering large proteins by means of the powder suspension type of MDIs. This may be because of the issues of solid state powder stability described above, or it may be due simply to lack of interest in this dosage form for protein delivery. However, as described above, KOS Pharmaceuticals is currently in Phase II clinical trials with a mealtime inhaled insulin delivered from an MDI incorporating breath actuation and a spacer (97).

PROTEINS FORMULATED FOR PULMONARY DELIVERY AND SPECIFIC CASE HISTORIES

In this section, we discuss some case studies in which proteins were formulated for aerosol delivery. The first involves the development of a protein for topical delivery to the lung using liquid formulations with nebulizers; the second describes development of a protein powder as a DPI for systemic delivery; and the third involves the formulation of a peptide as an MDI for systemic delivery.

Unfortunately, the third product has had a rather checkered development history and, as a result of the phase-out of CFC propellants, was abandoned as a commercial product. However, because of its unique place in protein/peptide formulation development, it is cited as a case study.

Formulation Development and Characterization of rhDNase [Pulmozyme® (Dornase Alpha)]

Cystic fibrosis patients, as a result of persistent bacterial infections, have large concentrations of host-neutrophil–derived DNA in their airway sections. This increased concentration contributes to an increased viscosity of the mucosal secretions in these patients. rhDNase can alter the viscoelastic properties of human sputum and was, therefore, developed for delivery targeted to the human airways. Pulmozyme was developed as a liquid inhalation product to be delivered by jet nebulizers. Details on the clinical use of DNase (bovine and human), formulation, and characterization of the aerosols are available in a recent review (100). The salient features of that review are presented here.

Formulation Development

As discussed earlier, osmolality, pH, and buffering agents can promote adverse reactions in the lungs, especially in patients with compromised pulmonary function. Thus, an isotonic liquid formulation was developed without any buffer components or preservatives. Although buffering agents were not used in Pulmozyme formulations, it was demonstrated that at 1 mg/mL, the protein itself provides sufficient buffering capacity to control the pH of the solution. It was shown that calcium is required for stability of this protein, and thus sufficient $CaCl_2$ was added to maintain activity during storage. Since this formulation did not contain any preservative, it was designed for single-use administration in nebulizers.

Many inhalation products are manufactured for single use, using blow-filled seal technology such as that employed by Automatic Liquid Packaging Inc. The plastic vials manufactured by this process are made from low-density polyethylene, which is permeable to water vapor and gases. Thus suitable foil packaging also had to be developed for this product. This type of packaging provides great convenience to patients besides allowing increased throughput for the manufacture of the product. In addition, the plastic surface resulted in a lower pH of storage for the product than is obtainable upon storage in glass vials. This actually resulted in longer shelf life for the product, since the major degradation route for Pulmozyme is deamidation of an asparagine residue that results in lower potency of the product.

Aerosol Characterization

As already mentioned, the distribution of the size of droplets in an aerosol is critical in determining deposition in the airways. An ideal combination of device

Table 3 1 mg/mL rhDNase Delivery by Jet Nebulizers

Nebulizer	n^a	FPF (%)	Nebulizer efficiency (%)	Delivery (%)
Marquest customized Respirigard II	8	51 ± 3	48 ± 6	24 ± 4
Marquest Acorn II	8	50 ± 4	55 ± 6	27 ± 3
Hudson T Up-Draft II	8	51 ± 4	49 ± 5	25 ± 2
BaxterAirlife Misty	7	46 ± 3	44 ± 2	21 ± 1

[a]Table values are a result of *n* independent measurements using *n* individual nebulizers, given as the mean value and standard deviation.
Abbreviations: rhDNase, recombinant human deoxyribonuclease; FPF, fine particle fraction.
Source: Adapted from Ref. 21.

and formulation will convert most of the solution into an aerosol that contains a majority of the droplets within the FPF. Since most cystic fibrosis patients are familiar with the use of jet nebulizers, these devices were used to deliver aerosol, and the in vitro behavior of these systems was characterized by techniques already discussed and reviewed (21,26,100). Initially, a cascade impactor was used to characterize four different nebulizers; the results are summarized in Table 3. The essential findings were that there were no significant differences in the performance of the four jet nebulizers (FPF or efficiency, E) and that the overall efficiency of delivery was ~25% [Eq. 4]. This type of performance is similar to that observed for other drugs delivered by jet nebulization. The more rapid analysis via laser light diffraction was later used and gave results comparable to those determined by the more labor-intensive analysis using cascade impactors.

The quality of rhDNase before and after nebulization was determined. It was shown that there was no reduction in activity, no generation of aggregates, and no significant alteration in rhDNase tertiary and secondary structures. Altogether, the four jet nebulizers that were tested were essentially equivalent in their ability to deliver respirable doses of intact, fully active, nonaggregated rhDNase (21,26,42).

Formulation Development and Characterization of Inhaled Insulin (Exubera®)

Insulin is the primary treatment for Type I diabetes and an important part of the therapy for Type 2 patients. Furthermore, physicians are using insulin earlier in the treatment of the disease state. Insulin is currently delivered with multiple (short and long acting) daily subcutaneous injections. However, patients usually still suffer from poor glucose control, due to needle phobia, lack of dosing flexibility, and poor compliance. Therefore, pulmonary delivery of insulin would provide a potential improvement of a suboptimal dosing regimen for a lifelong therapy (101).

Formulation Development

Insulin was formulated as an amorphous powder containing 60% insulin in a buffered matrix of mannitol, glycine, and citrate (101,102). Amorphous solids, termed glasses, do not melt but exhibit a T_g that depends on the chemical composition, processing history, and time scale of observation (103,104). Furthermore, when the formulation was exposed to various temperatures, no crystallization was observed. The T_g is a temperature at which a transition begins from a mechanical solid phase to a viscoelastic supercooled rubber phase. The T_g for the spray dried insulin formulation was determined to be between 78°C and 95°C at the specified moisture content (101). Water acts as a plasticizer and lowers the T_g of amorphous solids (105). Increasing the moisture content decreases the T_g and increases the rate of insulin degradation. When the T_g remains above pharmaceutically acceptable storage conditions, adequate stability is obtained. In order to maintain an appropriate distance between the T_g and the storage temperature, a moisture content of less than 5% was required. Spray dried insulin powders with ~2% water content result in a T_g of ~80°C (101).

In general, labile proteins exhibit kinetic stability and acceptable shelf life when stabilized in an amorphous glass stored below their T_g (104). When amorphous solids are stored well below their T_g, molecular motions and chemical reactivity are slowed (103). A strong glass exhibits mobility that obeys an Arrhenius relationship over a wide temperature range and has a weaker temperature dependence relative to fragile glasses of molecular mobility above the onset of the glass transition. In contrast, fragile glasses deviate from Arrhenius behavior and exhibit rapid increases in molecular mobility above the T_g (101,104). The spray dried insulin formulation is specifically termed a strong glass, with a temperature range of 20°C between the extrapolated onset and end point of the glass transition (106).

Spray dried insulin powders were characterized for chemical and physical stability. The rate of insulin degradation was determined to fit Arrhenius kinetics, confirming the presence of a strong glass (101,107). The structural stability was determined to be α-helical, and unchanged by processing conditions (101).

Spray drying was selected for its ability to produce homogenous powders within the desired particle size range (<5 μm), low moisture content, and high drug purity. The insulin solution is first atomized to form droplets. As the droplets cool, moisture is removed from the system, creating a particle skin, and shrinkage occurs to form raisin-shaped particles (5% wt/vol solids) (101,108). The true and bulk densities are 1.46 and 0.2 g/cm^3, respectively (101).

Aerosol Characterization

The pulmonary delivery system is a reusable DPI that was designed to deliver insulin to the small airways and alveoli for systemic absorption (101). Particle size distribution, from the inhaler, was predicted by the properties of the powder, specifically primary particle size (101). Product performance was characterized by ED, particle size distribution, and FPD. Aerosol delivery across a range of flow rates (5–60 L/min) and flow volumes (400–1400 mL) did not affect the aerosol performance (101). Furthermore, device priming and orientation (0–270°) showed little dependence.

Formulation Development and Characterization of LHRH MDI

Leuprolide is a 1 kDa nanopeptide possessing potent LHRH agonist properties. It is indicated in diseases such as endometriosis and prostate cancer. The molecule is hydrophilic, and it is not absorbed from the gastrointestinal tract. Leuprolide is currently administered by subcutaneous injection.

Formulation Development

Adjei and coworkers (109,110) showed that systemic delivery of leuprolide acetate was possible via the lung, and a series of papers document various aspects of their MDI formulations. Additionally, taste-masking studies with aspartame and menthol were performed (110). The solution MDI was formulated using ethanol as the solvent and CFC 12 as the propellant. The suspension aerosol utilized a blend of CFC 11 and CFC 12, the leuprolide acetate being suspended as a micronized powder. All the MDI formulations showed acceptable stability over three months. No literature is available to confirm a longer shelf life.

Proof of concept was demonstrated in beagle dogs using intratracheal administration of 1 mg/mL saline solutions of leuprolide acetate (111). Volumes of solution, based on body weight, were instilled at fixed distances from the epiglottis, and bioavailability was measured as dose-corrected areas under the curve relative either to subcutaneous or to intravenous injection. The results of these experiments clearly showed that as the drug was deposited more distally, bioavailability increased: at a distance of 25 cm, corresponding to the bifurcation at the base of the trachea, the drug was essentially 100% bioavailable. (*Note:* Leuprolide has both *C*- and *N*-terminals blocked, and hence is not easily metabolized in the lung.)

Aerosol Characterization

Following proof of concept, human studies were carried out using MDI formulations (111). Solution- and suspension-metered dose aerosols were compared in 23 normal volunteers (109). The human data confirmed the high bioavailability determined in dogs and demonstrated a correlation between FPF, determined by means of cascade impaction methods (112), and bioavailability. The FPF for the suspension aerosol was four times higher than for the cosolvent, presumably because of the poor evaporation kinetics of the spray (114). Relative bioavailabilities of 6.6% for the solution aerosol and 27.9% for the suspension aerosol were obtained. When corrected for the difference in FPFs, the bioavailabilities were similar: that is, 66% for the solution aerosol and 73% for the suspension.

These data, coupled with the inferred stability, demonstrate the technical feasibility of the MDI approach for leuprolide. However, all the formulations described above used CFC propellants and, as a result of the CFC phase-out, cannot be commercialized. Issues relating to the formulation of leuprolide in HFA propellants are yet to be discussed in the literature.

CHOOSING BETWEEN DRY POWDER AND LIQUID AEROSOL DELIVERY APPROACHES

This chapter has tried to summarize the merits and problems encountered in developing solid state and liquid formulations of proteins for pulmonary delivery. As already discussed, the difficulties of developing both liquid and powder formulations can be numerous and varied. In the case of powder preparations, formulation development is complicated by the need for biochemical stability of the macromolecule, physical stability of the powder, and a powder that readily lends itself to dispersion and aerosolization. In the case of liquids, the constraints are a little more relaxed, with the major issues being solution stability and survival during a nebulization process. However, the price paid for a solution's early versatility can be a less convenient product, which requires refrigeration during storage and a more cumbersome delivery system.

So where does a formulator begin? As with all development projects, the formulator must begin with the timelines and end product requirements in mind. If the delivery system needs to be readily available for administration when symptoms develop, or if dosing frequency is high, a handheld device coupled with a formulation that is stable at room temperature is highly desirable. Generally, liquid formulations of proteins will not be room-temperature stable and will not have the stability of a dry powder. Hence, when there is a need for maximum patient convenience, a dry powder product is more desirable. However, powders can sometimes require more "up-front" formulation development, and there is the major issue of device specificity in formulation development. In fact, device availability can be one of the main obstacles to carrying out early clinical work with powder formulations. Currently, devices that deliver liquid aerosols (i.e., nebulizers) are readily available, whereas "generic" dry powder devices are not. Also, nebulizer performance is, in general, formulation independent, whereas powder formulations, again, tend to be device specific. It can therefore be advantageous to begin clinical development and carry out "proof of concept"-type studies with early solution formulations and nebulizer devices with the intention of utilizing the dry powder approach for later studies and as the commercial product form. With this approach, however, two formulations must be developed, and after "proof of concept," the quickest time to market may sometimes be obtained by continuing development of the solution form. This can lead to an inferior product, but earlier commercial returns.

Although the development of a liquid formulation can often be less challenging, and hence quicker than work involving a dry powder, the requirements for the inclusion of a preservative to produce a multiuse formulation may complicate both the development and the approval processes. The number of preservatives used in inhalation products is limited, and the potential for direct interaction between protein and preservative is always present. It may be possible to limit these problems by selecting appropriate excipients, but this tactic may add considerable time to the formulation development process. The obvious way around these issues is to use a device capable of administering individually packed unit doses. Another approach could be to use multidose systems with nonventing closures. However, even these systems may have to employ preservatives in order to ensure an acceptably low risk of microbial contamination.

It should be clear that there is no single approach to the development of a protein pulmonary delivery product. In general, formulation and device must be coupled early in the development cycle and must be developed as an integrated system. If there is no, or limited, access to a dry powder device, the development of a liquid formulation and the use of nebulizer devices may facilitate early clinical work. However, the ultimate choice between dry powder and liquid must be dictated by the disease, the patient population, the dosing regimen required, the market competition, and the formulation and device possibilities. Finally, and sometimes above all, timelines and the need to be first to market may be important factors that guide a formulator on this difficult decision.

SUMMARY AND CONCLUSIONS

The challenges for the successful delivery of protein pharmaceuticals by aerosol delivery to the lung is augmented by the need to couple a delivery device with the formulation. Thus, as in the case of liquid formulation, it is necessary not only to assure the typical two-year pharmaceutical shelf life, but also to prevent protein degradation due to the stresses that result from aerosolization. Development of dry powders adds complexity because of the need to have both biochemical and physical stability. The latter is required for ease of manufacture, filling into devices, and the requirement to attain a high FPF on delivery. This chapter has summarized some of the available devices and strategies for producing formulations that result in pharmaceutical shelf life and efficient delivery to the lung.

REFERENCES

1. Clarke S, Stephenson RC, Lowenson JD, Ahern TJ, Manning MC. Stability of protein pharmaceuticals. Part A. Chemical and physical pathways of protein degradation. In: Ahern TJ, Manning MC, eds. Pharmaceutical Biotechnology. Vol. 2. New York: Plenum Press, 1992:434.
2. Cleland JL, Powell M, Shire SJ. The development of stable protein formulations: a close look at protein aggregation, deamidation and oxidation. Crit Rev Ther Drug Carrier Sys 1994; 10(4):307–377.
3. Guidance for Industry; Nasal spray and inhalation solution, suspension and spray drug products—chemical, manufacturing and control documentation. Federal Drug Administration Center for Drug Evaluation and Research (CDER), July 2002. (http://www.fda.gov/cder/guidance/4234fnl.htm)

4. Gonda I. Physiochemical principles in aerosol delivery. In: Crommelin DJA, ed. Topics in Pharmaceutical Sciences. 1991. Stuttgard: Medpharm Scientific.

5. Byron PR. Aerosol formulation, generation, and delivery using nonmetered systems. In: Byron PR, ed. Respiratory Drug Delivery III. Boca Raton: CRC Press, 1990:143–165.

6. Rudolf G, Gebhart J, Heyder J, Scheuch G, Stahlhofen W. Mass deposition from inspired polydisperse aerosols. Ann Occup Hyg 1988; 32:919–938.

7. Proceedings of International Society for Aerosols in Medicine Focused Symposium. Towards Meaningful Laboratory Tests for Evaluation of Pharmaceutical Aerosols, Puerto Rico, Jamaica. 29–31, 1997; J Aerosol Med 11 (1998).

8. Ranucci J. Dynamic plume-particle size analysis using laser diffraction. Pharm Technol 1992; 16(10):109–114.

9. Gorman WG, Carroll FA. Aerosol particle-size determination using holography. Pharm Technol 1993; 17(2):34–37.

10. Jager PD, De Stefano GA, Mcnamara DP. Particle size measurement using right-angle light scattering. Pharm Technol 1993; 17(4):102–120.

11. Ranucci JA, Chen CF. Phase doppler anemometry: a technique for determining aerosol plume-particle size and velocity. Pharm Technol 1993; 17(6):62–74.

12. Niven RW. Aerodynamic particle size testing using a time-of-flight aerosol beam spectrometer. Pharm Technol 1993; 17(l):72–78.

13. Clark AR. The use of laser diffraction for the evaluation of the aerosol clouds generated by medical nebulizers. Int J Pharm 1995; 115:69–78.

14. Milosovich SM. Particle size determination via cascade impaction. Pharm Technol 1992; 16(9):82–86.

15. Atkins PJ. Aerodynamic particle size testing—impinger methods. Pharm Technol 1992; 16(8):26–32.

16. USP/NF. Physical tests and determinations: Aerosols, Vol. USO XVII. IUSPharmacopeia. Rockville, MD: United States Pharmacopeial Convention, Inc., 1992:3158–3178.

17. European Pharmacopeia. Section 2.9.18—Preparations for inhalation: aerodynamic assessment of fine particles. European Pharmacopeia. 5th ed. (suppl 1). Council of Europe, 67075 Strasbourg, France, 2005:2799–2811.

18. USP 28-NF 23. Physical tests and determinations: Aerosols. Rockville, MD, USA: United States Pharmacopeia, 1992:3298–2316.

19. Lundback HH, Bagger-Jorgensen H, Sandell D, Sundahl M. Effect of inherent variability of inhalation products on impactor mass balance limits. In RN Dalby, PR Byron and SJ Far (eds) Respiratory Drug Delivery Europe, Interpharm press, Buffalo Griove, 2005; 1:177–180.

20. Pritchard SE, Mc Robbie DW. Studies of human oropharyngeal airspaces using magnetic resonance imaging—II. The use of 3D gated MRI to determine the influence of mouthpiece diameter and resistance of inhaled delivery systems on the orophayngeal airspace geometry. J Aerosol Med 2003; 17(4):310–324.

21. Cipolla DC, Gonda I. Method for collection of nebulized proteins. In: Cleland JL, Langer R, eds. Formulation and Delivery of Proteins and Peptides. Washington: American Chemical Society, 1994:342–352.

22. Niven RW, Butler JP, Brain JD. How air jet nebulizers may damage 'sensitive' drug formulations. 11th Annual Meeting of the American Association for Aerosol Research, San Francisco, Oct 12–16, 1992.

23. Niven RW. Delivery of biotherapeutics by inhalation aerosols. Pharm Technol 1993; 17(7):72–81.

24. Cipolla DC, Clark AR, Chan HK, Gonda I, Shire SJ. Assessment of aerosol delivery systems for recombinant human deoxyribonuclease. STP Pharma Sci 1994; 4(1):50–62.

25. Newman SP, Steed KP, Reader JS, Hooper G, Zierenberg B. Efficient delivery to the lungs of flunisolide aerosol from a new portable hand-held multidose nebulizer. J Pharm Sci 1996; 85(9):960–964.

26. Cipolla D, Gonda I, Shire SJ. Characterization of aerosols of human recombinant deoxyribonuclease I (rhDNase) generated by jet nebulizers. Pharm Res 1994; 11:491–498.

27. Ivri E. Apparatus and methods for delivery of therapeutic liquids to the respiratory system. U.S. patent 5,586,500, 1996.

28. Keller M, Lintz FC, Walther E. Novel liquid formulation technologies as a tool to design the aerosols performance of nebulizers using air jet (LC plus) or a vibrating membrane principle (E-Flow). Drug Delivery to the Lungs, London, Dec 13–14, 2001.

29. Schuster J, Rubsamen R, Lloyd P, Lloyd J. The AERx aerosol delivery system. Pharm Res 1997; 14(3):354–357.

30. Newman SP, Steed K, Towse L, Zierenberg B. The BINEB (final prototype): a novel hand-held multidose nebuliser evaluated by gamma scintigraphy. Eur Respir J 1996; 9:441S.

31. Weston TE, King AW, Dunne ST. Atomising devices and methods. World Intellectual Property Organization. Dunne Miller Weston Ltd., 1991. Patent WO 91/14468.

32. Fine JM, Gordon T, Thompson JE, Sheppard D. The role of titratable acidity in acid aerosol-induced broncho-constriction. Am Rev Respir Dis 1987; 135:826–830.

33. Balmes JR, Fine JM, Christian D, Gordon T, Sheppard D. Acidity potentiates bronchoconstriction induced by hypoos-molar aerosols. Am Rev Respir Dis 1988; 138:35–39.

34. Beasley R, Rafferty P, Holgate ST. Adverse reactions to the non-drug constituents of nebulizer solutions. Br J Clin Pharmacol 1988; 25:283–287.

35. Desager KN, Van Bever HP, Stevens WJ. Osmolality and pH of anti-asthmatic drug solutions. Agents Actions 1990; 31:225–228.

36. Snell NJC. Adverse reactions to inhaled drugs. Resp Med 1990; 84:345–348.

37. Sant'Ambrogio G, Andersen JW, Sant'Ambrogio FP, Mathew OP. Response to laryngeal receptors to water solutions of different osmolality and ionic composition. Respir Med 1991; 85(A):57–60.

38. Gonda I, Kayes JB, Groom CV, Fildes FJT. Characterization of hygroscopic inhalation aerosols. In: Stanley-Wood NG, ed. Particle Size Analysis. New York: Wiley Heyden, 1982:31–34.

39. Gonda I, Phipps PR. Some consequences of instability of aqueous aerosols produced by jet and ultrasonic nebulizers. In: Masuda S, Takahashi K, eds. Aerosols. New York: Pergamon Press, 1991:227–230.

40. Godden DJ, Borland C, Lowry R, Higgenbottam TW. Chemical specificity of coughing in man. Clin Sci 1986; 70:301–306.

41. Auffarth B, De Monchy JG, van der Mark TW, Postma DS, Koeter GH. Citric acid cough threshold and airway responsiveness in asthmatic patients and smokers with chronic airflow obstruction. Thorax 1994; 46:638–642.

42. Cipolla D, Gonda I, Meserve KC, Weck S, Shire SJ. Formulation and aerosol delivery of recombinant deoxyribonucleic acid derived human deoxyribonuclease I. In:

Cleland JL, Langer R, eds. Formulation and Delivery of Proteins and Peptides. ACS Symposium Series 567.Washington: American Chemical Society, 1994:322–342.

43. Tanford C. The Hydrophobic Effect. New York: Wiley, 1980.

44. Oeswein JQ et al. Aerosolization of protein pharmaceuticals. Proceedings of the Second Respiratory Drug Delivery Symposium, University of Kentucky, 1991.

45. Hora MS, Rana RK, Wilcox CL, Katre NV, Hirtzer P, Wolfe SN, Thomson JW. Development of a lyophilized formulation of interleukin-2. Dev Biol Stand 1992; 74:295–306.

46. Lam XM, Patapoff TW, Nguyen TH. The effect of benzyl alcohol on recombinant human interferon-gamma. Pharm Res 1997; 14(6):725–729.

47. Beasley R, Rafferty P, Holgate S. Benzalkonium chloride and bronchoconstriction [letter]. Lancet 1986; 2(8517):1227.

48. Berg OH, Henriksen RN, Steinsvag SK. The effect of a benzalkonium chloride-containin nasal spray on human respiratory mucosa in vitro as a function of concentration and time of action. Pharmacol Toxicol 1995; 76(4):245–249.

49. Klaustermeyer WB, Hale FC, Prescott EJ. Reproducibility of the response to diluent challenge in adult asthma. Ann Allergy 1979; 43(2):84–87.

50. Wright W, Zhang YG, Salome CM, Woolcock AJ. Effect of inhaled preservatives on asthmatic subjects. I. Sodium metabisulfite. Am Rev Respir Dis 1990; 141(6):1400–1404.

51. Zhang YG, Wright WJ, Tam WK, Nguyen-Dang TH, Salome CM, Woolcock AJ. Effects of inhaled preservatives on asthmatic subjects, II. Benzalkonium chloride. Am Rev Respir Dis 1990; 141:1405–1408.

52. Stadtman R. Metal ion-catalyzed oxidation of proteins: biochemical mechanism and biological consequences. Free Radical Biol Med 1990; 9(4):315–325. (Erratum published in Free Radical Biol Med 1991; 10:3–48.)

53. Patton JS, Platz RM. Aerosol insulin—a brief review. In: Byron PR, Dalby RN, Farr SJ, eds. Respiratory Drug Delivery IV. Buffalo Grove: Interpharm Press, 1994:65–74.

54. Chang, J Y H, Milby T, Oeswein JQ. Effects of metals and phenol on hGH degradation. Pharm Res 1994; 11(10):S81.

55. Lyda JC. FDA proposes sterility requirement of Inhalation solution drug products. PDA Let 1997:5–6.

56. Thipphawong J, Toulana B, Clauson P, Okikawa J, Farr SF. Pulmonary insulin administration using the AERx insulin diabetes system. Diabetes Technol Ther 2002; 4:499–504.

57. Leung FK, Li J, Song Y, et al. Glucose responses of pulmonary-delivered alveair formulated with unmodified chinese-made analog insulin in normal rats. ADA 65th Scientific Session, 2005, 2058-PO.

58. Exubera® [Package insert]. New York: Pfizer Inc, 2006.

59. Lee VHL, Dodda-Kashi S, Grass GM, Rubas W. Oral route of peptide and protein drug delivery. In: Lee VHL, ed. Peptide and Protein Drug Delivery. New York: Marcel Dekker, 1991:891.

60. Eljamal M, Nagarajan S, Patton JS. In situ and in vivo methods for pulmonary delivery. Pharm Biotechnol 1996; 8:361–374.

61. Rudolph G, Kobirch R, Stahlhofen W. Modeling and algebraic formation of regional aerosol deposition in man. J Aerosol Sci 1990; 21(1):S306–S406.

62. Woiszwillo JE, Fothstein F. Method for isolating biomolecules from a biological sample with linear polymers. US Patent 5,599,719, 1997.

63. Steiner S, Pfutzner A,Wilson BR, Harzer O, Heinemann L, Rave K. Technosphere/insulin proof of concept study with a new insulin formulation for pulmonary delivery. Exp Clin Endocrinol Diabetes 2002; 110:1721.

64. Edwards DA, Hanes J, Caponetti G, et al. Large porous particles for pulmonary drug delivery. Science 1997; 276(5320):1868–1871.

65. Dellamary LA, Tarara TE, Smith DJ, et al. Hollow porous particles for inhalation. Pharm Res 2000; 17:168–174.

66. Lechuga-Ballesteros D, Charan C, Liang Y, Stults CLM, Vehring R, Kuo M. Designing stable and high performance respirable particles of pharmaceuticals. In: Dalby R, Byron P, eds. Respiratory Drug Delivery IX. Buffalo Grove: Interpharm Press, 2004:565–568.

67. Chan HK, Clark A, Gonda I, Mumenthaler M, Hsu C. Spray dried powders and powder blends of recombinant human deoxyribonuclease (rhDNase) for aerosol delivery. Pharm Res 1997; 14:431.

68. Visser J. An invited review: van der waals and other cohesive forces affecting powder fluidization. Powder Technol 1989; 58:1–10.

69. Lechuga-Ballesteros D, Charan C, Stults CLM, et al. Trileucine improves aerosol performance and stability of spray-dried powders for inhalation. J Pharm Sci. in-press.

70. Clark AR, Dasovich N, Gonda I, Chan HK. The balance between biochemical and physical stability for inhalation protein powder: rhDNase as an example. In: Dalby R, Byron P, eds. Respiratory Drug Delivery V. Buffalo Grove: Interpharm Press, 1996:167.

71. Carpenter JF, Prestrelski SJ, Anchordoquy T, Arakawa T. Interactions of stabilizers with proteins during freezing and drying. In: Cleland JL, Langer R, eds. Formulation and Delivery of Proteins and Peptides. ACS Symposium Series 567. Washington: American Chemical Society, 1994:134–147.

72. Franks F, Hatly RH. Storage of materials. US patents 5,098,893, 1992.

73. Levine H, Slade L. Another view of trehalose for drying and stabilizing biological materials. Biopharm 1992; 5:36.

74. French DL, McAuly AJ, Chang B, Niven RW. Moisture induced state changes in spray-dried trehalose/protein formulations. Pharm Res 1995; 12(9):S-83.

75. Lechuga-Ballesteros D, Abdul-Fattah A, Stevenson CL, Bennett DB. Properties and stability of a liquid crystal form of cyclosporine: the first reported naturally occurring peptide that exists as a thermotropic liquid crystal. J Pharm Sci 2003; 92:1821–1831.

76. Lechuga-Ballesteros D, Bennett D, Cabot K,et al. Aerosol drug development of a spray dried liquid crystalline formulation of cyclosporine. In: Dalby R and Byron P, eds. Respiratory Drug Delivery VII. Rayleigh: Serentec Press, 2000:147–256.

77. Quan C, Wu S, Hsu C, Canova-Davis E. Title. Ninth Symposium of the Protein Society, Boston, 1995.

78. Wolf R. The safety of inhaled therapeutic proteins. J Aerosol Med 1998; 11(4):197–219.

79. Anderson SD, Brannan J, Spring J, et al. A novel bronchial provocation test (BPT) using a respirable dry powder of mannitol. Australian and New Zealand thoracic Society ASM. Perth, Australia, 1996.

80. Platz R. Improved process for preparing micronized polypeptide drugs. WO93/13752, 1993.

81. Chang BS, Fischer NL. Development of an efficient single-step freeze-drying cycle for protein formulations. Pharm Res 1995; 12:831–837.

82. Mumenthaler M, Hsu CC, Pearlman R. Feasibility study on spray-drying protein pharmaceuticals: recombinant human growth hormone and tissue type plasminogen activator. Pharm Res 1994; 11:12–20.

83. Maa YF, Nguyen PAT, Hsu SW. Spray drying of air-liquid interface sensitive recombinant human growth hormone. J Pharm Sci 1998; 87:152–159.

84. Yang B, Lesikar D, Tan MM, Ramchandran S, Stevenson CL. Formulation of human growth hormone for pulmonary delivery. AAPS PharmSci 2002; 4:W5073.

85. Nguyen XC, Herberger JD, Burke PA. Protein powders for encapsulation: a comparison of spray-freeze-drying and spray drying of darbepoetin alfa. Pharm Res 2004; 21:507–514.

86. Vanbever R, Mintzes JD, Wang J, et al. Formulation and physical characterization of large porous particles for inhalation. Pharm Res 1999; 16:1735–1742.

87. Chan HK, Clark AR, Feeley JC, et al. Physical stability of salmon calcitonin spray dried powders for inhalation. J Pharm Sci 2004; 93:792–804.

88. Owens DR, Zinman B, Bolli GB. Insulins today and tomorrow. Lancet 2001; 358:589–597.

89. Pfuztner A, Mann AE, Steiner SS. Technosphere/insulin—a new approach for effective delivery of human insulin via the pulmonary route. Diabetes Technol Ther 2002; 4:589–593.

90. Heinemann L, Pfutzner A, Heise T. Alternative routes of administration as an approach to improve insulin therapy: update on dermal, oral, nasal and pulmonary insulin delivery. Curr Pharm Des 2002; 7:1327–1352.

91. Pfutzer A, Flacke F, Pohl R, et al. Pilot study with technosphere/PTH (1-34)—a new approach for effective pulmonary delivery of parathyroid hormone (1-34). Horm Metab Res 2003; 35:319–323.

92. Niven RW, Lott FD, Ip AY, Crips KM. Pulmonary delivery of powders and solutions containing recombinant human granulocyte colony-stimulating factor (rhG-CSF) in the rabbit. Pharm Res 1994; 11:1101–1109.

93. Shahiwala A, Misra A. A preliminary pharmacokinetic study of liposomal leuprolide dry powder inhaler: a technical note. AAPS PharmSciTech 2005; 6:E482–E486.

94. Bosquillon C, Preat V, Vanbever R. Pulmonary delivery of growth hormone using dry powders and visualization of its focal fate in rats. J Control Release 2004; 96:233–244.

95. Maa YF, Nguyen PA, Andya JD, et al. Effect of spray drying and subsequent processing conditions on residual moisture content and physical/biochemical stability of protein inhalation powders. Pharm Res 1998; 15:768–775.

96. Maa YF, Nguyen PA, Sweeney T, Shire SJ, Hsu CC. Protein inhalation powders: spray drying vs spray freeze-drying. Pharm Res 1999; 16:249–254.

97. Hausmann M, Bellweg S, Heinemann L, Buchwald A, Heise T. Add-on therapy with Kos inhaled insulin dosed using a metered dose inhaler is as efficacious as add-on therapy with Lantus in poorly controlled type 2 diabetic patients treated with sulfonylureas or metformin. ADA 65th Scientific Sessions, 417-OR, 2005.

98. Brown R, Slusser JG. Propellant-driven aerosols of functional proteins as potential therapeutic agents in the respiratory tract. Immunopharmacology 1994; 28:241–257.

99. Brown R, Pickrell JA. Propellant-driven aerosols for delivery of proteins in the respiratory tract. J Aerosol Med 1995; 8(1):43–57.

100. Shire SJ. Stability characterization and formulation development of recombinant human deoxyribonuclease I [Pulmozyme®, (Dornase Alpha)] In: Pearlman R,

Wang J, eds. Formulation, Characterization, and Stability of Protein Drugs. New York: Plenum Press, 1996.

101. White S, Bennett DB, Cheu S, et al. Exubera: pharmaceutical development of a novel product for pulmonary delivery of insulin. Diabetes Technol Ther 2005; 7:896–906.

102. Skyler SJ, Cefalu WT, Kourides IA, et al. Efficacy of inhaled human insulin in type 1 diabetes mellitus: a randomized proof-of-concept study. Lancet 2001; 357:331–335.

103. Hancock BC, Shablin SL, Zografi G. Molecular mobility of amorphous pharmaceutical solids below their glass transition temperatures. Pharm Res 1991; 12:799–806.

104. Franks F, Hatley RHM, Mathias SF. Materials science and the production of shelf-stable biologicals. Pharm Technol Int 1991; 3:24–34.

105. Hancock BC, Zografi G. The relationship between glass transition temperature and the water content of amorphous pharmaceutical solids. Pharm Res 1994; 11:471–477.

106. Kajiwara K, Franks F, Echlin P, Greer AL. Structural and dynamic properties of crystalline and amorphous phases in raffinose-water mixtures. Pharm Res 1999; 16:1441–1448.

107. Strickley RG, Anderson BD. Solid state stability of human insulin. II. Effect of water on reactive intermediate partitioning in lyophilizes from pH 2-5 solution stabilization against covalent dimer formation. J Pharm Sci 1997; 86:645–653.

108. Tahl K, Claesson M, Lilliehorn P, Linden H, Backstrom K. The effect of process variables on the degradation and physical properties of spray dried insulin intended for inhalation. Int J Pharm 2002; 233:227–237.

109. Adjei A, Sundberg D, Miller J, Chun A. Bioavailability of leuprolide acetate following nasal and inhalation delivery to rats and healthy humans. Pharm Res 1992; 9:244–249.

110. Zheng JY, Fulu MY, Lee DY, Barber T, Adjei AL. Pulmonary peptide delivery: effect of taste-masking excipients on leuprolide suspension metered dose inhalers. Pharm Dev Tech 2001; 6:521–530.

111. Adjei L, Garren J. Pulmonary delivery of peptide drugs: effect of particle size on bioavailabiltiy of leuprolide acetate in healthy male volunteers. Pharm Res 1990; 7(6):565–569.

112. Adjei L, Carrigan PJ. Pulmonary bioavailability of LH-RH analogs: some biopharmaceutical guidelines. J Biopharm Sci 1992; 3(1/2):247–254.

113. Clark AR. MDIs: the physics of aerosol formation. J Aerosol Med 1996; 9(1): S19–S25.

11

Using Needle-Free Injectors for Parenteral Delivery of Proteins

Stephen J. Farr and Brooks Boyd

Zogenix, Inc., Emeryville, California, U.S.A.

Paul Bridges

Genentech, Inc., San Francisco, California, U.S.A.

Lawrence S. Linn

Zogenix, Inc., Emeryville, California, U.S.A.

NEEDLE-FREE INJECTION TECHNOLOGIES: OVERVIEW AND CURRENT APPROACHES

Why Needle-Free Injectors? The Patient Perspective

Needle-free injectors deliver drugs through the skin without the use of a needle. The market drivers for NFI are improved patient sentiment, enhanced compliance, improved safety, and marketing advantages. In addition, some devices present significant ease-of-use benefits.

In November 2000, the U.S. Congress passed the Needlestick Safety and Prevention Act that subsequently resulted in Occupational Safety and Health Administration regulations requiring health-care facilities to select safer delivery devices as they become available (among other requirements). Therefore, the safety argument for needle-free delivery for the patient and care-worker are easily understood and well documented (1,2).

Some clinical data suggest reduced sensation from needle-free devices; however, it should be noted that patient sentiment is not solely associated with pain scores: the benefit of needle-free technology is more significant than the pain scores suggest, and is likely to be much more associated with reduced fear of (and the elimination of) the needle-stick itself. Needle-free injection (NFI) is therefore particularly attractive to patients who dislike needles, particularly those not used to self-inject, those who require multiple injections over time, those at risk of needle-stick injuries, and children.

Prefilled and disposable NFI designs, particularly, have the ability to enhance patient compliance by making self-administration quicker and easier. The current trend in improving existing needle-based pen technology includes increased use of disposable pen systems, improved safety and handling, and smaller needles plus the arrival of more sophisticated devices with automatic needle insertion and injection. If a scale of device type versus patient conve-nience existed, then needle- and syringe–based lyophilized formulations would be at the low end of the scale and disposable prefilled autoinjectors would be at the upper end (3). The progression toward the upper end suggests continuous improvements in patient convenience, where single-use disposable needle-free devices could offer the ultimate in convenience for transcutaneous administration whilst also mitigating needle phobia or fear that can lead to significant problems with compliance (4).

Historical Development, Successes, and Failures

The general principle of the NFI system is to accelerate the drug formulation in order to penetrate the skin and deliver it to the subcutaneous (SC), intradermal, or potentially even the intramuscular space.

First patented in the 1930s, early generation NFIs developed in the 1940s were bulky and primarily practitioner-used for insulin delivery and mass vacci-nation in the military. These early devices were not designed with low sensation and patient reassurance in mind, sometimes causing more bruising and bleeding than a needle and syringe and were often impractical, requiring disinfection after use.

Later, more refined and portable NFI technology was developed, for exam-ple the Biojector® 2000 (B2000) by Bioject in the late 1990s. However, this technology has not been associated with any great success.

Some companies, such as Valeritas and Zogenics, have been developing more user friendly and convenient, prefilled, disposable capable NFI systems such as the Mini-Ject and Intraject®.

Like NFI devices, pen injectors have also been developed that are either reusable or disposable. The majority of pen injectors use prefilled containers such as cartridges or prefilled syringes. One big benefit of prefilled pens is that the exact dose required is contained within the device (vs. up to 25% overage in a vial), which is particularly important for delivering expensive biotechno-logically derived therapeutics. This benefit is no different for a prefilled NFI technology.

The downside of developing such convenient NFI drug delivery systems is that the device itself, together with the often dedicated manufacturing equipment, tends to be more complicated than that for the typical autoinjector. Some of the challenges include the following:

- The power source required for jet injection, its manufacture, storage, and reliability thereof. Power sources are typically chemical, gas, or spring.
- Related to the above is the energy during the delivery event itself. The forces exerted on components can result in customized triggering mechanisms and drug reservoirs.
- As a result of the novel components and energy source, the resulting manufacturing and assembly processes are also often customized, potentially requiring more development time than is typical for an autoinjector.

At the same time as tackling an often lengthy development process, companies developing needle-free devices need to offer patented, leading-edge technologies in order to avoid the risk of external super cession by competitors (5).

So why has needle-free delivery not been a significant success so far? This can be partly attributed to the "bad press" of the earlier, bulky systems that could be more painful than injection with a needle. Another significant reason is likely to be that at one time, the technology developers promised to meet many drug-delivery, patient compliance issues with devices that were really no more convenient to use than existing reusable autoinjectors. NFI development companies clearly expected pharmaceutical companies to actively pursue the "no-needle" benefits of NFI; however, being needle-free by itself is clearly not enough. The poor market performance of the less convenient, "reusable" NFI is perhaps a demonstration of this. More likely are the significant cost and time hurdles to developing the technology, including intellectual property management, R&D, and manufacturing, as well as quality and regulatory considerations. Perhaps the largest "mental" barrier is present with combination products, where the challenges of customized primary containers must also be overcome in partnership with the pharmaceutical client.

The combination of overpromising on what the technology could achieve and then sometimes underdelivering on execution of those promises has undermined the position of needle-free in the minds of pharmaceutical and biotech companies.

NFI products need to be designed and engineered to ensure they are easy and convenient for patients to use. Only when these factors are present will the advantage of the absence of a needle be realized by both patients and their doctors. Reusable NFI devices have categorically failed to improve patient convenience. In fact, it could be argued that many such devices on the market have made drug injections more cumbersome and complicated than using a needle and syringe or an autoinjector. Prefilled, disposable NFI systems should be able to meet these requirements, but what is needed is a change in the mindset of the developers of NFI devices in order to embrace the regulatory and clinical challenges in the development of successful drug–device combination products.

NFI General Theory of Operation

Most NFI devices pressurize the drug container in order to force the aqueous formulation to exit an orifice at high velocity.

The key physical requirements to a successful NFI are (*i*) being able to penetrate the skin and (*ii*) maintaining a jet velocity sufficient to deliver the drug formulation into the intended space without unintended tissue damage or pain. For this reason, some injectors employ a specific delivery profile design to modulate the flow rate over time, usually starting with a peak jet velocity to pierce the skin, then slowing in order to deliver the drug formulation through the hole without damaging tissue whilst maintaining enough pressure to deliver all of the fluid drug formulation successfully.

Recent research has revealed that the fundamental mechanism of skin penetration with an NFI is similar to that of a needle: In this case, it is a liquid jet that forms and opens a hole in the skin. Experiments have demonstrated that a sharp-tipped punch model can be used to determine the pressure required to penetrate the skin with a liquid jet (6).

Most devices complete delivery in a fraction of a second. The short time is partly determined by the mechanics of maintaining enough container pressure and subsequent jet velocity in order to deliver a successful injection across all patient skin types. The short time also mitigates the opportunity for the patient to disturb the delivery process. The exception to this is (*i*) the Avant technology, which delivers most of the dose at a rate similar to that of a traditional needle-based autoinjector, utilizing a vacuum to maintain the nozzle position against the skin and (*ii*) the Anesiva (formerly Corgentech and AlgoRx) and PowderMed (Pfizer, New York, U.S.A.) technologies, which are not liquid based: They accelerate tiny solid particles to a sufficient velocity to deliver them through the dermis, and these devices are only capable of delivering to the intradermal space.

For liquid delivery, the different NFI energy sources have their own advantages and disadvantages, with varying associated costs of development, often dependent on whether the device needs to be either reusable or light weight and disposable.

Current NFI Technological and Regulatory Approaches

An increased force required to inject with a needle and syringe is usually attributable to higher protein concentrations, with concomitant increases in viscosity and hydrodynamic flow resistance (7). As a result, patient comfort and therefore compliance has previously had to be compromised to practically deliver higher-concentration protein formulations. This is one area where some NFI devices can offer a clear benefit. Due to (*i*) the large injection forces already generated to successfully deliver under the skin and (*ii*) the generally much shorter fluidic path length, even the most viscous biologic drugs can potentially be delivered via existing NFI technology in a very short period of time as described below in the section titled Effect of Viscosity on Injection.

The complexity of design, development, and manufacture of NFI means that the cost per unit dose delivered is more expensive than that delivered by traditional routes, and that will be the case at least until the manufacturing technology is established at large volumes. One of the challenges is reducing the cost of the manufacture of the devices' energy source. For these reasons, the first therapeutics to be delivered, particularly via prefilled NFI, are likely to be more expensive and therefore in the more specialist therapeutic areas.

In the future, it would appear that the most successful NFI devices will be (*i*) those developed to accept prefilled drug containers and (*ii*) those that are the most simple and intuitive to use. This will address some key reasons for slow adoption so far: The constant need to tailor the technology to each therapeutic area and the complexity of those first (mainly reusable) devices on the market.

A comprehensive list of NFI products on the market or in development pipelines is shown in Table 1. All of the devices listed in Table 1 have been used in a publicized clinical trial. The table includes information on the technology and formulations, as well as the approach to regulatory approval, if known. Additional information on the delivery systems as it applies to a specific therapeutic protein is also described in the section titled Clinical Experience with NFI of Proteins.

NFI for Intradermal or Intramuscular Delivery

PowderMed, Bioject, and Valeritas all claim capability in delivering to either or both intradermal and intramuscular space. It should be noted that most clinical data to date have been generated with NFI delivery to the SC space.

FORMULATION CONSIDERATIONS FOR NFI DELIVERY

The principles of protein formulation for NFIs are the same as for other delivery methods. The formulation must be compatible with the container-closure materials of the delivery system (8,9). As for conventional needle-syringe systems, the injection system must function reliably and maintain a protective barrier between the formulation and the environment. For needle-free systems, however, the functional considerations are arguably more rigorous due to the high pressures and velocities involved in penetrating the skin without using a needle (6). Shearing may occur during delivery and its effect on the integrity of the protein and formulation should also be considered, although data to date suggest that this is not an issue (4,10).

Lyophilized, Aqueous, and Powder Formulations

Formulations for needle-free delivery are generally either aqueous solutions or made into aqueous solutions just prior to delivery. Most proteins have traditionally been formulated as a lyophilized powder contained in a glass vial for reconstitution at the time of injection. This has changed in recent years as more aqueous protein

Table 1 Needle-Free Injector Products, Currently Marketed or Under Development

Company name	Technology name	Home/ practitioner intended use	Power source (chemical, gas, spring)	Pre-/ patient filled	Variable/ fixed dose	Single/ multiple use	Delivery space (SC, IM, ID)	Maximum volume (mL)	Notes/ regulatory approval
AdvantaJet	GentleJet/ Activa	Home		Patient	Variable	Multiple	SC	0.5	For insulin. Class II approved
Anesiva (previously Corgentech)	Zingo™	Pract.	Gas	Prefilled	Fixed	Multiple	ID		Powder delivery of lidocaine as topical local anesthetic. Rebranded PowderJect technology
Antares	Medi-Jector VISION®	Home	Spring	Patient	Variable	Multiple	SC	0.5	For insulin and growth hormone. Specific orifices per skin type. Other platforms developed
Avant Medical	Avant Guardian 101	Home	Spring	Patient	Variable	Multiple	SC	1.0	Vacuum controls tip/ skin interface during slow injection. 510K approved. Other platforms

Company	Product	Setting	Power	Fill	Dose	Use	Route	Vol (mL)	Comments
Biojector	Biojector® 2000	Pract.	Gas	Patient	Variable	Multiple	All	1.0	Some 510K approved. Vitaject for insulin and as Cool.click™ for growth hormone
	Vitaject	Home	Spring	Patient	Variable	Multiple	All	1.0	
	Iject	Home	Gas	Prefilled	Variable	Single	All	1.0	
Crossject		Both	Chem.	Prefilled	Fixed	Single			Multinozzle design
Felton	HIS 500	Pract.	Gas	N/A	Fixed	Multiple	SC	0.5	High workload, 6 injections/min
Injex-Equidyne	Injex 30	Home	Spring	Patient	Fixed	Multiple	SC	0.3	510K approved
	Injex 50	Home	Spring	Patient	Fixed	Multiple	SC	0.5	510K approved
National Medical Products	J-Tip	Home	Gas	Patient	Variable	Single	SC	0.5	All plastic molded design
PowderMed	PMED®	Home	Gas	Patient	Single	Multiple	ID		Not liquid—propels particles into epidermis. Same rebranded PowderJect technology
The Medical House	GH1	Home	Spring	Patient		Multiple	SC		Partnered in 2004 with Serono for growth hormone
Valeritas (Biovalve)	Mini-Ject	Home	Chem.	Prefilled	Fixed	Single	SC	1.3	IM and ID capable to lower volumes
Zogenics	Intraject®	Home	Gas	Prefilled	Fixed	Single	SC	0.5	Previously developed by Weston Medical then Aradigm Corp.

Abbreviations: IV, intravenous; SC, subcutaneous; IM, intramuscular; ID, intradermal; PMED, particle-mediated epidermal delivery.

formulations have entered the market either as prefilled syringes, in glass vials, or in ampoules. There are now many marketed products in which a protein is formulated as a liquid (e.g., rhIFN-α-2a, rhIFN-α-2b, rhG-CSF, rhGH, rhInsulin) as well as a large number of aqueous protein formulations currently in development (11). The manufacture of lyophilized formulations for reconstitution, the associated device design requirements, and associated operational steps for patient or clinician administration are considerably more complicated, expensive, and time consuming than for aqueous formulations.

All of the approved needle-free systems currently on the market were approved as 510K or premarket approval devices that are used with existing formulations (Table 1). While these can be used with either reconstituted, lyophilized powder, or liquid formulations, the liquified formulation must be loaded into the device prior to injection. These 510K-approved devices have not gained wide acceptability, likely because they are more inconvenient to use than a needle and syringe or a pen-injector.

Liquid formulations offer the simplest solution from the perspective of a patient, for NFI formulation. A liquid formulation eliminates the need for reconstitution, thus making the product more convenient. A prefilled liquid formulation is even better. Although in many cases, liquid formulations can offer additional formulation stability challenges, the effort is worthwhile, and required, to compete with other autoinjectors in the market place.

The formulation of powders for needle-free delivery can be significantly more challenging than for liquid delivery owing to the fact that solid particles must be robust enough to penetrate the skin yet still be stable for the life of the product. Anesiva has developed a powder formulation and delivery system currently pending regulatory approval called Zingo™ for topical anesthesia for cannulation procedures (12). They produce small solid particles of lidocaine that are accelerated towards the skin at a rate that will enable penetration after which the lidocaine will dissolve to provide analgesia. Although this topical delivery method has been investigated for proteins, the SC delivery, formulation, and stability challenges are substantial. There are not any protein formulations currently being developed for this technology.

PowderMed is developing the particle-mediated epidermal delivery (PMED®) technology for vaccine delivery. The formulation consists of elemental gold particles of 1 to 3 μm in diameter onto which a DNA vaccine is precipitated. There are similar delivery, formulation, and stability challenges for this technology as for Zingo.

Effect of Shear on Macromolecular Integrity

Proteins are widely believed to be susceptible to shear forces. Under certain process conditions, extensive or extreme shear forces can be detrimental to protein integrity. In a drug delivery setting, shear forces are normally transient, and proteins have generally been stabilized by formulation excipients, making shear less problematic, e.g., as described in Ref. (13). A much greater problem is the generation of a large amount of air–liquid interfacial area at which proteins can unfold. The creation of an interfacial area is often associated with shear. The kinetics of protein diffusion to the surface and subsequent unfolding are proportional to the time the interface exists, the extent of the interfacial area, and the relative distance between the bulk formulation and the interfacial area.

NFI is generally achieved by forcing liquid through a small orifice, which produces a high-velocity liquid jet. During the bulk solution's formation into a jet, it sustains transient shear forces as a result of its acceleration. The effect, however, is generally very limited in duration (<1 second or, in some cases, <0.1 seconds). Because the needle-free device is, of necessity, held in close proximity to the skin and a high jet velocity is created in order to penetrate the skin, there is a minimal surface area created before the liquid jet penetrates the skin, and it is created for a very short duration. During skin penetration itself, the shear forces can generally be expected to be equal to or less than those created during jet formation and of a similar duration.

One example of a prefilled, single-use, disposable, needle-free delivery system, Intraject, has been tested using a variety of protein formulations, and there has been little or no adverse effect on protein integrity during delivery (rhGH, erythropoietin (EPO), G-CSF, rIFN-α-2b, and a monoclonal antibody (mAb)) (4,10). For this system, delivery is achieved in <0.1 seconds. The lack of damage to proteins is also evident from pharmacokinetic (PK) data as shown, for example, in Figures 1 and 2 in which NFI has nearly identical PKs to needle and syringe injection.

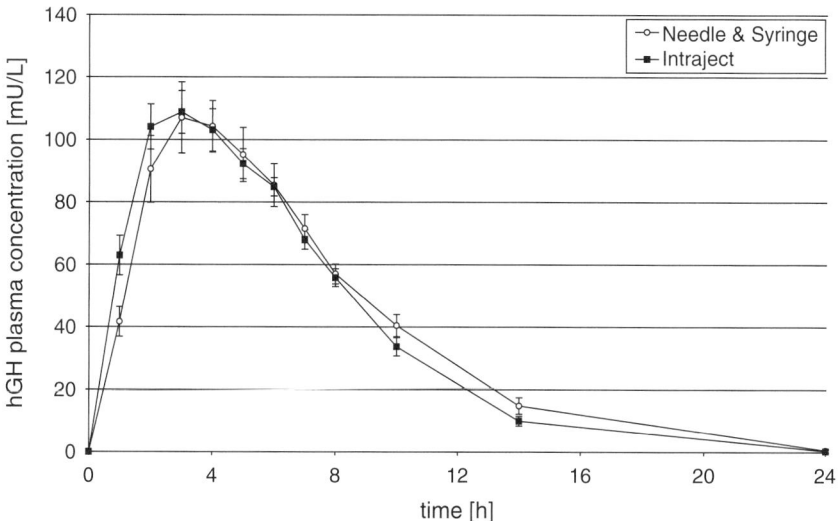

Figure 1 Concentration-time profiles of single-dose rhGH administered by Intraject and needle and syringe (healthy subjects).

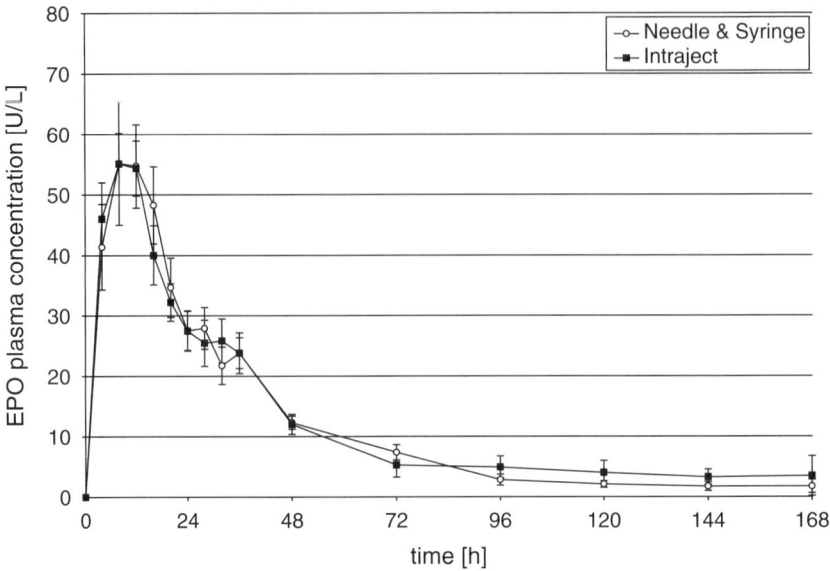

Figure 2 Concentration-time profiles of single-dose rhEPO administered by Intraject and needle and syringe (healthy subjects).

Effect of Viscosity on Injection

Because only a fixed volume, usually less than 1.5 mL, can be administered in a single injection, it is often necessary to deliver viscous formulations. This is because a protein solution can be highly concentrated to achieve the desired dose in a single injection, as in the case of mAbs, or a controlled release vehicle is used, which will provide a sustained effect as in the case of a viscous polymer solution. In these instances, NFI can provide a distinct advantage over needle and syringe or pen injectors by delivering viscous formulations over significantly shorter time (Fig. 3). The data in Figure 3 were generated using a 23-gauge needle. For smaller-diameter needles (higher gauge), the rate of delivery through the needle will be even slower. To date, there has not been a significant amount of information published on the delivery of viscous formulations by NFI; however, this is an area in which needle-free delivery can offer a distinct advantage.

Materials of Construction and Compatibility

Protein formulation for NFIs is determined in part by the materials used in the delivery device container-closure system. This is true both for prefilled disposable systems in which long-term storage of the formulation is required and for patient-filled reusable systems in which the formulation is in contact with device materials for a relatively short period of time. In either case, extensive charac-

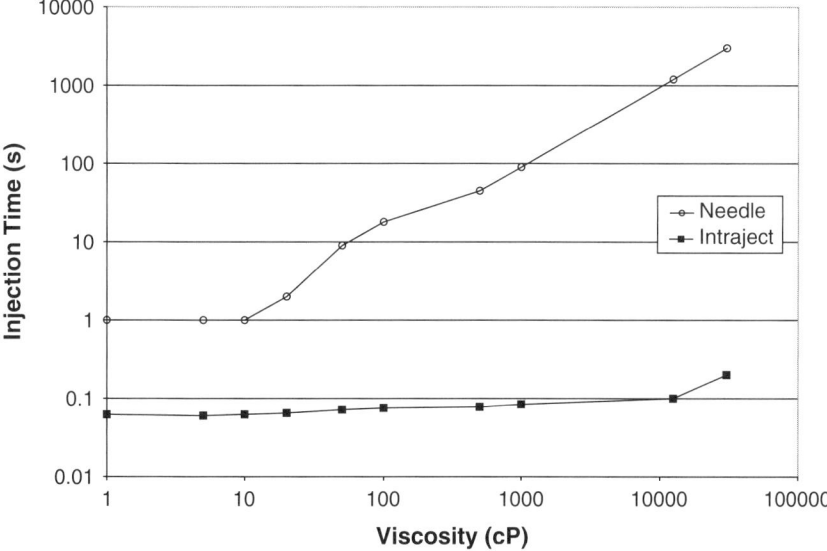

Figure 3 Injection time as a function of viscosity for Intraject NFI and 23g needle. *Abbreviation*: NFI, needle-free injection.

terization of the formulation-contact materials is required with similar, or identical, requirements to conventional needle-syringe or pen-injector systems. From a formulation perspective, this is generally assessed initially using stressed or accelerated studies and, as development progresses, as part of long-term stability studies. In the United States, the requirements for assessing material interaction with the formulation is generally determined by delivery route and formulation type (Table 2). As for any injectable, the container-closure system must be compatible with the formulation; it must also provide a sterile seal against the outside environment. As required by the formulation, it must also provide an oxygen barrier and light protection. For a needle-free delivery system, as for a prefilled syringe, the container-closure must also be compatible with the functionality of the injection system. This includes accommodating moving parts, high pressures during delivery, and the generation of a liquid jet, typically through an orifice.

Prefilled Disposable Systems

Among prefilled, single-use, disposable systems, Intraject is furthest advanced in development. The primary formulation contact material used in Intraject is U.S. Pharmacopoeia Type I borosilicate glass, an industry standard. By using this type of glass and other similarly standard materials, there are few limitations on the type of formulation that may be used in the system as compared with conventional prefilled systems. The challenging part of delivery system design then becomes ensuring that the formulation container is compatible with device function.

Table 2 Container-Closure Assessment by Delivery Route and Formulation Type

Degree of concern	Likelihood of packaging component-dosage form interaction		
	High (Solutions/Suspensions)	Medium (Powders)	Low (Tablets/capsules)
Highest			
Pulmonary	Inhalation aerosols and solutions	Sterile powders and powders for injection	
Parenteral	Injections and injectable suspensions	Inhalation powders	
High			
Ophthalmic	Ophthalmic solutions and suspensions		
Transdermal	Transdermal ointments and patches		
Nasal	Nasal aerosols and sprays		
Low			
Topical	Topical solutions and suspensions; topical and lingual aerosols	Topical powders; oral powders	
Oral	Oral solutions and suspensions		Oral tablets and oral (hard and soft gelatin) capsules

Source: From Ref. 8.

For Intraject, this meant improving the strength of the glass to make it capable of withstanding the pressures necessary to create a liquid jet that can reliably penetrate the skin and produce NFI.

Other systems have used different approaches. For example, the Crossject system (Crossject Medical Technology, Paris, France) has a glass container in which the protein formulation is stored and a plastic chamber into which the formulation is dispensed prior to delivery. This separates the performance requirements from the container-closure requirements of long-term storage, simplifying the engineering design problem (at least from a materials selection standpoint). The delivery system, however, is less convenient for end users because of the additional steps required to move the liquid from one container to another prior to delivery.

As often occurs in drug delivery technology, the delivery system is developed independently of the formulation, and the protein formulation must be made compatible with the system. This is frequently true for needle-free delivery systems because of the extensive amount of engineering work that goes into creating a reliable and easy-to-use, needle-free delivery system. As a result, it is clearly advantageous to design the system using materials that are of standard construction to minimize the amount of compatibility work required from the formulation perspective.

Patient-Filled Durable Systems

These systems have much shorter contact times with the injected formulation but require a complicated (sometimes a highly complicated) series of steps by the patient for use. They are often approved by regulatory agencies as durable devices not associated with a specific drug or formulation for delivery (i.e., a 510K filing). From a consumer perspective, the filling requirements combined with the fact that NFI generally has similar pain and tolerability to needle injection means that these systems have not gained wide acceptance.

CLINICAL EXPERIENCE WITH NFI OF PROTEINS

Clinical Considerations for Needle-Free Delivery of Proteins

Despite its well-known drawbacks, injection remains a highly appealing route of delivery of therapeutics from a clinical perspective, especially for proteins, for a variety of compelling reasons. These include: (*i*) bypass of the degradative effects of enterohepatic cycling first-pass metabolism of susceptible molecules, (*ii*) partial or complete avoidance of gastrointestinal sequelae, (*iii*) ease, speed, and convenience of delivery, and (*iv*) the favorable absorption and distribution kinetics associated with SC capillary tissue uptake. The great majority of therapeutic proteins are delivered subcutaneously, intravenously, or intramuscularly, with current portfolios and pipelines dominated by the product classes of cytokines [e.g., EPOs, interferons (IFNs), CSF, interleukins], hormones (e.g., insulin, growth hormone, fertility hormones), and other therapeutic proteins including mAbs and vaccines. Table 3 includes representatives of each of these classes, some of which have already been developed into NFI products or are strong candidates for this technology. Many of these indications—e.g., anemia, rheumatoid arthritis, multiple sclerosis (MS), chronic hepatitis, and diabetes—require a chronic injection-dosing regimen, a prospect highly unpleasant for most patients.

The challenge of offering solutions for patients apprehensive of needles while at the same time optimizing treatment administration for chronic conditions that require this route of delivery, coupled with the alarming worldwide incidence of sharps-associated blood-borne pathogen transmission, has long prompted the development of NFI methods. Because prospective needle-free products are frequently developed on the back of legacy needle and syringe products (insulin and

Table 3 Injectable Proteins and Routes of Delivery

Trade (generic name)	Manufacturer	Indication(s)	Route
Activase (alteplase) [glycoprotein]	Genentech	Acute myocardial infarction	IV
		Acute ischemic stroke	
		Pulmonary embolism	
Aranesp (darbepoetin alfa)	Amgen	Anemia	SC/IV
Avonex (interferon beta-1a)	Biogen Idec	Multiple sclerosis	IM
Avastin (bevacizumab)	Genentech	Colon or rectal carcinoma	IV
Betaseron (interferon beta-1b)	Berlex	Multiple sclerosis	SC
Enbrel (etanercept)	Amgen and Wyeth	Rheumatoid arthritis	SC
		Psoriatic arthritis	
		Ankylosing spondylitis	
		Plaque psoriasis	
Herceptin (trastuzumab)	Genentech	Breast cancer	IV
Humalog (insulin lispro)	Lilly	Diabetes	SC
Humira (adalimumab)	Abbott	Rheumatoid arthritis	SC
		Psoriatic arthritis	
		Ankylosing spondylitis	
Humulin (human insulin)	Lilly	Diabetes	SC
Intron A (interferon alfa-2b);	Schering	Malignant melanoma	SC, IM, IV, intralesional
		Hepatitis B/C	
		Condyloma acuminata	
		Follicular/non-Hodgkin's lymphoma	
		AIDS-related Kaposi's sarcoma	
		Hairy cell leukemia	

PEG-Intron (PEG-interferon alfa-2b)	Schering	Chronic hepatitis C	SC
PEG-Intron (PEG-interferon alfa-2b)			SC
NeoRecormon (epoetin beta)	Roche	Anemia	SC
Neulasta (pegfilgrastim)	Amgen	Chemotherapy-associated neutropenia	SC
Neupogen (filgrastim)	Amgen	Chemotherapy associated neutropenia	SC, IV
Nutropin (somatropin)	Genentech	Growth hormone deficiency (adult and children)	SC
		Growth failure	
		Short stature (Turner's syndrome and idiopathic)	
Novolog/Novorapid (insulin aspart, human insulin analog)	Novo Nordisk	Diabetes	SC
Pegasys (peginterferon alfa-2a)	Roche	Chronic hepatitis B/C	SC
Procrit, Eprex (epoetin alfa)	Ortho Biotech and Janssen-Ortho (Johnson & Johnson)	Anemia	SC/IV
Raptiva (efalizumab)	Genentech	Plaque psoriasis	SC
Rebif (interferon beta-1a)	Serono and Pfizer	Multiple sclerosis, relapsing	SC
Remicade (infliximab)	Centocor	Rheumatoid arthritis Chron's disease Psoriatic arthritis Ankylosing spondylitis Plaque psoriasis Ulcerative colitis	IV
Rituxan/Mabthera (rituximab)	Genentech/Biogen idec	Non-Hodgkin's lymphoma Rheumatoid arthritis	IV
TNKase (tenecteplase) [glycoprotein]	Genentech	Acute myocardial infarction	IV
Xolair (omalizumab)	Genentech	Asthma	SC

Abbreviations: IV, intravenous; SC, subcutaneous; IM, intramuscular.

growth hormone are good examples), an additional challenge frequently encountered in NFI development is to ensure that bioequivalence or biosimilarity to a predicate product is achieved.[a]

Regardless, though, of whether a prospective NFI product is an innovator or is being developed along a 510K pathway, the safety and efficacy of an NFI product depends to a large extent on how well it mimics the reliability and performance of benchmark needle and syringe injection; underdosing or not achieving comparable PKs (e.g., in T_{max}, C_{max}) may lead to an altered or unacceptable safety or efficacy profile. Achieving this benchmark, however, can pose challenges on numerous fronts.

Clinical Validation of NFI Technology

Early clinical validation of an NFI technology, whether in developing a protein therapeutic or other drug class, comes primarily from standard PK/pharmacodynamic (PD) studies, typically to benchmark key PK parameters and assess bioavailability and dosimetry relative to other routes of delivery [e.g., a dosage form expansion from IV to SC], assess bioequivalence or biosimilarity relative to needle and syringe, and to evaluate adequacy of biomarker response. Imaging studies also provide a highly useful tool to show precise tissue deposition of the injectate and to determine if the intended SC, intramuscular, or intradermal injection has been achieved. Device design parameters can be configured during early development to produce the desired depth of injection, primarily by varying injection force (the pressure of compressed gas or spring compression) and nozzle size or other parameters to achieve an optimal combination sufficient to penetrate the epidermis and dermis, drill into fatty tissue, and deposit injectate into the SC space, but with the injectate stopping short of muscle if an SC injection is indicated.

Figure 4 shows data from a preclinical model dosed with a therapeutic mAb using the Intraject NFI system. In this study, there were no significant differences in performance between Intraject and conventional SC injection with respect to PKs and bioavailability. Furthermore, it was demonstrated in a separate study that the mAb structure was not altered during injection. Figure 5 shows an Intraject injection to the SC space of the thigh (human subject).

Data from both pharmacologic and imaging studies are combined to enable an assessment of product feasibility and can also be used to determine if any additional refining to the device configuration is needed.

Liquid NFI Performance and the Patient–Device Interface

How subject variables such as skin type, body mass index (BMI), or ethnicity interact with device design to potentially influence the performance of a device

---

[a] In the case of vaccines specifically, seroconversion rates of an NFI product is an appropriate endpoint for demonstrating equivalence. For other biologics, demonstrating biosimilarity is appropriate for products manufactured by recombinant DNA technology to show clinical similarity to a reference product (such products are also referred to as biogenerics, similar biological products, or follow-ons).

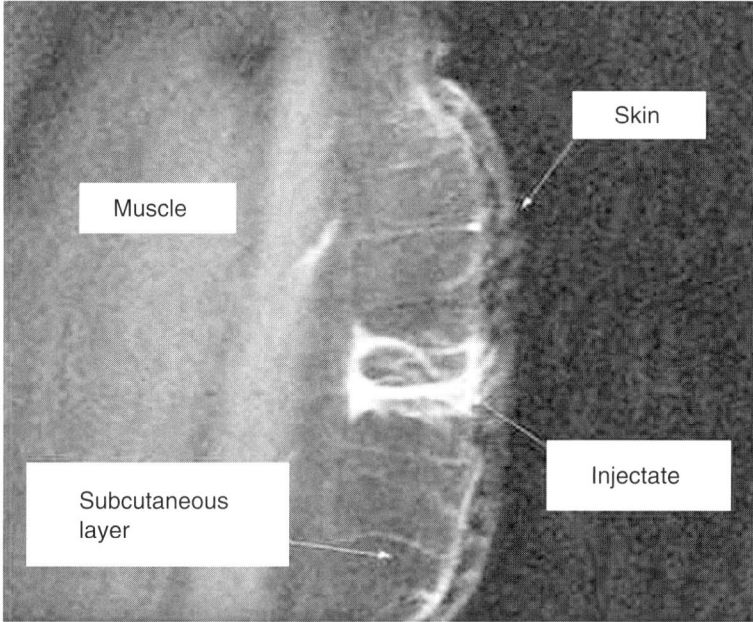

Figure 4 MR image of Intraject subcutaneous delivery of a mAb (animal model). *Abbreviations*: MR, magnetic resonance; mAb, monoclonal antibody.

should also be considered, since safety and efficacy are increasingly being recognized by regulatory authorities as being inextricably linked to device performance (14). For liquid formulations, NFI performance may be measured in terms of the observed frequency of incomplete or "wet" injections, i.e., an injection that disperses a portion or all of the injectate outside of the injection site (disregarding the normal "ooze" of injectate from the injection hole). Injection performance should be assessed across a broad array of individuals of varying body types, ethnicities, skin types, and ages. These factors are of potential importance, since varying tissue composition of SC fat, collagen, elastin, sebum, or other skin constituents may interact at the biology–device interface to potentially influence injection performance.

There are currently little published data clarifying the relationship between subject variables and skin properties and injection performance, but several mechanical and biophysical properties of human skin relevant to transcutaneous drug delivery have been studied, including the physics of forces applied to the skin surface (15–17), the stress, strain, and stiffness characteristics of varying skin composition (proportion of collagen and elastin) (18), the effects of varying physicochemical environments on stress–strain variables (19), and racial differences in how the skin is affected by chemical insult (20). From our own early multivariate analyses of several dozen Intraject prototypes of varying configurations (14),

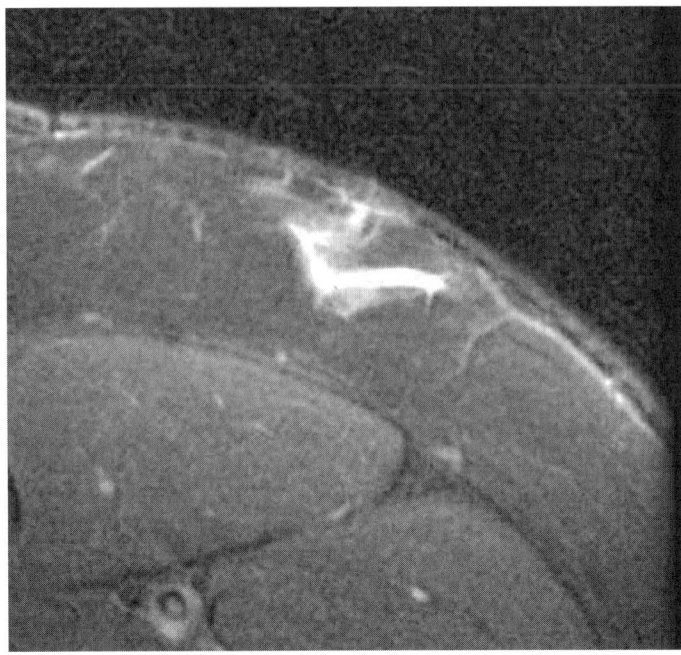

Figure 5   MR image of Intraject subcutaneous delivery (human thigh). *Abbreviation*: MR, magnetic resonance.

it was apparent that two key device design features, orifice size and injection pressure, and two subject variables, subject BMI and site of injection (skin thickness), interact to mediate injection performance with this device. Up to specific limits, greater pressure (gas mass) and a larger orifice size lead to improved injection performance, while a BMI greater than an extreme minimum threshold and injection sites having relatively thinner skin lead to optimal performance (i.e., individuals having more than just minimal SC fat and injection sites where there is relatively thin skin, such as the lateral abdomen). Device optimization of Intraject has led to injection performance of >98%. No relationship was found between injection performance and subjects' Fitzpatrick skin type, sex, or age (14).

Immunogenicity

Immunogenic response, i.e., antibody production, is almost certain to occur to some extent when using any parenteral protein, and the skin is a particularly effective immunologic barrier organ that has evolved to be highly responsive to frequent environmental insults as it has a rich concentration of dendritic and other antigen-presenting cells (APCs)—so it is perhaps not surprising that SC delivery of proteins generally provokes a greater response compared to IV delivery. Immunogenic response, however, varies dramatically both across and within parenteral protein products in terms of the observed incidence between individuals, resulting antibody titer levels, and clinical significance (21). Further, immunogenic potential is generally less for proteins structurally closer to self-proteins (e.g., insulin, growth hormone) as opposed to novel proteins, and greater for proteins of increasing size and complexity, when they become aggregated[b], or are repetitively administered (21). Antibody production can affect both safety and efficacy of the product. Antibodies may be neutralizing or nonneutralizing and cross-reactive or non-cross-reactive to corresponding endogenous proteins. Cross-reactive neutralizing proteins probably present the most significant threat to safety and efficacy (21). Although the great majority of protein therapeutic products delivered parenterally are associated with antibody response, it should be emphasized that serious adverse events, such as life-threatening hypersensitivity reactions, remain rare.

The evidence is mixed regarding immunogenic response differences between NFI delivery versus needle and syringe for the same treatment; occasionally, NFI is associated with greater immunogenic response compared to needles (22–25), perhaps due to the more diffuse pattern of injectate deposition among APCs with NFI, but often there is less antibody production or no apparent difference compared to needle and syringe (26,27).

NFI Therapeutic Protein Products and the Development Pipeline

The therapeutic protein market has been dominated by recombinant DNA products, most notably the EPOs, IFNs, and insulin, and to a lesser extent growth hormone, blood factors, and interleukins. Recombinant DNA products will continue to dominate in the long term in terms of total sales, but are regarded as maturing markets owing to their level of market saturation and consequent slowing forward growth. The area of greatest near- to intermediate-term accelerating growth is expected to be for fusion inhibitors, a class of antiretroviral agents used to treat viral infections, such as HIV, by blocking the virus from fusing with and entering a cell, and CSFs (28) to treat neutropenia following chemotherapy or for other reasons. Examples in this class include Neupogen (filgrastim) and Neulasta (pegfilgrastim). Rapid growth is also expected for mAbs, which are poised to at least double and perhaps triple in market value over the next five to six years, led by treatments for oncology indications as well as for arthritis and immune and inflammatory disorders.

Monoclonal antibodies are also being developed for other indications including respiratory, cardiovascular, and ophthalmology indications.

As already mentioned, the great majority of therapeutic proteins, whether in rapidly growing product classes or in moderating or maturing product classes, require chronic dosing regimens using SC or IV delivery. From the patient perspective, as awareness of the availability of currently marketed NFI products and technology grows, demand for application of this technology to the burgeoning

---

[b] As previously noted, needle-free delivery of various proteins does not result in aggregation which could render them immunogenic.

therapeutic proteins arena is likely to grow as well. Following is a brief survey of currently marketed NFI therapeutic protein products as well as products that are currently under development. Additional comparative information for NFI technologies is provided in Table 1.

Insulin

NFI of insulin has a long history and probably dates back to as early as 1947 when its originator (an engineer) stumbled on the idea when an accidental explosion embedded oil droplets in a victim's skin without marking his skin (29). The idea was eventually developed, and "Hypospray" (30) for administering insulin was introduced into the market in the early 1960s (31) but was discontinued in 1971. The vital need to offer a needle-free option to the many diabetics and for other important patient populations remained and was one of the key factors driving the evolution of NFI technology. Another benefit may lie in evidence that jet injection of insulin has been associated with lower antibody production and less PK variability than when delivered by needle and syringe (26).

Bioject's Vitajet® 3 NFI system (also licensed by The Medical House under the name "mhi-500") is designed for SC self-injection of insulin. Like Bioject's B2000, the Vitajet has a reusable injector and a disposable nozzle and uses a spring-based power source. The Vitajet 3 received Food and Drug Administration (FDA) marketing approval for SC injections of insulin in 1996.

Antares Pharma (Antares Pharma, Inc., New Jersey, U.S.A.) is the developer and manufacturer of the Medi-Jector VISION® needle-free insulin delivery system. Antares, formerly known as Medi-Ject Corporation, was founded in 1979 and changed its name to Antares Pharma, Inc. in 2001. Medi-Jector devices were introduced for insulin injections in 1979 and are powered by metal springs.

The J-Tip Injector was developed and manufactured by National Medical Products, Inc. (California, U.S.A.) It is a single use, presterilized, disposable syringe incorporating a carbon dioxide power source and is filled in a similar manner to needle and syringe. A J-Tip Transporter allows insulin to be taken from the bottle and filled into J-Tip Injectors using a J-Tip Adapter, which is attached to an insulin bottle, allowing the transporter to be filled with the patient's daily-anticipated insulin dosage requirements. It is provided as a sterile disposable, and is discarded when the medication bottle is emptied. A J-Tip Cartridge Transfer System is an accessory incorporating the prepackaged Novo insulin cartridge, allowing the patient to easily fill the injector from the Novo insulin vials. This unique product accessory is designed to enhance use of the Squibb Novo insulin cartridge design used in the Squibb Novo pen.

Visionary Medical Products Corporation (VMPC) (California, U.S.A.) established PenJet® Corporation to manufacture its patented PenJet needle-free drug delivery systems. VMPC was founded in 1993. Among the devices that VMPC has licensed to some of the world's largest medical product manufacturers are its insulin pens, insulin pens with memory, and pocket-size insulin pens combined with a blood glucose meter. PenJet is a small, inexpensive, needle-free, disposable jet injector that is prefilled with a single dose of medication and powered by self-contained compressed inert gas. PenJets are available for SC, intradermal, and some intramuscular injections. Dosage sizes range from 1.0 to 0.1 mL. Standard-size PenJet ampoules hold 0.5 mL. Prefilled PenJet ampoules are inserted into a sterile PenJet housing immediately after filling. If desired, the self-contained PenJet gas canister can be inserted into the device just prior to its use.

Growth Hormone

There are now at least six NFI human growth hormone (hGH) products marketed globally, a few of which come from the same manufacturer and use essentially the same underlying NFI technology platform, but are branded under different names for different indications or for distribution in different regions worldwide. Growth hormone is produced by recombinant DNA technology and is referred to as somatropin, or rhGH, to distinguish it from endogenous hormone. It is administered by daily SC injection to children and adults diagnosed with growth hormone deficiency (GHD/AGHD) small for gestational age, Prader-Willi syndrome, chronic wasting syndrome associated with advanced HIV disease, Turner's syndrome, or short-bowel syndrome (although not all of these indications have NFI products currently available).

Cool.click™ is an NFI system designed for delivery of Serono's Saizen® rhGH and is a customized version of Bioject's Vitajet 3 NFI system for children with GHD. It uses an internal spring to power injections. Cool.click was the first needle-free delivery system for hGH injection approved by FDA in June 2000. The system includes customized dosage features to deliver variable doses of Saizen and was designed to make the injector attractive and nonthreatening to children. Pediatric patients indicate an overall preference for the Cool.click system over syringes and it creates less discomfort (32).

Bioject's SeroJet™ NFI system is designed for delivery of Serostim®, Serono's high-dose rhGH formulation and is indicated for wasting syndrome associated with HIV disease. Serostim was approved by FDA in 2001.

Antares Pharma, Inc. has partnered with specialty biopharmaceutical companies to use its Medi-Jector NFIs as the platform, branded under various trade names, for the delivery of growth hormone as well as insulin. Ferring Pharmaceuticals BV currently markets the ZOMAJET® 2 VISION in Europe, Ferring's trade name for the Medi-Jector VISION device loaded with Ferring's hGH product ZOMACTON®. The Twin-Jector® EZ II with hGH is marketed by JCR Pharmaceuticals Co. Ltd. for distribution in Japan, and the SciTojet2™ injector is marketed by SciGen Ltd. (Gateway East, Singapore) in Asia.

BioPartners GmbH (Baar, Switzerland), a privately owned biopharmaceuticals company, and the Medical House PLC (Attercliffe, Sheffield, U.K.) have established a long-term relationship for use of The Medical House's GH1 reusable, spring-powered, needle-free delivery system for delivery of BioPartners' rhGH product, Valtropin® (codeveloped with LG Life Sciences of Korea) for Turner's syndrome and GHD in children. Marketing approval for the European

Union (EU) from the European Commission for Valtropin has been granted; it is only the second biosimilar product to receive EU marketing approval from the European Commission. Biopartners has in-licensed Valtropin from LG Life Sciences and has commercialization rights for Valtropin in Europe, Japan, and other parts of Asia.

Single-dose administration of rhGH was studied in the Intraject injector in 13 healthy males for PK, safety, and tolerability comparison to conventional needle and syringe (3.84 mg somatropin in 0.48 mL in both injectors) (33). Injections from both devices were found to be safe and well tolerated, and Intraject achieved bioequivalence relative to the same dose delivered by conventional needle and syringe (Fig. 1).

Fuzeon

Fuzeon (enfuvirtide), currently under development by Roche Pharmaceuticals and Trimeris, Inc., is a linear 36-amino acid synthetic peptide inhibitor of the fusion of HIV with CD4⁺ cells. Unlike other HIV drugs that work after HIV has entered human immune cells, Fuzeon works outside the CD4⁺ cell, blocking HIV from entry. It is intended for patients who have developed resistance to other HIV treatments, though it is acknowledged that resistance to Fuzeon could also develop. It is given twice daily, subcutaneously. Roche and Trimeris, Inc., are developing the drug for use with the B2000. The FDA recently acknowledged that similar blood levels had been achieved compared to standard needle–syringe. However, FDA indicated that a small number of certain adverse events related to administration with the B2000 device (hematomas and nerve pain) warrant review of additional information in order to better characterize the incidence of events. Some of these events were associated with use of B2000 to deliver Fuzeon either in close proximity to bone joints or into scar tissue.

Erythropoietin

There are currently no marketed NFI-EPO products, although daily SC EPO injections via NFI in dialysis patients have been reported (34). Single-dose administration of recombinant epoetin alpha given by the Intraject NFI versus conventional needle and syringe was studied in an open-label, crossover study in 14 healthy males subjects to compare PK profiles, relative bioavailability, PDs (reticulocyte count), and safety and tolerability between the two types of injectors (35). The PKs of EPO administered by Intraject were found to be similar to that of EPO administered by conventional needle and syringe (Fig. 2). Injections from both systems were safe and well tolerated, and a transient increase in the average absolute reticulocyte count was observed with both injection systems.

Interferons

There are currently no marketed IFN-containing NFI products. IFN-α has been studied in jet injection for treatment of Palmar and plantar warts (36,37), and IFN-γ has been studied for use in lepromatous leprosy (38). IFN-β-1b is an approved treatment for use in MS, but promising NFI product development efforts for this indication have at least been temporarily thwarted by various regulatory and clinical hurdles. IFN-β-1b is self-administered every other day by SC injection for MS.

An additional therapeutic area that is promising for NFI product development which requires chronic SC or intramuscular dosing with IFNs is treatment of chronic viral hepatitis. IFN-α-2a (Roferon-A), IFN-α-2b (Intron-A), and IFN alfacon-1 (Infergen) are approved for the treatment of adults with chronic hepatitis C and treatment is administered for six months to two years. Less-frequent weekly dosing is available with pegylated versions of IFN, pegylated IFN-α-2b (PEG-Intron), and pegylated IFN-α-2a (Pegasys). IFN-α-2b is effective in the treatment of adults with chronic hepatitis B virus infection and treatment is for 16 weeks or longer.

Vaccines

Perhaps NFI is best recognized for its long history of association with vaccination. Gun-type jet injectors were ideally suited to providing quick and efficient mass immunization from bulk multidose vials to the military and civilian populations, but the practice was generally abandoned after several decades of use due to concerns over rare instances of cross-contamination (39,40).[c] The need to provide safe, large-scale immunization capabilities is greater now than ever before as acutely heightened concerns over bioterrorism, pandemic disease, and disease eradication are forefront in the consciousness of the health officials, the media, and the public. Also, the minimum number of immunizations recommended for children and adolescents in the United States has risen from 8 in 1989 (41) to now 14 in 2006 (42).

Most vaccines are currently delivered using needle and syringe, are in the 1 mg or less dosage range, and do not require precision in dose titration. As previously discussed, the skin is a good immunologic organ having a rich concentration of dendritic and other APCs. Some reports have shown that NFI produces equal or greater seroconversion rates for some vaccines compared to needle and syringe (22–25), perhaps because of a more diffuse distribution of formulation achieved in the SC tissue with NFI. Other reports, however, suggest no difference in immunogenicity between injection methods when delivering plasmid DNA vaccines in nonhuman primates (27).

There is considerable ongoing developmental activity in NFI vaccines, especially in the areas of influenza, chronic viral infections, and oncology. PowderMed's immunotherapy program uses their proprietary DNA-PMED technology as previously described. DNA vaccine is delivered directly to the immunologically rich area of active APCs in the epidermis. Data would suggest that significantly lower

---

[c] The latest generation of high-speed jet injectors currently under development and designed for large-scale vaccination initiatives now eliminate the potential for cross-contamination. Examples include the Lectraject® (DCI, Syracuse, New York, U.S.A.), which uses a disposable nozzle and an injection technology developed by Felton International (Kansas City, Kansas, U.S.A), which uses a shield to prevent backflow of blood into the injector, should bleeding occur.

doses of vaccine are required using their technology when compared with intramuscular injection for an equivalent immune response (43,44).

PowderMed is developing therapeutic vaccines (a new type of vaccine that uses the individual's immune system to mount an immunogenic response and thereby produce the therapeutic effect) in the areas of annual and pandemic influenza, chronic viral diseases, and cancer. The influenza vaccines are in clinical development for prophylaxis of seasonal and avian flu. In addition, PowderMed has several other development programs, some in partnership, to test vaccines against genital herpes (herpes simplex type 2), hepatitis B, genital warts (human papilloma virus), HIV/AIDS, and lung cancer (non–small cell lung cancer).

The B2000 is being used in clinical investigations of melanoma vaccines being developed by Memorial Sloan-Kettering and in AIDS. Completed preclinical work has shown that the B2000 showed improved immunogenicity in melanoma. In another study recently presented by the Karolinska Institute of Sweden at the AIDS Vaccine 2006 International Conference, the use of the B2000 showed promise in an initial HIV vaccine clinical study, which included intramuscular and intradermal injections in 40 patients. The B2000 is also being utilized by the National Institutes of Health in human trials of plasmid vaccines for HIV and the Ebola virus (45).

Tolerability, Patient Acceptance, and Reliability

Tolerability and Patient Acceptance

Tolerability of injection is generally measured along the dimensions of pain (Visual Analog Scale) and soreness, and local injection site reactions ("reactogenicity") such as swelling, erythema, bleeding, and bruising, while patient acceptance is generally measured in terms of preference versus needle and syringe. NFI should not imply pain free, and some sensation will be perceived by most patients. NFI has similar levels of discomfort compared to needle systems in children and young adults (46), although in general, injection into SC tissues is not usually a highly painful experience (47). As pain and local injection site reaction can be expected to vary with the type of needle or pen injector used, based primarily on the size of the needle, similarly, they can be expected to vary with the NFI device used, based primarily on the size of the orifice as well as the velocity of the liquid jet during injection.

Pain and local site reactions can derive from the immediate tissue trauma and nociception caused by injection as well as by any subsequent immunogenic or inflammatory response to the formulation. There is some evidence that NFI delivery of proteins or other formulations, having greater tissue diffusion, is associated with a somewhat greater local injection site reaction compared with needle and syringe (24,48–50).

In one moderately large immunization study (22), the pain, safety, and immunogenicity of an influenza virus vaccine administered by NFI (Bioject) compared to needle and syringe was studied in 304 healthy young adults. Three different doses were administered via either of two Bioject needle-free devices or by needle and syringe. Higher pain levels at the time of vaccination were associated with female subjects and with NFI, and these factors were also associated with local injection site reactions following vaccination. Immune response did not vary significantly by dosage, but administration by one NFI device was associated with higher post-vaccination H1N1 antibody titers. Other studies have found no differences in pain ratings (or immunogenicity) between NFI and needle and syringe (51).

Needle-phobia is, not surprisingly, a particular concern in most children requiring parenteral treatment, and studies that have investigated children's attitudes for a needle-free option show their consistent preference for the needle-free mode of administration (49,52,53). Similarly, patients who use insulin jet injectors show that despite reported problems with the injectors, 70% still preferred taking insulin by jet injector (54). Pooled clinical study data on Intraject suggests an overwhelming preference for the NFI versus needle and syringe even though pain ratings between the two are equivocal (55).

Ultimately, patient acceptance (preference) for NFI does not appear to be primarily related to reduction in pain experienced, but rather, NFI is preferred over needle and syringe primarily for psychological reasons pertaining to needle-phobia or needle-aversion. None of the current devices are pain free, but technology that reduces pain or reduces the perception of pain is likely to increase compliance and product acceptance (56).

Clinical Reliability

The reliability of NFI delivery has evolved considerably from early injectors, which had as much as a 10% to 20% unsuccessful injection rate (57). There is little published data reporting specifically on the reliability of modern NFI systems, but this mode of delivery has been in gradually increasing clinical use for over 50 years, beginning with multiuse jet injectors and later with the development of single-use injectors or those using disposable cartridges, with no reports of major complications. Incomplete injection, i.e., delivery of less than the intended dose of injectate, can occur with jet injectors (47) but can also occur with standard needle and syringe. In one small study of 19 diabetic children using needle and syringe insulin administration, insulin loss at the delivery site was observed in 23% of all the injections (58).

The injection performance of Intraject has been studied during development to evaluate the relationship between frequency of wet or incomplete injections, device-related factors, and subject (physiological) variables (14). In all, 26 different device configurations, varying by nozzle size, gas mass, and other parameters, were used to deliver a total of 3211 SC injections into the abdomen of 302 healthy volunteers, and two complementary validated methods were used to determine completeness of each injection (defined as ≥90% dose delivery). The reliability of complete injections ranged from 59% to 98% among the various combinations of device configurations, with the two device parameters (nozzle size and gas mass) and two subject variables (BMI and skin thickness) showing a strong association with injection reliability. It appears that, at least with Intraject, while both device design and subject variables interact to ultimately determine the reliability of the NFI, the device can be designed in such a way as to overcome the subject variables, as the best performing configuration in the study had 98% complete injections.

CONCLUSIONS

Although NFI technology has a long history and can be traced back as far as the mid-1800s, until now, there have been few catalysts to establish this technology more firmly in the drug delivery landscape. There are several reasons to believe that the pipeline of NFI products will expand rapidly, the primary areas being the continued intense development in the recombinant, monoclonal, and other therapeutic protein technology areas for a wide array of oncologic, immune, inflammatory, endocrine, and infectious diseases that generally require chronic parenteral administration regimens. Patient convenience, improved compliance, and the critical need to sharply reduce the global hazard of infection from needle-stick injuries are also likely to play key roles in moving more formulations from needle and syringe into NFI platforms in the years ahead.

There are now a number of different NFI technologies under development. Concerns over the possible degradative effects of shearing forces generated by jet injection of liquid formulations containing macromolecules have not materialized. The amount of pain associated with NFI is consistent with that in a needle and syringe. NFI technology has solved the critical problems of needle-based injury and risk of cross-contamination, and offers those patients who are highly averse to needles, a less distressing option. The risk of cross-contamination that eventually halted widespread use of early generation gun-type multiuse jet injectors for mass immunization campaigns, although small, has now been overcome.

The reliability of NFI has greatly improved over early generation systems as well, with injection performance generally equivalent or superior to needle-based injection. Patient surveys across the various NFI technologies show that the great majority of individuals prefer NFI compared to conventional needle and syringe, despite the slightly higher rate of injection site sequelae associated with NFIs. As patients with chronic disease become increasingly aware of the availability and advantages of NFI products, and as the pipeline of therapeutic proteins continues to expand, there appears to be substantial opportunity for needle-free technologies to contribute to patients' quality of life as well as to help shape the future of the biopharmaceutical industry.

REFERENCES

1. Centers for Disease Control. Public health service statement on management of occupational exposure to human immunodeficiency virus, including considerations regarding zidovudine postexposure use. MMWR 1990; 39.

2. Henry K, Campbell S. Needlestick/sharp injuries and HIV exposure among health care workers. Minnesota Med 1995; 78:41–44.
3. Thompson I. New-generation auto-injectors: completing the scale of convenience for self-injection. PharmaVent Drug Deliv R Autumn/Winter 2005.
4. Hamilton JG. Needle phobia: a neglected diagnosis. J Family Practice 1995; 41:169–175.
5. Gorman S, Brown P. Drug delivery: past, present and future. PharmaVentures, Oxford: The Drug Delivery Companies Report Spring/Summer, 2004.
6. Shergold OA, Fleck NA, King TS. The penetration of a soft solid by a liquid jet, with applications to the administration of a needle-free injection. J Biomech 2006; 39:2593–2602.
7. Whitaker S. Introduction to fluid mechanics. The Physical and Chemical Engineering Sciences. Prentice-Hall International Series. Englewood Cliffs, N.J.: Prentice-Hall, 1968:457.
8. US FDA Guidance for Industry. Container Closure Systems for Packaging Human Drugs and Biologics, Chemistry Manufacturing and Controls Documentation, May 1999.
9. EMEA Guideline on Plastic Immediate Packaging Materials, CPMP/QWP/4359/03.
10. Hlodan R, Uddin S, King T, Edwards S, Varley P. Monoclonal antibody injection without a needle. Br J Pharmac 2000; 131:218.
11. Wang W. Instability, stabilization, and formulation of liquid protein pharmaceuticals. Int J Pharm 1999; 185(2):129–188.
12. Anesiva press release: http://investors.anesiva.com/releasedetail.cfm?ReleaseID=219712.
13. Maa YF, Hsu CC. Investigation on fouling mechanisms for recombinant human growth hormone sterile filtration. J Pharm Sci 1998; 87(7):808–812.
14. Linn L, Boyd B, Iontchev H, King T, Farr S. The effects of system parameters on in vivo injection performance of a needle-free injector in human volunteers. Pharm Res (in press).
15. Pereira JM, Mansour JM, Davis BR. The effects of layer properties on shear disturbance propagation in skin. J Biomechanical Eng 1991; 113:30–35.
16. Pereira JM, Mansour JM, Davis BR. Analysis of shear wave propagation in skin; application to an experimental procedure. J Biomech Eng 1990; 23:745–751.
17. Mridha M, Ödman S, Öberg PÄ. Mechanical pulse wave propagation in gel, normal and oedematous tissues. J Biomech Eng 1992; 25:1213–1218.
18. Manschot JFM, Brakkee AJM. The measurement and modelling of the mechanical properties of human skin in vivo. I. The measurement. J Biomech 1986; 19:511–515.
19. Wildnauer RH, Bothwell JW, Douglass AB. Stratum corneum biomechanical properties. I. Influence of relative humidity on normal and extracted human strateum corneum. J Invest Derm 1971; 56:72–78.
20. Berardesca E, Maibach HI. Is skin color expression of racial differences in skin function? Retrieved (Dec 2006) from www.ulb.ac.be/medecine/loce/espcr/b_iss/Dis-9b.htm).
21. Koren E, Zuckerman LA, Mire-Sluis AR. Immune responses to therapeutic proteins in humans—clinical significance, assessment, and prediction. Current Pharma Biotechnol 2002; 3:349–360.
22. Jackson LA, Austin G, Chen RT, et al. (Vaccine Safety Datalink Study Group). Safety and immunogenicity of varying dosages of trivalent inactivated influenza vaccine administered by needle-free jet injectors. Vaccine 2001; 19(32):4703–4709.

23. Kotloff KL. Needle-free vaccine delivery technologies in development. Center for Disease Control National Immunization Conference, Nashville, Tennessee, May 2004. (WebEx presentation, retrieved on December 12, 2006 at http://cdc.confex.com/cdc/nic2004/wrfredirect.cgi?paperid=5874).

24. Williams J, Fox-Leyva L, Christensen C, et al. Hepatitis A vaccine administration: comparison between jet-injector and needle injection. Vaccine 2000; 18(18):1939–1943.

25. Parent du Chatelet I, Lang J, Schlumberger M, et al. Clinical immunogenicity and tolerance studies of liquid vaccines delivered by jet-injector and a new single-use cartridge (Imule): comparison with standard syringe injection. Imule Investigators Group. Vaccine 1997; 15(4):449–458.

26. Jovanovic-Peterson L, Sparks S, Palmer JP, Peterson CM. Jet-injected insulin is associated with decreased antibody production and postprandial glucose variability when compared with needle-injected insulin in gestational diabetic women. Diabet Care 1993; 16:1479–1484.

27. Rao SS, Gomez P, Mascola JR, et al. Comparative evaluation of three different intramuscular delivery methods for DNA immunization in a nonhuman primate animal model. Vaccine 2006 24(3):367–373.

28. Belsey MJ, Pavlou AK. Leading therapeutic recombinant protein sales forecast and analysis to 2010. J Commercial Biotech 2005; 12(1):69–73.

29. Time Magazine. Shot Without Pain. October 27, 1947.

30. Spiess H. Insulin administration using hypospray injection. Arch Kinderheilkd [German] 1968; (suppl 58):13–14.

31. Weller C, Linder M. Jet injection of insulin vs the syringe-and-needle method. JAMA 1966; 195:844–847.

32. Murray FT et al. Bioequivalence and patient satisfaction with a growth hormone (Saizen®) needle-free device—results of clinical and laboratory studies. Today's Therapeut Trend 2000; 18(1):71–86.

33. Zogenix, Inc. Pilot study to compare the individual blood levels of growth hormone following conventional syringe and novel device administration. Clinical Study Report on file; Study 486/WM, December 1999.

34. Suzuki T, Takahashi I, Takada G. Daily subcutaneous erythropoietin by jet injection in pediatric dialysis patients. Nephron 1995; 69(3):347.

35. Zogenix, Inc. Comparison of pharmacokinetics of recombinant human erythropoietin (epoetin alpha) administered by subcutaneous injection using two different injection systems. Clinical Study Report on file; Study 528/WM, August 2000.

36. Brodell RT, Bredle DL. The treatment of palmar and plantar warts using natural interferon and a needleless injector. Dermatol Surg 1995; 21:213–218.

37. Gibson JR, Harvey SG, Kemmett D, Salisbury J, Marks P. Treatment of common and plantar viral warts with human lymphoblastoid interferon-alpha-pilot studies with intralesional, intramuscular, and Dermo Jet injections. Br J Dermatol 1986; (115 suppl):76–79.

38. Nathan CF. Local and systemic effects of intradermal recombinant interferon gamma in patients with lepromatous leprosy. N Eng J Med 1986; 315(1):6–15.

39. Centers for Disease Control. Hepatitis B associated with jet gun injection—California. MMWR 1986; 35(23):373.

40. Canter J et al. An outbreak of hepatitis-B associated with jet injections in a weight-reduction clinic. Arch Intern Med 1990; 150:1923–1927.

41. Weniger BG. The challenge to immunization practice by increasing numbers of parenteral vaccines. Abstract S26, The First Annual Conference on Vaccine Research, Washington, D.C., May 30, 1998.

42. American Academy of Pediatrics. 2006 Childhood Immunization Schedule and Catch-up schedule (retrieved on December 12, 2006 from http://www.cispimmunize.org/IZSchedule_2006.pdf).

43. Burkoth TL, Bellhouse BJ, Hewson G, Longridge DJ, Muddle AG, Sarphie DF. Transdermal and transmucosal powdered drug delivery. Crit Rev Ther Drug Carrier Syst 1999; 16:331–384.

44. Degano P, Sarphie DF, Bangham CRM. Intradermal DNA immunization of mice against influenza A virus using the novel PowderJect(r) system. Vaccine 1998; 16:394–398.

45. Sheets RL, Stein J, Manetz TS, et al. Biodistribution of DNA plasmid vaccines against HIV-1, Ebola, severe acute respiratory syndrome, or West Nile virus is similar, without integration, despite differing plasmid backbones or gene inserts. Toxicol Sci 2006; 91(2):610–619.

46. Schneider U, Birnbacher R, Schober E. Painfulness of needle and jet injection in children with diabetes mellitus. Euro J Paediatrics 1994; 153:409–410.

47. Bremseth D, Pass F. Delivery of Insulin by jet injection: recent observation. Diabetes Technol Ther 2001; 3:225–232.

48. Mathei C, Van Damme P, Meheus A. Hepatitis B vaccine administration: comparison between jet-gun and syringe and needle. Vaccine 1997; 15:402–444.

49. Dorr HG, Zabransky S, Keller E, et al. Are needle-free injections a useful alternative for growth hormone therapy in children? Safety and pharmacokinetics of growth hormone delivered by a new needle-free injection device compared to a fine gauge needle. J Pediatr Endocrinol Metab 2003; 16:383–392.

50. Levine M. Can needle-free administration of vaccines become the norm in global immunization? Nat Med 2003; 9:99–103.

51. Sarno MJ et al. Clinical immunogenicity of measles, mumps and rubella vaccine delivered by the Injex jet injector: comparison with standard syringe injection. Pediatr Infect Dis J 2000; 19:839–842.

52. Silverstein JH, Murray FT, Malasanos T, et al. Clinical testing results and high patient satisfaction with a new needle-free device for growth hormone in young children. Endocrine 2001; 15(1):15–17.

53. Verrips GH, Hirasing RA, Fekkes M, et al. Psychological responses to the needle-free Medi-Jector or the multidose disetronic injection pen in human growth hormone therapy. Acta Paediatr 1998; 87:154–158.

54. Denne JR, Andrews KL, Lees DV, Mook W. A survey of patient preference for insulin jet injectors versus needle and syringe. Diabetes Educ 1992; 18:223–227.

55. Zogenix, Inc. Pooled Clinical Study Data. Data on file.

56. Dumas H, Parker D, Pongpairochana V. Understanding and meeting the needs of those using growth hormone injection devices. BMC Endocr Disord 2006; 6:5.

57. King T. Needle-free drug delivery. In: Swarbrick J, Boylan JC, eds. Encyclopedia of Pharmaceutical Technology. 2nd ed. New York: Taylor & Francis, 2004:1–12.

58. Stewart NL, Darlow BA. Insulin loss at the injection site in children with Type 1 diabetes mellitus. Diabet Med 1994; 11:802–805.

12

Oral Delivery of Biopharmaceuticals Using the Eligen® Technology

Ehud Arbit, Shingai Majuru, and Isabel Gomez-Orellana

Emisphere Technologies, Inc., Tarrytown, New York, U.S.A.

INTRODUCTION

Today, most biopharmaceuticals are administered parenterally. Their oral bioavailability is negligible, due to poor absorption and rapid and extensive degradation in the gastrointestinal tract. Limited absorption stems mostly from their large molecular size (typically >1000 Da) and, in most cases, also from their low lipophilicity, which limits permeability across the gastrointestinal epithelium (1,2). In spite of these inherent difficulties, the development of oral biopharmaceuticals has been pursued for decades, since the discovery of the first biopharmaceuticals, insulin and heparin, nearly a century ago. Efforts were abandoned after initial limited success; however, the advent of biotechnology and the proliferation of biopharmaceuticals over the past two decades have led to a renewed interest in developing oral forms of peptides and proteins. A number of approaches have been explored, including coadministration with enzyme inhibitors, use of permeation enhancers, and encapsulation in nano- and microparticles (3–7). Some approaches have been abandoned and, to date, only few have shown commercial viability and progressed to clinical studies (8). Success factors in developing viable oral biopharmaceuticals vary from drug to drug but overall include reasonable bioavailability and variability, stability of final product, acceptable cost of goods and manufacturing, and limited side effects and toxicity.

The benefits of oral administration extend beyond patient convenience and facilitated compliance. Oral delivery offers other potential advantages such as availability to larger patient populations, improved therapies, and the possibility for expansion to additional therapeutic indications. This is the case, for instance, for oral calcitonin, which has been found to be a potential disease-modifying agent to treat osteoarthritis in addition to its known osteoporosis indications (9–11). In addition, oral delivery offers more possibilities than other forms of administration for achieving desired pharmacokinetic profiles and for combination therapies. For example, with the same drug, oral forms can be designed to generate rapid onsets with pulse-type pharmacokinetic profiles as well as sustained-release ones. The rapid onset and pulsatile pharmacokinetics that has been obtained with some oral delivery forms mimics the physiological release of hormones and can benefit treatments based on drugs such as parathyroid hormone, growth hormone, or insulin. Lastly, oral administration can be particularly advantageous for drugs that target the liver or benefit from entering the bloodstream via the portal circulation, as it appears to be the case for insulin, incretins, and the so-called "gut hormones" (12). On the other hand, oral administration may not be a suitable approach for drugs with adverse high first-pass metabolism. In these cases, administration in sublingual tablets or using fast-dissolve technologies that avoid portal delivery by targeting absorption from the buccal cavity may be a plausible alternative.

FACTORS INFLUENCING ORAL BIOAVAILABILITY

Rapid and extensive degradation in the gastrointestinal tract is one of the key factors determining the low oral bioavailability of proteins and peptides. Early studies indicated that insulin could be absorbed to some extent through the gastrointestinal tract if protected from enzymatic degradation (13–16). In addition, it has been shown that pegylated proteins that have improved resistance to proteolytic degradation also have higher oral bioavailability than unconjugated proteins (17,18). This observation is somehow puzzling given the large molecular size and hydrodynamic radius of pegylated proteins and has raised speculations about possible specific mechanisms involved in their absorption process. Nevertheless, the oral bioavailability of pegylated proteins is extremely low, typically less than 0.5%, and peglylation alone does not appear to be a viable approach to enable oral absorption.

Peptides and proteins with low susceptibility to gastrointestinal degradation, such as cyclic peptides and peptides containing D amino acids, have some oral bioavailability. For example, apolipoprotein A1 mimetics, D-4F and reversed-D4F, which are made mostly of D amino acids, have a reported oral bioavailability of approximately 4% (19,20). Desmopressin and cyclosporine, two relatively small peptides with low gastrointestinal degradation, have been available in oral forms for over a decade. Their low degradation greatly facilitated the development of oral forms using microemulsion-based formulations. Unfortunately, this formulation approach has not met with much success when applied to other proteins and peptide drugs that have larger molecular size and are more susceptible to presystemic metabolism.

Overall, these examples indicate that proteins and peptides may cross the gastrointestinal epithelium naturally, on their own if protected from degradation, although the extent of this absorption is extremely low. Thus, although gastrointestinal degradation is a major challenge for the oral delivery of biopharmaceuticals, low permeability across the intestinal epithelium has proven to be just as challenging, if not more. Interestingly, most efforts and successful strategies in oral protein delivery have aimed at improving absorption rather than focusing on minimizing degradation. Perhaps this success is due to the fact that the methods used to improve absorption may indirectly contribute to reduce degradation; for example, some absorption enhancers have been shown to inhibit enzymatic degradation. In addition, approaches that improve the rate and extent of absorption also contribute to reduce residence time in the gastrointestinal tract and, consequently, losses through degradation.

There are three main absorption pathways for oral drugs (*i*) passive diffusion, which can be transcellular, through the cell membrane or paracellular, through the intercellular spaces, (*ii*) transcytosis and (*iii*) facilitated transport, which requires the presence of endogenous transporter molecules and, typically, takes place either transcellularly or via transporter-mediated endocytosis. Most drugs of molecular weight lower than 500 to 700 are readily absorbed by passive diffusion, either transcellularly or paracellularly. Transcellular passive diffusion is arguably the most common mechanism of drug absorption. It requires that the drug be sufficiently hydrophobic or lipophilic to partition spontaneously into the cell membrane of the epithelial cells. Transcellular absorption via passive diffusion can be improved by modifying the characteristics of the drug, making it more lipophilic or by modifying the cell membrane with surfactants and/or absorption enhancers capable of increasing membrane permeability. Hydrophilic molecules that cannot partition into cell membranes may be absorbed paracellularly; however, their bioavailability is often low, mainly because the overall surface area of the paracellular pathway is rather limited. Efforts to increase paracellular absorption have typically focused on identifying agents that can open the tight junctions between adjacent cells. This not only increases the overall surface area but also enables the absorption of larger molecules.

Facilitated transport requires drugs to have a chemical moiety that acts as a recognition site for endogenous transporter molecules that can shuttle the drug across the intestinal epithelium. Some transporter-mediated mechanisms have broad specificity with regard to the characteristics of the molecules that they shuttle and can transport relatively large molecules of various physicochemical characteristics. The main challenges of this approach include possible limited or variable distribution of the transporter in the gastrointestinal epithelium and the need to modify the drug molecule to attach the chemical moiety that enables recognition by the transporter.

STRATEGIES TO DEVELOP ORAL BIOPHARMACEUTICALS

The strategies developed over the years to enable oral absorption of biopharmaceuticals have aimed both at circumventing the challenges and at utilizing the different natural mechanisms of drug and nutrient absorption. The main strategies explored to date can be grouped in four categories (*i*) drug conjugates, (*ii*) use of permeation enhancers, (*iii*) micro- and nanoparticles, and (*iv*) coadministration with delivery agents. These strategies are not mutually exclusive; on the contrary, they are often used in combination. In addition, they may be combined with other approaches, primarily coadministration with enzyme inhibitors capable of reducing presystemic degradation and/or formulation with mucoadhesive polymers that can improve absorption by increasing residence time and proximity to the intestinal epithelium.

Drug conjugates aim at improving oral bioavailability through the attachment of a chemical moiety that either increases the lipophilicity and permeability of the macromolecule (21–23) or enables its absorption via a transporter protein or receptor-mediated endocytosis (24). Conjugates must retain biological activity or release the attached molecule after absorption. So far, this approach has met with limited success, in part due to the low oral bioavailability achieved and the manufacturing challenges and cost. Examples of drug conjugates include fatty acid conjugates of insulin, deoxycholic acid conjugates of heparin, and hexyl–insulin monoconjugate 2, a modified insulin conjugated to an amphililic polymer that has shown efficacy in type I and type II diabetic patients (25–28). A similar approach has been applied to the oral delivery of calcitonin (29).

In other cases, drug conjugates have been designed to take advantage of transporter-mediated absorption by attaching to the biopharmaceutical a chemical moiety that is recognized by an endogenous transport system. The glucose transporter, the di- and tripeptide transporters, the bile acid transporter, the vitamin B12 receptor, and the Fc receptor are examples of cellular transport systems that have been explored to enhance the oral absorption of biopharmaceuticals. Some transporters have limitations with regard to the size and characteristics of the molecules that they are able to carry. However, others appear to have broad specificity. For example, it has been reported that the vitamin B12 transporter is able to transport a molecule as large as erythropoietin (30–32).

Permeation enhancers are typically surfactants, bile salts, or medium-chain fatty acids capable of altering the permeability of cell membranes and/or opening tight junctions between adjacent cells, thus increasing absorption across the intestinal epithelium (33–38). Examples of this approach are oral formulations of calcitonin and parathyroid hormone consisting of a combination of absorption enhancers and enzyme inhibitors that are currently being evaluated in clinical trials (39). Absorption enhancers have been applied also to the development of oral forms of an antisense therapeutic (40,41) and heparin (42). In addition, an oral spray that targets absorption from the buccal cavity has been applied to deliver insulin. This product has been approved in Ecuador and is now advancing through clinical studies in other countries (43,44). Many absorption enhancers increase both paracellular and transcellular absorption and often cause permanent opening of tight junctions, which can cause irritation and increase the risk of potential toxicities. Identifying agents that can open tight junctions in a reversible manner, without causing permanent damage to the intestinal epithelium has been challenging (45–50). In addition, recent findings indicating that leaky tight junctions might be linked to the development of type 1 diabetes and other autoimmune diseases have raised concerns about this approach (51–53). Natural polymers such as chitosans have drawn interest as potential agents to enhance the oral absorption of biopharmaceuticals, mainly because of their mucoadhesive properties, ability to increase paracellular absorption, and possibilities for drug encapsulation in micro- and nanoparticles (36,54,55). Results from their application to develop oral calcitonin and oral insulin are encouraging (56,57). Micro- and nanoparticle formulations offer the advantage of protection from degradation in the gastrointestinal tract and possibilities for sustained release in the bloodstream. Nanoparticles may be absorbed paracellularly or through the gut-associated lymphoid tissue. This approach has been investigated primarily for the development of oral vaccines (58–61). Loading capacity, material compatibility, and manufacturing challenges are important aspects to evaluate when considering this approach.

Specific drug delivery agents that interact weakly and reversibly with macromolecules, increasing their lipophilicity and enabling transcellular absorption, were introduced in the mid-1990s (62–66). This approach forms the basis of the Eligen® technology, which has been successfully applied to the development of oral forms of several protein and peptide drugs and has advanced into clinical trials of more biopharmaceuticals than any other approach to date. These include insulin, calcitonin, parathyroid hormone, growth hormone, unfractionated heparin, and low-molecular-weight heparin (3,10,11,67–75). The success stems in part from the versatility of this technology (it has been applied to drugs as diverse as calcitonin and heparin), its scalability, relatively low cost of goods, and uncomplicated formulation and manufacturing. The safety of this technology has been demonstrated in over 100,000 human doses, including studies lasting up to three months. No serious adverse events related to the delivery agents have been reported to date.

CASE STUDY: ORAL INSULIN

Therapeutic Advantages Conferred by Route of Administration

The oral route of administration is regarded as the most convenient, safe, and economical for drug administration. A drug given by this route is absorbed into the mesenteric veins of the gastrointestinal tract to drain into the portal-hepatic venous system bound for the liver. Depending on the drug metabolic pathway or site of action, the effects of the oral-portal route of absorption may vary. Systemic bioavailability of a drug may be significantly reduced if the drug undergoes extensive first-pass metabolism in the liver. Alternatively, for drugs that exert a desirable effect on the liver, the initial exposure of the liver to the drug may confer considerable advantages, a case in point being insulin. In healthy individuals, insulin is secreted by the pancreatic β cells directly into the hepatoportal circulation in response to a glucose load or meal ingestion. Up to 80% of the insulin secreted is cleared by the liver on first pass and directly binds to hepatic insulin receptors to activate signaling pathways (76). The liver, situated in a strategic junction between the stomach, the pancreas, and the peripheral circulation plays an important role in carbohydrate metabolism. It buffers the entry of ingested glucose from the portal vein into the systemic circulation, minimizing plasma glucose excursions and concurrently functioning as a storage depot for glucose in the form of glycogen. In response to a glucose load and subsequent insulin secretion, the liver responds by simultaneous suppressing glucose production and increasing hepatic glucose uptake. The insulin that has not been utilized by the liver (20–50%) is shunted to the systemic circulation where it stimulates glucose utilization by peripheral tissues, predominantly muscle (77–80). The signals that activate the liver to stop glucose output and increase glucose uptake include (*i*) the "portal–peripheral insulin gradient" created when insulin is secreted into the portal vein to reach two- to threefold higher insulin concentrations than in the peripheral circulation; (*ii*) the "portal signal," i.e., the signal induced by the presence of high glucose concentration in the portal vein as compared to the periphery, following meal ingestion, and (*iii*) the distinctive biphasic pattern of insulin secretion consisting of an early phase (5–10 minutes) and a late phase (two to three hours) response (81–85). In variance, subcutaneous insulin, as well as the recently approved inhaled insulin, is delivered into the peripheral circulation resulting in a reversed portal–peripheral gradient, with muscle and fat tissue exposure to higher insulin levels than the liver. As a consequence, the liver is deprived of appropriate activation signals. In addition, peripheral hyperinsulinemia predisposes to hypoglycemia, is thought to be linked to weight gain, and is implicated in increased insulin resistance. Support for the importance of direct insulinization of the liver for attaining a better glycemic and overall metabolic control (e.g., lipids, cortisol) with low risk of hypoglycemia is available but scarce. Studies in humans where insulin was administered directly into the portal vein or into the peritoneal cavity (from which ~ 50% of insulin is absorbed through the portal vein) and from follow-up studies of islet cell transplant recipients support this hypothesis (86).

Oral Insulin Developed Using the Eligen® Technology

The Eligen drug delivery technology relies on the coadministration of biopharmaceuticals with low molecular weight compounds termed delivery agents or carriers. These delivery agents interact weakly and reversibly with the coadministered drug, increasing its lipophilicity and enabling transcellular absorption (65). Versatility and low manufacturing cost are among the main strengths of this approach.

The oral insulin developed using the Eligen technology was viewed from the beginning not only as an opportunity to improve patient convenience and quality of life but also as a possibility for developing improved treatments for diabetes that would benefit from the delivery of insulin into the portal circulation, restoring physiological portal/systemic concentration ratios. Early studies in rodents and nonhuman primates showed that the pharmacokinetic profile obtained with liquid and solid oral insulin formulations mimicked the physiological profile of the early phase of insulin secretion. This raised interest in developing an oral insulin form to treat primarily type 2 diabetes in which the early insulin secretion phase is inhibited but there is still some degree of β-cell function and various levels of insulin secretion in the second phase. The potential of oral insulin to slow down the progression of type 2 diabetes by restoring the early phase of insulin secretion has been speculated and this certainly warrants further studies. The first studies in healthy volunteers were done using hard gelatin capsules filled with a physical blend of insulin and delivery agent. The results confirmed the ability of this oral form to mimic the early phase of insulin secretion and demonstrated concomitant decrease of blood glucose and c-peptide levels, confirming that the exogenous insulin that was administered orally was bioactive and capable of lowering blood glucose and suppressing endogenous insulin secretion (Fig. 1) (3,71). Similar results were observed in subsequent studies in type 2 diabetic patients. Tablet formulations were developed and tested for their ability to control post-meal blood glucose excursions in diabetic patients. These studies demonstrated the effectiveness of oral insulin tablets when taken with meal or 10 minutes earlier. A multidose, 14-day study in which patients received oral insulin four times a day, 10 minutes before meals and at bedtime, evaluated the safety, efficacy, and tolerability of oral insulin compared to placebo in type 2 diabetics treated with diet alone. Oral insulin was well tolerated and led to improvements in postprandial glucose excursions, both under oral glucose tolerance test and standardized meal (Fig. 2). Improvements in fasting glucose levels and insulin sensitivity were observed as well. No hypoglycemic events were observed. Additional studies are ongoing, including a 90-day study in type 2 diabetic patients treated with oral agents to evaluate the effect of oral insulin on hemoglobin A1C levels—a marker of long-term stability of glucose levels—as well as weight gain.

The results from clinical and preclinical studies indicate that the oral insulin developed using the Eligen technology rectifies two important features lacking in parenteral insulin administration: (*i*) It restores the physiological pathway of endogenous insulin secretion into the portal venous system to facilitate the liver's priming and engagement in the process of glycemic control and (*ii*) With its unique pharmacokinetic profile of rapid onset and nondepot effects, it simulates the early phase of insulin secretion. The latter may be particularly beneficial for patients with type 2 diabetes, which typically have a blunted early-phase response and a progressive loss of secondary-phase response. The loss of the early-phase insulin response is normally the first step in the development of the disease and has significant metabolic implications. Its correction may actually normalize hepatic response to insulin and possibly slow down the progression of the disease. These findings led to focus efforts on developing oral insulin formulations capable of generating pharmacokinetic profiles similar to the early insulin phase. Nevertheless, the Eligen technology is not limited to this type of pharmacokinetic profiles and efforts are under way to develop oral formulations capable of

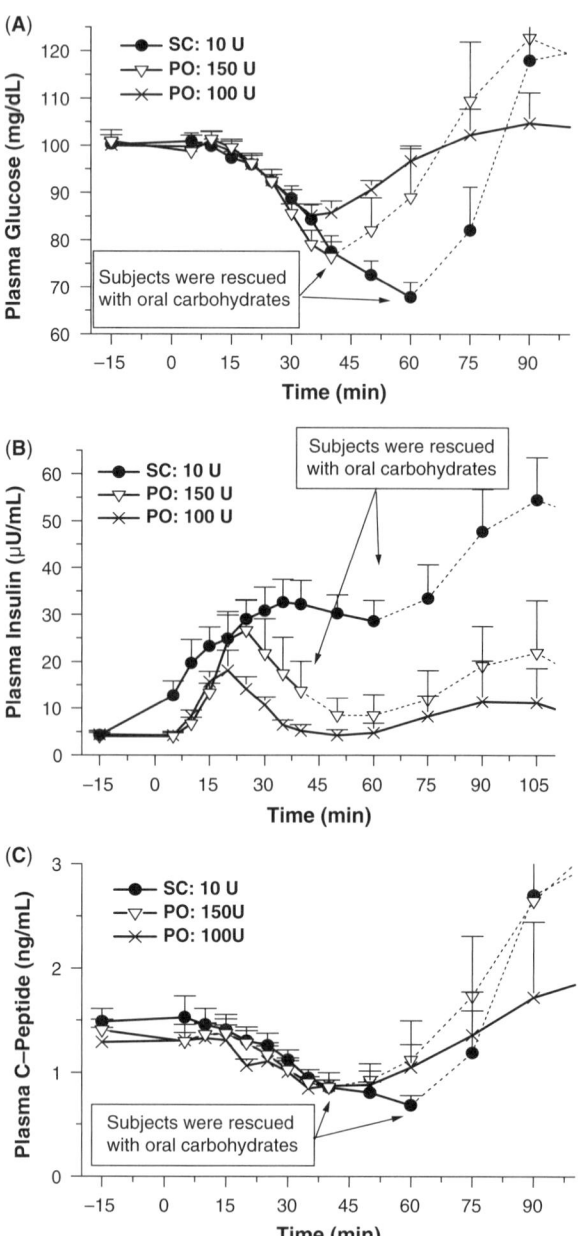

Figure 1 Effect of Eligen® oral insulin on fasting plasma glucose (**A**), insulin (**B**), and c-peptide (**C**) levels in a clinical study in healthy volunteers (*n* = 8 per group). Subjects in the subcutaneous insulin group and in the high oral insulin group had to be rescued with oral carbohydrates (*dotted lines*) because their glucose levels dropped by more than 30%, which was the safety level established by the study protocol. No changes in insulin, glucose, or c-peptide levels were observed on the placebo groups that received oral doses of insulin alone or vehicle alone.

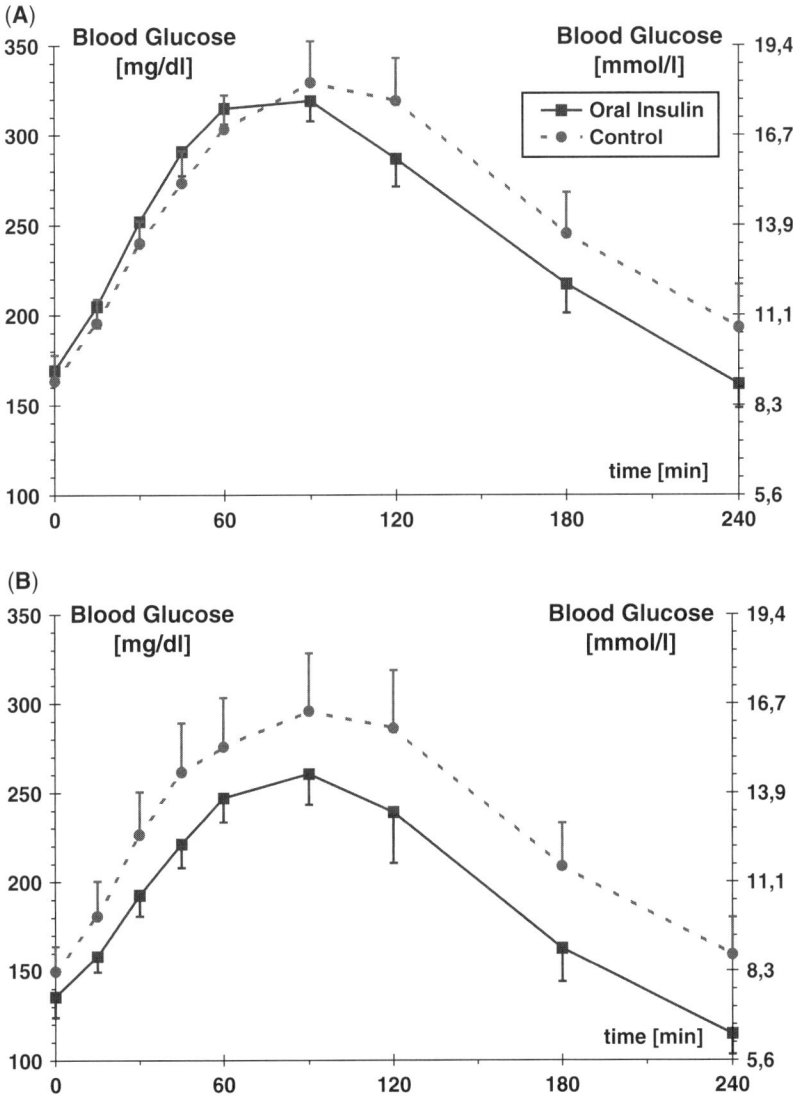

Figure 2 Impact on blood glucose excursions of a two-week treatment with oral insulin in type 2 diabetic patients controlled with dietary treatment only. The study included 13 patients, 7 treated and 6 placebo. Each subject in the treatment group received 300 IU insulin and 160 mg of delivery agent four times a day (administered as two tablets 10 minutes before the main meals and before bedtime) over 14 days. Subjects in the placebo groups followed the same dosing regimen but their tablets contained only the vehicle, 200 mg of delivery agent. The graphs show the blood glucose profiles during an OGTT (**A**) at day 0, before the treatment and (**B**) at day 14, end of the two-week study. A significant reduction of glucose levels were observed after the study in the treated group, whereas no significant changes were observed in the placebo. *Abbreviation*: OGTT, oral glucose tolerance test.

generating sustained insulin levels similar to those of the second phase of secretion. These examples illustrate the advantages of oral delivery with regard to the possibilities for a variety of pharmacokinetic profiles using the same delivery technology.

FORMULATION CONSIDERATIONS

As mentioned earlier, the success of oral delivery approaches depends not only on their ability to enable or increase oral drug bioavailability but also on their ability to generate a commercially viable product. This depends in part on the technology, or approach itself, and part on manufacturing- and formulation-related aspects, such as scalability, manufacturing costs, and stability and tolerability of final formulations. The Eligen drug delivery technology has been shown to be amenable to formulation in a variety of dosage forms, including liquid formulations, tablets, and capsules (73,87). Formulation development, scale-up, and commercial manufacturing can be done without investment in manufacturing equipment unique to the technology. In this regard, formulation development with the Eligen technology involves preformulation characterization, excipient compatibility testing, formulation development, stability studies, and scale-up.

Preformulation

As indicated in the Chapter 5 an understanding of the physical and chemical properties of the biopharmaceutical and the appropriate delivery agent is necessary for the development of an adequate formulation and ultimately a robust commercial product. In the absence of this information, formulation development can become an exercise in trial and error. In the case of the Eligen technology, preformulation is the process of optimizing the combination biopharmaceutical and delivery agent through the determination of those physical and chemical properties that are important to the development of a stable, effective, and safe dosage form. Preformulation studies are conducted on the drug and the delivery agent separately and in combination. Special studies are conducted depending on the type of dosage form, the drug molecule, and the delivery agent selected. These studies yield valuable information for formulation development and optimization. Additionally, the solubility of the drug and delivery agent in liquid formulation vehicles needs to be determined as well. These studies form the basis of dissolution media development for the dosage form.

Solid-State Characterizations Studies

The solid-state properties of the biopharmaceutical and the delivery agent are important for the development of scalable formulations. The properties evaluated include thermal properties, crystallinity, morphology, moisture sorption–desorption, moisture content, particle size, and flow properties. Thermal properties are usually determined using thermogravimetric analysis (TGA) and differential scanning calorimetry (DSC). The crystallinity of the material is determined using a powder X-ray diffractometer and the morphology is determined using a scanning electron microscope. Evaluation of moisture sorption–desorption involves equilibrating the material at a given relative humidity created using saturated salt solutions as described by Greenspan (88) and weighing the material to obtain its water uptake.

The particle size and flow properties of the drug and delivery agent are relevant to ensuring the final product quality in pharmaceutical solid dosage form production and, as such, are determined at the preformulation stage of development. It has been adequately demonstrated that poor flow can result in dosage forms possessing unacceptably high weight variation. When good flowing formulations are tableted, compression and ejection force levels are more uniform and weight variations decrease. In tableting applications, powders with a high degree of flowability offer several advantages including minimizing air-pocket formation, increase in weight uniformity, and increase in reproducibility and feed parameters resulting in consistent tablet hardness, friability, and dissolution rate (89). A variety of techniques including angle of repose, compressibility, and flow through an orifice are used to estimate the flow potential of powders. Particle size can be determined using sieve analysis or laser light scattering equipment. Particle size is relevant to dosage form content uniformity and powder flow. It is important to note that the preformulation studies conducted to support formulation development using the Eligen technology are the same as typically conducted to support formulation development for other solid dosage forms that do not use this technology. This is a significant advantage of this oral drug delivery technology because no additional equipment and expertise unique to the technology is required for preformulation characterization.

Excipient Compatibility

Drug stability is important for successful development and commercialization of a dosage form; as such; studies to understand drug stability are necessary early in the development process. The acceleration of drug development and optimization of drug stability are some of the major goals of drug development programs. Since excipients are an integral part of pharmaceutical dosage forms, an early program to explore drug-excipient reactions (drug-excipient compatibility) is critical to avoid unexpected formulation stability problems in Phase II and III clinical trials. Several methods are used for the evaluation of drug-excipient compatibility since there is no one single generally accepted method. Studies can be performed using mixtures or prototype formulations. The amount of excipient added should be at least what is anticipated to be in the formulation. Studies are typically performed at elevated temperatures and the addition of water to powder mixtures is used to accelerate any reactions. To support compatibility studies, samples are analyzed by a variety of techniques including DSC, TGA, high-performance liquid chromatography, X-ray diffraction, and scanning electron microscope while nuclear magnetic resonance and liquid chromatography–mass spectrometry can be used for the identification of degradation products. The extent of excipient compatibility testing to support development activities with Eligen technology is similar to that performed to support development of formulations that do not use the technology and more details can be found in Chapter 5.

Polymorph Mapping

The development of formulations using Eligen technology involves using a carrier proprietary carrier molecule (delivery agent) in the formulation. The delivery agent is a single entity small organic molecule and for successful formulation development it needs to be well characterized and its polymorphic form needs to be controlled. Polymorphism in material science is the ability of a solid material to exist in more than one form or crystal structure. Together with polymorphism, the complete morphology of a material is described by other variables such as crystal habit, amorphous fraction, or crystallographic defects. For pharmaceutical solids, the criteria for identification of different polymorphs of a material are nonidentical powder X-ray diffraction patterns. Other techniques used for polymorph identification include thermal analysis, Fourier transform infrared, and scanning electron microscopy. Different polymorphs of the same drug or delivery agent can have different dissolution rates that could lead to differences in bioavailability. In preformulation studies, the goal is the identification of potential polymorphs and to determine the most stable form that is suitable for development. The effect of processing such as wet granulation, milling, drying temperature, and compression is also evaluated during the preformulation stage.

Formulation Development

Once preformulation work has been completed, formulation development activities begin. Early formulation attempts using Eligen technology with heparin were constrained by the large amount of delivery agent required per dose and thus focused on an oral solution as the dosage form. Formulation improvements concentrated mainly on taste enhancement. A taste masked oral heparin formulation with a delivery agent was developed and used in clinical studies as previously reported (67,68).

Solid dosage formulation strategies that are compatible with the Eligen technology include solid dosage formulations filled into hard gelatin capsules, simple blends of drug and delivery agent filled into hard gelatin capsules, liquid or semisolid formulations filled into hard or soft gelatin capsules, and granulation using standard techniques followed by blending with standard excipients and then compressing in to tablets. These techniques have been successfully applied to the development of oral solid dosage forms of several biopharmaceuticals, including heparin, low-molecular-weight heparin, insulin, parathyroid hormone, calcitonin, and growth hormone, using the Eligen drug delivery technology (90). In the case of heparin for example, formulation techniques have yielded a solid dosage form that delivers heparin to an extent comparable to the liquid formulation used in the early Phase III study, yet utilizing less delivery agent and heparin. In the case of insulin and other peptides, formulation development efforts have led to improved bioavailability and have yielded tablets that are stable for more than 12 months at room temperature (RH) packaged in high density polyethylene (HDPE) bottles with induction seal or other packaging that provide protection from moisture (Fig. 3). For formulations based on Eligen technology stability studies need to be conducted in the container closure system as per International Committee for Harmonization stability guidelines for drug product. Typical storage conditions stability conditions will be based on the stability of the protein or macromolecular drug and may include −20°C, 2°C to 8°C, and 25°C/60% RH. For products that have to be stored at 20°C or 2°C to 8°C, the RH stability data will provide information on storage of the drug after dispensing to the patient; as such, stability of one to three months at RH will be adequate. If the product is stable at RH, then the typical stability storage conditions will be 25°C/60% RH, 30°C/65% RH, and 40°C/75% RH. Since stability is not always linear with temperature, the shelf-life of the drug product will be based on real time stability at the storage condition.

CONCLUSIONS

Oral administration is regarded as the preferred form of drug intake, mainly because of its convenience—particularly in chronic indications—potential for availability to large patient populations, and cost effectiveness. In addition, oral delivery can offer added advantages and lead to improved therapeutics in the case of drugs that target the liver or benefit from entering the bloodstream via the portal circulation, as is the case of insulin, incretins, and the so-called gut hormones. Although interest in developing oral proteins and peptides dates back to almost a century ago, the advent of biotechnology and proliferation of biopharmaceuticals have brought about a renewed interest in oral biopharmaceuticals in recent years. Several oral proteins and peptides are currently in clinical trials and,

Figure 3 Example of an insulin tablet developed using the Eligen® drug delivery technology.

hopefully, they will be available to patients in a not too distant future. The success of oral delivery approaches and technologies will depend not only on their ability to enable or increase oral bioavailability but also on their ability to generate commercially viable products of reasonable manufacturing costs and appropriate stability and tolerability.

REFERENCES

1. Woodley JF. Enzymatic barriers for GI peptide and protein delivery. Crit Rev Ther Drug Carrier Syst 1994; 11:61–95.
2. Lipka E, Crison J, Amidon GL. Transmembrane transport of peptide type compounds: prospects for oral delivery. J Control Release 1996; 39:121–129.
3. Goldberg M, Gomez-Orellana I. Challenges for the oral delivery of macromolecules. Nat Rev Drug Discov 2003; 2:289–295.
4. Sood A, Panchagnula R. Peroral route: an opportunity for protein and peptide drug delivery. Chem Rev 2001; 101:3275–3303.
5. Shah RB, Ahsan F, Khan MA. Oral delivery of proteins: progress and prognostication. Crit Rev Ther Drug Carrier Syst 2002; 19:135–169.
6. Gomez-Orellana I, Paton DR. Advances in the oral delivery of proteins. Exp Opin Ther Patents 1998; 8:223–234.
7. Gomez-Orellana I, Paton DR. Advances in the oral delivery of proteins. Update. Exp Opin Ther Patents 1999; 9:247–253.
8. Gomez-Orellana I. Strategies to improve oral drug bioavailability. Exp Opin Drug Delivery 2005; 2:419–433.
9. Bagger YZ, Tanko LB, Alexandersen P, et al. Oral salmon calcitonin induced suppression of urinary collagen type II degradation in postmenopausal women: a new potential treatment of osteoarthritis. Bone 2005; 37:425–430.
10. Tanko LB, Bagger YZ, Alexandersen P, et al. Safety and efficacy of a novel salmon calcitonin (sCT) technology-based oral formulation in healthy postmenopausal women: acute and 3-month effects on biomarkers of bone turnover. J Bone Miner Res 2004; 19:1531–1538.
11. Buclin T, Cosma Rochat M, Burckhardt P, Azria M, Attinger M. Bioavailability and biological efficacy of a new oral formulation of salmon calcitonin in healthy volunteers. J Bone Miner Res 2002; 17:1478–1485.
12. Arbit E. The physiological rationale for oral insulin administration. Diabetes Technol Ther 2004; 6:510–517.
13. Laskowski M Jr, Haessler HA, Miech RP, Peanasky RJ, Laskowski M. Effect of trypsin inhibitor on passage of insulin across the intestinal barrier. Science 1958; 127:1115–1116.
14. Danforth E Jr, Moore RO. Intestinal absorption of insulin in the rat. Endocrinology 1959; 65:118–123.
15. Fujii S, Yokoyama T, Ikegaya K, Sato F, Yokoo N. Promoting effect of the new chymotrypsin inhibitor FK-448 on the intestinal absorption of insulin in rats and dogs. J Pharm Pharmacol 1985; 37:545–549.
16. Yamamoto A, Taniguchi T, Rikyuu K, et al. Effects of various protease inhibitors on the intestinal absorption and degradation of insulin in rats. Pharm Res 1994; 11:1496–1500.

17. Mansoor S, Youn YS, Lee KC. Oral delivery of mono-PEGylated sCT (Lys18) in rats: regional difference in stability and hypocalcemic effect. Pharm Dev Technol 2005; 10:389–396.

18. Jensen-Pippo KE, Whitcomb KL, DePrince RB, Ralph L, Habberfield AD. Enteral bioavailability of human granulocyte colony stimulating factor conjugated with poly(ethylene glycol). Pharm Res 1996; 13:102–107.

19. Li X, Chyu K-Y, Neto JRF, et al. Differential effects of apolipoprotein A-I-mimetic peptide on evolving and established atherosclerosis in apolipoprotein E-null mice. Circulation 2004; 110:1701–1705.

20. Reddy ST, Anantharamaiah GM, Navab M, et al. Oral amphipathic peptides as therapeutic agents. Expert Opin Investig Drugs 2006; 15:13–21.

21. Asada H, Douen T, Waki M, et al. Absorption characteristics of chemically modified-insulin derivatives with various fatty acids in the small and large intestine. J Pharm Sci 1995; 84:682–687.

22. Lee Y, Nam JH, Shin HC, Byun Y. Conjugation of low-molecular-weight heparin and deoxycholic acid for the development of a new oral anticoagulant agent. Circulation 2001; 104:3116–3120.

23. Liu J, Pervin A, Gallo CM, et al. New approaches for the preparation of hydrophobic heparin derivatives. J Pharm Sci 1994; 83:1034–1039.

24. Swaan PW. Recent advances in intestinal macromolecular drug delivery via receptor-mediated transport pathways. Pharm Res 1998; 15:826–834.

25. Clement S, Dandona P, Still JG, Kosutic G. Oral modified insulin (HIM2) in patients with type 1 diabetes mellitus: results from a phase I/II clinical trial. Metabolism 2004; 53:54–58.

26. Clement S, Still JG, Kosutic G, McAllister RG. Oral insulin product hexyl-insulin monoconjugate 2 (HIM2) in type 1 diabetes mellitus: the glucose stabilization effects of HIM2. Diabetes Technol Ther 2002; 4:459–466.

27. Kipnes M, Dandona P, Tripathy D, Still JG, Kosutic G. Control of postprandial plasma glucose by an oral insulin product (HIM2) in patients with type 2 diabetes. Diabetes Care 2003; 26:421–426.

28. Wajcberg E, Miyazaki Y, Triplitt C, Cersosimo E, Defronzo RA. Dose-response effect of a single administration of oral hexyl-insulin monoconjugate 2 in healthy nondiabetic subjects. Diabetes Care 2004; 27:2868–2873.

29. Chin CM, Gutierrez M, Still JG, Kosutic G. Pharmacokinetics of modified oral calcitonin product in healthy volunteers. Pharmacotherapy 2004; 24:994–1001.

30. Russell-Jones GJ. Use of vitamin B12 conjugates to deliver protein drugs by the oral route. Crit Rev Ther Drug Carrier Syst 1998; 15:557–586.

31. Russell-Jones GJ, Westwood SW, Farnworth PG, Findlay JK, Burger HG. Synthesis of LHRH antagonists suitable for oral administration via the vitamin B12 uptake system. Bioconjug Chem 1995; 6:34–42.

32. Russell-Jones GJ, Westwood SW, Habberfield AD. Vitamin B12 mediated oral delivery systems for granulocyte-colony stimulating factor and erythropoietin. Bioconjug Chem 1995; 6:459–465.

33. Eaimtrakarn S, Rama Prasad YV, Ohno T, et al. Absorption enhancing effect of labrasol on the intestinal absorption of insulin in rats. J Drug Target 2002; 10:255–260.

34. Onuki Y, Morishita M, Takayama K, et al. In vivo effects of highly purified docosa-hexaenoic acid on rectal insulin absorption. Int J Pharm 2000; 198:147–156.

35. Suzuki A, Morishita M, Kajita M, et al. Enhanced colonic and rectal absorption of insulin using a multiple emulsion containing eicosapentaenoic acid and docosahexaenoic acid. J Pharm Sci 1998; 87:1196–1202.

36. Thanou M, Verhoef JC, Junginger HE. Oral drug absorption enhancement by chitosan and its derivatives. Adv Drug Deliv Rev 2001; 52:117–126.

37. Ziv E, Kidron M, Raz I, et al. Oral administration of insulin in solid form to nondiabetic and diabetic dogs. J Pharm Sci 1994; 83:792–794.

38. Ziv E, Lior O, Kidron M. Absorption of protein via the intestinal wall. A quantitative model. Biochem Pharmacol 1987; 36:1035–1039.

39. Mehta NM. Oral delivery and recombinant production of peptide hormones. Part I: making oral delivery possible. BioPharm International June 2004. www.biopharminternational.com/biopharm/article/articleDetail.jsp?id=10228.

40. Raoof AA, Chiu P, Ramtoola Z, et al. Oral bioavailability and multiple dose tolerability of an antisense oligonucleotide tablet formulated with sodium caprate. J Pharm Sci 2004; 93:1431–1439.

41. Raoof AA, Ramtoola Z, McKenna B, Yu RZ, Hardee G, Geary RS. Effect of sodium caprate on the intestinal absorption of two modified antisense oligonucleotides in pigs. Eur J Pharm Sci 2002; 17:131–138.

42. Nissan A, Ziv E, Kidron M, et al. Intestinal absorption of low molecular weight heparin in animals and human subjects. Haemostasis 2000; 30:225–232.

43. Guevara-Aguirre J, Guevara M, Saavedra J, Mihic M, Modi P. Beneficial effects of addition of oral spray insulin (Oralin) on insulin secretion and metabolic control in subjects with type 2 diabetes mellitus suboptimally controlled on oral hypoglycemic agents. Diabetes Technol Ther 2004; 6:1–8.

44. Modi P, Mihic M, Lewin A. The evolving role of oral insulin in the treatment of diabetes using a novel RapidMist System. Diabetes Metab Res Rev 2002; 18(suppl 1): S38–S42.

45. Daugherty AL, Mrsny RJ. Regulation of the intestinal epithelial paracellular barrier. Pharm Sci Technol Today 1999; 2:281–287.

46. Madara JL. Modulation of tight junctional permeability. Adv Drug Deliv Rev 2000; 41:251–253.

47. Ward PD, Tippin TK, Thakker DR. Enhancing paracellular permeability by modulating epithelial tight junctions. Pharm Sci Technol Today 2000; 3:346–358.

48. Salama NN, Eddington ND, Fasano A. Tight junction modulation and its relationship to drug delivery. Adv Drug Deliv Rev 2006; 58:15–28.

49. Fasano A, Uzzau S. Modulation of intestinal tight junctions by Zonula occludens toxin permits enteral administration of insulin and other macromolecules in an animal model. J Clin Invest 1997; 99:1158–1164.

50. Yen WC, Lee VH. Role of Na+ in the asymmetric paracellular transport of 4-phenylazobenzyloxycarbonyl-L-Pro-L-Leu-Gly-L-Pro-D-Arg across rabbit colonic segments and Caco-2 cell monolayers. J Pharmacol Exp Ther 1995; 275:114–119.

51. Sapone A, de Magistris L, Pietzak M, et al. Zonulin upregulation is associated with increased gut permeability in subjects with type 1 diabetes and their relatives. Diabetes 2006; 55:1443–1449.

52. Watts T, Berti I, Sapone A, et al. Role of the intestinal tight junction modulator zonulin in the pathogenesis of type I diabetes in BB diabetic-prone rats. Proc Natl Acad Sci USA 2005; 102:2916–2921.

53. Meddings JB, Jarand J, Urbanski SJ, Hardin J, Gall DG. Increased gastrointestinal permeability is an early lesion in the spontaneously diabetic BB rat. Am J Physiol 1999; 276:G951–G957.

54. Prego C, Fabre M, Torres D, Alonso MJ. Efficacy and mechanism of action of chitosan nanocapsules for oral peptide delivery. Pharm Res 2006; 23:549–556.

55. Prego C, Torres D, Alonso MJ. The potential of chitosan for the oral administration of peptides. Expert Opin Drug Deliv 2005; 2:843–854.

56. Bernkop-Schnurch A. Chitosan and its derivatives: potential excipients for peroral peptide delivery systems. Int J Pharm 2000; 194:1–13.

57. Guggi D, Kast CE, Bernkop-Schnurch A. In vivo evaluation of an oral salmon calcitonin-delivery system based on a thiolated chitosan carrier matrix. Pharm Res 2003; 20:1989–1994.

58. Koping-Hoggard M, Sanchez A, Alonso MJ. Nanoparticles as carriers for nasal vaccine delivery. Expert Rev Vaccines 2005; 4:185–196.

59. Webster DE, Gahan ME, Strugnell RA, Wesselingh SL. Advances in oral vaccine delivery options. Am J Drug Deliv 2003; 1:227–240.

60. van der Lubben IM, Verhoef JC, Borchard G, Junginger HE. Chitosan and its derivatives in mucosal drug and vaccine delivery. Eur J Pharm Sci 2001; 14:201–207.

61. Shakweh M, Ponchel G, Fattal E. Particulate uptake by Peyer's patches: a pathway for drug and vaccine delivery. Exp Opin Drug Deliv 2004; 1:141–163.

62. Leone-Bay A, Paton DR, Weidner JJ. The development of delivery agents that facilitate the oral absorption of macromolecular drugs. Med Res Rev 2000; 20:169–186.

63. Malkov D, Wang HZ, Dinh S, Gomez-Orellana I. Pathway of oral absorption of heparin with sodium N-[8-(2-hydroxybenzoyl)amino] caprylate. Pharm Res 2002; 19:1180–1184.

64. Stoll BR, Leipold HR, Milstein S, Edwards DA. A mechanistic analysis of carrier-mediated oral delivery of protein therapeutics. J Control Release 2000; 64:217–228.

65. Malkov D, Angelo R, Wang HZ, Flanders E, Tang H, Gomez-Orellana I. Oral delivery of insulin with the *eligen*® technology: mechanistic studies. Curr Drug Deliv 2005; 2:191–197.

66. Ding X, Rath P, Angelo R, et al. Oral absorption enhancement of cromolyn sodium through noncovalent complexation. Pharm Res 2004; 21:2196–2206.

67. Baughman RA, Kapoor SC, Agarwal RK, Kisicki J, Catella-Lawson F, FitzGerald GA. Oral delivery of anticoagulant doses of heparin. A randomized, double-blind, controlled study in humans. Circulation 1998; 98:1610–1615.

68. Berkowitz SD, Marder VJ, Kosutic G, Baughman RA. Oral heparin administration with a novel drug delivery agent (SNAC) in healthy volunteers and patients undergoing elective total hip arthroplasty. J Thromb Haemost 2003; 1:1914–1919.

69. Gonze MD, Manord JD, Leone-Bay A, et al. Orally administered heparin for preventing deep venous thrombosis. Am J Surg 1998; 176:176–178.

70. Gonze MD, Salartash K, Sternbergh WC III, Baughman RA, Leone-Bay A, Money SR. Orally administered unfractionated heparin with carrier agent is therapeutic for deep venous thrombosis. Circulation 2000; 101:2658–2661.

71. Kidron M, Dinh S, Menachem Y, et al. A novel per-oral insulin formulation: proof of concept study in non-diabetic subjects. Diabet Med 2004; 21:354–357.

72. Leone-Bay A, Sato M, Paton D, et al. Oral delivery of biologically active parathyroid hormone. Pharm Res 2001; 18:964–970.

73. Majuru S. Advances in the oral delivery of heparin from solid dosage forms using Emisphere's *eligen®* oral drug delivery technology. Drug Deliv Technol 2004; 4:84–89.
74. Money SR, York JW. Development of oral heparin therapy for prophylaxis and treatment of deep venous thrombosis. Cardiovasc Surg 2001; 9:211–218.
75. Pineo GF, Hull RD, Marder VJ. Orally active heparin and low-molecular-weight heparin. Curr Opin Pulm Med 2001; 7:344–348.
76. Meier JJ, Veldhuis JD, Butler PC. Pulsatile insulin secretion dictates systemic insulin delivery by regulating hepatic insulin extraction in humans. Diabetes 2005; 54:1649–1656.
77. Ferrannini E, Bjorkman O, Reichard GA, et al. The disposal of an oral glucose load in healthy subjects. A quantitative study. Diabetes 1985; 34:580–588.
78. Galassetti P, Shiota M, Zinker BA, Wasserman DH, Cherrington AD. A negative arterial-portal venous glucose gradient decreases skeletal muscle glucose uptake. Am J Physiol Endocrinol Metab 1998; 275:E101–E111.
79. DeFronzo RA, Ferrannini E, Hendler R, Wahren J, Felig P. Influence of hyperinsulinemia, hyperglycemia, and the route of glucose administration on splanchnic glucose exchange. Proc Natl Acad Sci USA 1978; 75:5173–5177.
80. Myers SR, McGuinness OP, Neal DW, Cherrington AD. Intraportal glucose delivery alters the relationship between net hepatic glucose uptake and the insulin concentration. J Clin Invest 1991; 87:930–939.
81. Caumo A, Luzi L. First-phase insulin secretion: does it exist in real life? Considerations on shape and function. Am J Physiol Endocrinol Metab 2004; 287:E371–E385.
82. Del Prato S. Loss of early insulin secretion leads to postprandial hyperglycaemia. Diabetologia 2003; 46(suppl 1):M2–M8.
83. Cherrington AD, Sindelar D, Edgerton D, Steiner K, McGuinness OP. Physiological consequences of phasic insulin release in the normal animal. Diabetes 2002; 51(suppl 1):S103–S108.
84. Tolic IM, Mosekilde E, Sturis J. Modeling the insulin-glucose feedback system: the significance of pulsatile insulin secretion. J Theor Biol 2000; 207:361–375.
85. Grubert JM, Lautz M, Lacy DB, et al. Impact of continuous and pulsatile insulin delivery on net hepatic glucose uptake. Am J Physiol Endocrinol Metab 2005; 289: E232–E240.
86. Shishko PI, Kovalev PA, Goncharov VG, Zajarny IU. Comparison of peripheral and portal (via the umbilical vein) routes of insulin infusion in IDDM patients. Diabetes 1992; 41:1042–1049.
87. Majuru S. Development of an oral insulin solid dosage formulation using Emisphere's *eligen®* technology. Drug Deliv Technol 2005; 5:74–78.
88. Greenspan L. Humidity fixed points of binary saturated aqueous solutions. J Res NBS 1977; 81A:89–96.
89. Gioia A. Intrinsic flowability: a new technology for powder-flowability classification. Pharm Technol 1980; 65–68.
90. Singh B, Majuru S. Oral delivery of therapeutic macromolecules: a perspective using *eligen®* technology. Oral Deliv Technol 2003; 3(4):1–5.

13

Setting Specifications and Expiration Dating for Biotechnology Products

Carol Hasselbacher

Stability, Corporate Product Quality, Amgen Inc., Longmont, Colorado, U.S.A.

Wassim Nashabeh

Regulatory Affairs, Genentech Inc., San Francisco, California, U.S.A.

Anthony Mire-Sluis

Corporate Product Quality, Amgen Inc., Thousand Oaks, California, U.S.A.

INTRODUCTION

A specification is defined as a list of tests, references to analytical procedures, and appropriate acceptance criteria that are numerical limits, ranges, or other criteria for the tests described (1). As such, a specification defines a quality standard to confirm the quality of products and establishes the set of criteria to which a drug substance, drug product, or materials at other stages of its manufacture should conform in order to be considered acceptable for its intended use (2). "Conformance to specification" means that the drug substance and drug product, when tested according to the listed analytical procedures, will meet the acceptance criteria. Specifications are critical quality "standards" that are proposed and justified by the manufacturer and approved by regulatory authorities as conditions of approval. Specifications not only apply at the release of drug substances and drug products, but are also required as part of stability studies that occur during product development (to assure material used in the clinic remains safe and efficacious for the lifetime of the studies) as well as ongoing annual stability studies that are required post-approval.

Specifications are not the only controls used to assure product quality. Product quality is assured through a range of programs during product development, quality systems, and controls. Examples include thorough product characterization during development, upon which many of the specifications are based, adherence to Good Manufacturing Practices, a validated manufacturing process, raw materials testing, in-process testing, stability testing, etc. In-process controls (IPCs) that are associated with rejection limits are similar to lot-release specifications, except that they are tests that are carried out during the manufacturing process. Many IPCs are associated with action or alert limits and differ from specifications regarding the action required should limits be exceeded. Specifications also apply to the raw materials used in the manufacturing process and must be considered if assurance of product quality is to be maintained.

Since specifications are only one aspect of the quality assurance process even if specifications are met, if other quality assurance activities fail to be met then investigations have to be carried out to assure the product is safe and effective.

The setting of specifications is an activity that occurs throughout the life-cycle of a product, from early development, licensure, and postapproval. The setting of specifications for biotechnology products is not straightforward, and many data services need to be taken into account in order to ensure that appropriate specifications are developed. These include, although not exclusively, analytical, preclinical, and clinical data, data derived from previous experience with similar molecules, and published data.

The selection of which tests to include in a specification from the wide range available to characterize biological products is dependent on the level of understanding of the criticality of a specific quality attribute, i.e., does a product characteristic have an impact on safety or efficacy? The establishment of criticality through nonclinical studies and clinical trials is essential for meaningful specifications. In addition to nonclinical and clinical considerations, additional factors in selecting which quality parameters to specify include process development and validation data, physicochemical and biological product characterization data, pharmaceutical development studies including product stability data, and manufacturing experience along with cGMP regulatory requirements. Acceptance criteria for specifications should also be established and justified based on data obtained from lots used in nonclinical and/or clinical studies, in addition to data from lots used for demonstration of manufacturing consistency and data from stability studies, and relevant process development data. The use of proper statistical approaches applied to the historical data should be considered a key contributing factor in setting the acceptable ranges or limits, but this is not the sole factor, as the potential nonclinical and clinical implications or lack thereof of any quality parameter should be considered the predominant factor in any decision-making process.

Stability studies not only support product characterization and setting of specifications, but are also required to define the expiration dating or shelf-life for the product. An appropriate expiration date is one that assures product safety and efficacy through expiry. Setting an appropriate expiry date relies on a comprehensive understanding of the degradation pathways of the product and how degradation impacts product quality, safety, and efficacy.

SPECIFICATIONS—TESTS AND ACCEPTANCE CRITERIA

Product characterization is the foundation of setting specifications. In order to set appropriate specifications, a comprehensive understanding of the physico-chemical and biological characteristics of a product is required, from which a subset are selected that most appropriately reflect the critical quality attributes of that product. Selecting which characteristics to specify should also take into account the capability of the assays to detect protein variants.

Advances in analytical sciences now allow most protein products to be characterized extensively in terms of their identity, heterogeneity, and impurity profile (3). Current analytical methods (both physicochemical and biological) can characterize the primary, secondary, and, to some extent, higher-order structure of proteins. These methods allow for the identification and characterization not only of the desired protein component, but also of many product-related substances. In addition, techniques are powered sufficiently to detect product-related impurities and process-related impurities present in the drug product. An appropriate selection of analytical and biological tools allows for the evaluation of both the physicochemical and functional characteristics of a product.

Only after the characteristic profile of a product is thoroughly understood can the analysis of material that has been studied in nonclinical and clinical studies, as well as material manufactured throughout product development, allow for the criticality of specific product quality characteristics to be established.

The recent focus of the concept of Quality by Design (QbD) by regulatory authorities also has direct impact on the setting of specifications (4). The use of QbD in protein design can reduce specific, undesired product quality attributes (e.g., a deamidation site) that would have required control through a specification test. In addition, the knowledge gained through the studies described in the following sections can provide a "design space" around which specification tests and acceptance criteria can be built. This involves the understanding of which product variants are critical quality attributes that could affect safety and/or efficacy. The greater this knowledge of your product, the wider acceptance limits can become (i.e., no longer tied simply to clinical exposure and process variability) and the number of tests can be reduced accordingly.

Product Characterization to Define Product Quality Attributes and Their Criticality

Demonstrating the Physicochemical Attributes of a Product

Proteins exhibit primary, secondary, tertiary, and, dependent on the product, quaternary structure. In addition, most protein products undergo some degree of post-translational modification, such as *N*- and *C*-terminal amino acid heterogeneity, oxidation, deamidation, acetylation, *N*-formylation, proteolytic processing, glycosylation, and glutamic acid γ-carboxylation. Such modifications can occur through intracellular activities during fermentation/bioreaction or be caused by the manufacturing process. The combinatorial heterogeneity from posttranslational modification as well as protein degradation results in a wide range of protein species whose complexity can challenge the limits of analytical technology. Even when an extensive physicochemical characterization is undertaken, it is rare to have a complete understanding of the relationship between protein structure and function. In order to understand whether any particular physicochemical property is important for safety, efficacy, or consistency, it is necessary to thoroughly characterize, and thus identify, as many relevant characteristics of a product as is practically possible (3).

In general, primary and secondary structure can be adequately assessed with peptide mapping coupled with mass spectrometric technology in addition to other orthogonal tests such as sequencing and sodium dodecyl sulfate polyacrylamide gel electrophoresis. Tertiary structure (and in some cases, quaternary structure) is often difficult to adequately define using current physicochemical analytical technology. Techniques such as analytical ultracentrifugation are available for characterization studies but do not readily lend themselves to routine lot release. Techniques such as X-ray crystallography and multidimensional nuclear magnetic resonance spectroscopy can define tertiary protein structure and, to varying extents, quaternary structure; these technologies, however, require considerable effort and expertise, and not all proteins are amenable to these techniques. Therefore, a biological assay is likely to be required to characterize the functionality of protein products and may confirm the presence of a necessary higher-order structure.

The desired product can also be a mixture of post-translationally modified forms (e.g., glycoforms). It is necessary, depending on the quantity of any particular modification, to gain an understanding of its biological activity to obtain an understanding of the potential impact on safety and/or efficacy of the product by the variant. The pattern of heterogeneity of the desired product should be developed, and consistency of lots used in preclinical and clinical studies should be demonstrated. Even if the impact on safety or efficacy cannot be demonstrated for each and every product variant, a consistent level of heterogeneity reduces the burden placed on illustrating the criticality of a product attribute.

The tests required to detect and characterize where on the protein these modifications occur are specific to the effect that the modification has on the physicochemical properties of the protein. Most of the described modifications can be identified through mass spectrometry, when changes in mass occur, either of the whole protein or from peptide maps. Other techniques can be used for modifications that result in changes in charge, size, hydrodynamic radius, or hydrophobicity (see Methods in Table 1). However, it must be taken into account that some modifications may be buried within the protein and not readily accessible to nondisruptive

Table 1 Physicochemical Characterization Tests

Quality parameter	Analytical tests
Identity	
Primary structure	*Amino acid composition analysis*
	Primary amino acid sequence (e.g., *N*-terminal sequencing, *C*-terminal sequencing, peptide mapping and sequencing after enzymatic or chemical cleavage)
	Disulfide linkage
Higher-order structure	*Secondary and tertiary structure evaluation*
	Circular dichroism
	X-ray crystallography
	Magnetic resonance spectroscopy immunoreactivity with conformation-dependent antibodies
	Biological assay
Size, size distribution	Mass spectrometry
	Hydrodynamic methods (e.g., size exclusion chromatography equilibrium density gradient centrifugation)
	Light scattering
	SDS-PAGE, CE-SDS
Charge properties	IEF gel electrophoresis or capillary IEF
	CE
	Ion-exchange chromatography
Hydrophobic characteristics	Hydrophobic interaction HPLC
	Reverse phase HPLC
Immuno-reactivity	Immunoblotting
	ELISA
For glycosylated products: glycosylation	*Monosaccharide analysis*
	Mass spectrometry
	Spectrometry
	HPAEC-PAD
	Reverse phase HPLC
	CE
	Oligosaccharide analysis
	Residue-specific enzymatic degradation
	HPAEC-PAD
	Reverse phase HPLC
	CE
	Identification of glycosylation sites
	Peptide mapping followed by mass spectrometry
	Glycosylation occupancy analysis
	Peptide mapping
	CE-SDS

(Continued)

Table 1 Physicochemical Characterization Tests *(Continued)*

Quality parameter	Analytical tests
Determination of purity (product-related substances or impurities)	
Hydrophobic characteristics	Hydrophobic interaction HPLC
	Reverse phase HPLC
Charge characteristics	Ion exchange chromatography
	Capillary zone electrophoresis
	IEF (gel or capillary formats)
Size, size distribution	Size exclusion HPLC
	Analytical ultracentrifugation
	Light scattering Field flow fractionation
	SDS-PAGE or CE-SDS
Protein content	
Protein content	Reference methods
	Gravimetric analysis
	Total nitrogen (Kjeldahl)
	Quantitative amino acid analysis (other methods such as UV absorption, HPLC, or dye-binding may be validated against a reference method)

Abbreviations: SDS-PAGE, sodium dodecyl sulfate polyacrylamide gel electrophoresis; CE-SDS, capillary electrophoresis-sodium dodecyl sulfate; IEF, isoelectric focusing; HPLC, high performance liquid chromatography; ELISA, enzyme-linked immunosorbent assay; HPAEC-PAD; UV, ultraviolet.

tests, or may only exist on a proportion of the test material and could be masked within a mixture.

Assessing Biological Activities and Potency

The complex nature of the biological products produced by the biotechnology industry has required the use of a variety of physicochemical and biological tests to adequately characterize them. However, only a biological assay can estimate the biological activity or potency of the product. A bioassay is defined as a functional assay where the product induces some form of biological response in a test system (5). For some products, binding assays can suffice as a potency assay, but this requires validation and thorough comparison to a suitable bioassay (6). It has been generally accepted that bioassays are a quality issue and that they should not necessarily need to be designed to predict or reflect any clinical efficacy per se, as this is the purpose of clinical trials and "mimicking the biological activity in the clinical situation is not always necessary" (1). A correlation between the expected clinical response and the activity in the biological assay should be established in pharmacodynamic or clinical studies (5).

If data do suggest that the structure of biologically active proteins may contain different areas devoted to exercising different biological activities, it would be valid to have more than one test for bioactivity characterization studies, depending on the intended use of the molecule. For example, a monoclonal antibody can have both ligand-binding activity of the Fab portion of the molecule and functionality of the Fc portion of the molecule (e.g., complement binding).

However, since a bioassay used for lot-release specifications is required to demonstrate batch-to-batch consistency, unless data is available to prove otherwise, typically a single assay format is suitable, as long as the assay selected is relevant, precise, and robust. Whilst a biological reaction may be used in order to compare the strength of two preparations, one with respect to the other, it cannot be used by itself to define the potency of one preparation alone. Assays vary in response between assay types and from day to day, and therefore, it is not appropriate to use some characteristic of the assay itself (e.g., ED50) to establish potency.

In order to define the potency of a biological material, the assay must be calibrated using a suitable reference standard. This is most often carried out by comparing the biological response of a series of dilutions made from a reference standard to those of test samples.

The potency of any biological product should be expressed relative to a well-defined reference preparation. This is the concept of "relative potency." Therefore, the selection and validation of a suitable (well-characterized, stable, etc.) reference preparation is vital both for assay validation and for specification setting. Careful consideration must be given to validate the comparison of reference to sample. Some aspects of validity can be assessed statistically, and assays should be designed so that this can be achieved to as high a degree as is appropriate.

Following appropriate assay development and monitoring of clinical batches, specification limits must be applied to both the potency of the product (limits around the mean stated potency) and the assay [fiducial limits of the assay (see discussion below)]. Although guidelines exist [European pharmacopoeia (EP) and United States pharmacopoeia (USP)] on how to set such limits, it is the data that dictate what is sensible. Often, multiples of standard deviations are quoted, such as the use of the static 3SD approach or the dynamic tolerance intervals at the 95/99% confidence limits that are adjusted using a life-cycle management approach to specifications.

Since potency assays for lot release primarily control for batch-to-batch consistency, specification limits are designed to restrict the differences in potency between batches to a level that reflects the consistency of the manufacturing process and the "quality" of the product.

The major difference between setting limits for physicochemical assays and bioassays is the perceived variability of bioassays. Therefore, some measure of control of assay variability is required and has led to the requirement of setting fiducial limits of the bioassay in addition to those for the potency value. In general, fiducial limits equal confidence intervals and are usually set at 95%.

There are no hard and fast rules when choosing at what time to set limits on potency assays, although this usually occurs during the latter stages of Phase

III trials, where qualification lots of material are being produced and the bioassay validation has been completed.

Data used for setting bioassay specifications should only be derived from assays that are well controlled and found to be consistent. It is sensible to look at the data of a number of clinical production lots to check for trends and variability over time. It is often best to follow the *principles* of the pharmacopoeias unless the data suggest otherwise.

Characterization of Immunochemical Properties

When binding to a ligand is part of the activity of the product, the manufacturer should use analytical tests to characterize the product in terms of this specific property (e.g., if binding to a receptor is inherent in protein function, then this property needs to be measured). Various methods such as surface plasmon resonance, microcalorimetry, or classical Scatchard analysis can provide information on the kinetics and thermodynamics of binding, which can be related to biological activity and higher-order structure characterization.

When the product is an antibody, its immunological properties should be fully characterized. Binding assays of the antibody to purified antigens and defined regions of antigens should be performed, as feasible, to determine affinity, avidity, and immunoreactivity (including cross-reactivity). In addition, the target molecule bearing the relevant epitope should be biochemically defined and the epitope itself defined, when feasible.

Protein molecules are often examined using immunochemical procedures (e.g., enzyme-linked immunosorbent assay, Western blot) utilizing antibodies that recognize different epitopes of the protein molecule. The binding of antibodies, especially those recognizing tertiary or quaternary structure, can be a valuable tool in assessing the correct folding of the product. Immunochemical properties of a protein may serve to establish its identity, homogeneity, or purity, or serve to quantify it.

Defining the "Desired Product" Through Product Characterization

Analytical characterization of the product allows for the definition of the "desired product" and identifies variants of the product that are either biologically similar in activity, defined as product-related substances, or have altered activity, defined as product-related impurities. Differences in other nonclinical or clinical attributes [e.g., pharmacokinetic (PK) or bioavailability] can also define whether a particular variant is a related substance or an impurity (1).

The fact that most biological products are heterogeneous in nature results in the inability to define purity as an absolute measure from a single assay, i.e., purity is demonstrated through a variety of orthogonal tests (e.g., size exclusion to show molecular weight variants, isoelectric focusing to show charge variants).

Process-related impurities are often present in medicinal products and encompass those that are derived from the manufacturing process, i.e., cell substrates (e.g., host cell proteins, host cell DNA), cell culture (e.g., antibiotics, enzymes, or media components), or from downstream purification (e.g., chemicals used in refolding, column purification).

Since product-related impurities and process-related impurities can impact safety and efficacy, it is necessary to develop specifications for those that either are known to have such impact or are at a level that may have an impact. For certain process-related impurities (such as DNA, host cell impurities, protein A) specifications may not be required if process-validation studies along with manufacturing consistency data can demonstrate consistent removal of these impurities to appropriate levels. For product-related substances, the use of action limits rather than specifications should be considered as an alternative means to assure product consistency from lot to lot.

Lastly, biological products can contain contaminants that include all adventitiously introduced materials not intended to be part of the manufacturing process, such as chemical and biochemical materials (e.g., microbial proteases, raw material contaminants), and/or microbial species. Specifications are required to control such contaminants (e.g., sterility or bioburden), unless otherwise justified (e.g., through process validation).

Stability Studies

Stability studies form a vital component of the understanding of product characteristics. Protein products degrade over time, the extent of which is product dependent. The nature of the degradation pathways must be well characterized to identify if the pathways involve either an increase in existing product variants identified in newly manufactured material or the creation of additional variants (such as truncations, chemical modifications, higher molecular weight species, etc.). An understanding of the exposure to such variants during nonclinical and clinical studies as well as an assessment of their impact on safety and efficacy becomes part of the specification-setting process—not only for lot release, but during stability studies, where specifications also have to be met to assure continued product quality throughout the expiry of the product. Therefore, it is valuable to gain an understanding of the age of material to which patients have been exposed during clinical studies.

Nonclinical and Clinical Characterization

The "gold standard" of product characterization occurs in the clinical setting, where patient exposure provides the most holistic data on the safety and efficacy of a product. However, it is rare that direct links between product quality attributes and safety or efficacy are made unless they are specifically designed into the clinical program. This is due to relatively few lots of material being placed into clinical studies and that often safety signals of biological products are associated with the active component themselves rather than product or process impurities. Tracking of which patients receive a given lot can be carried out to provide some relationship of differing levels of quality attributes between lots to clinical efficacy and/or safety. However, it is often the case that a single patient receives more than one lot through the course of treatment, making the assessment complicated. However, even with the above limitations, clinical trials can detect gross differences that may be associated with a given product variant such as those related to pharmacodynamic properties (e.g., glycosylation, charge variants).

It is extremely valuable to gain an understanding of the impact on the physicochemical and biological properties of the product when injected into humans or animals. One can maximize nonclinical and clinical studies by taking blood samples at various times postinjection and monitoring the effect on the product. There have been cases where the glycosylation of a product have been shown to be irrelevant to potency, PK, or bioavailability—effectively removing the need for a specification on a lot-to-lot basis. In addition, it has been found that some products get immediately processed or chemically modified (e.g., reduced or deamidated), thus making the control of such a product variant at lot release less critical. Nonclinical studies can be further enhanced by studying purified variants in vivo to gain an understanding of the impact on potency, PK, or bioavailability.

Development of Lot-Release Specifications

Selecting Product Quality Attributes and Tests for Specifications

The development of specifications is a life-cycle process, starting from the initial application to requesting regulatory permission to enter clinical trials, through product development, and the regulatory licensing application review and post-approval changes. The focus of setting specifications is different at each stage of development, as more process data become available, methods improve, and the ability to correlate analytical data to clinical exposure increases. Specifically, lot-release specifications are a set of test and acceptance criteria used to disposition (release) lots of material after they have been produced—they are required for release of both drug substance and drug product (Tables 2 and 3).

The tests selected for lot release are dependent on the characterization of the product and the understanding of the biochemical and biological profile of the product in relation to safety, efficacy, and consistency. It is not surprising that any quality attribute that directly links to safety and efficacy should be controlled with a specification test (e.g., potency, quantity). However, not all quality attributes are critical to safety or efficacy, and where such attributes have been shown not to impact potency or PK/bioavailability, it may not be necessary to set a specification. Yet, there may be attributes that do not impact safety or efficacy, but are markers for consistency of product lot-to-lot and may be considered a reflection for other variations not easily detectable by specification testing alone (e.g., certain glycosylation parameters); therefore, these require either inclusion in specification testing or control through internal action limits that,

Table 2 Test Procedures and Acceptance Criteria for Biotechnological/Biological Drug Substance Specifications

Appearance and description	A qualitative statement describing the physical state (e.g., solid, liquid) and color of a drug substance should be provided.
Identity	The identity test(s) should be highly specific for the drug substance and should be based on the unique aspects of its molecular structure and/or other specific properties. More than one test (physicochemical, biological, and/or immunochemical) may be necessary to establish identity. The identity test(s) can be qualitative in nature.
Purity and impurities	The absolute purity of biotechnological and biological products is difficult to determine and the results are method dependent. Consequently, the purity of the drug substance is usually estimated by a combination of methods. The choice and optimization of analytical procedures should focus on the separation of the desired product from product-related substances and from impurities.
Process-related impurities	Process-related impurities in the drug substance may include cell culture media, host cell proteins, DNA, monoclonal antibodies, or chromatographic media used in purification, solvents, and buffer components. These impurities should be minimized by the use of appropriate, well-controlled manufacturing processes.
Product-related impurities	Product-related impurities in the drug substance are molecular variants with properties different from those of the desired product formed during manufacture and/or storage. For the impurities, the choice and optimization of analytical procedures should focus on the separation of the desired product and product-related substances from impurities. Individual and/or collective acceptance criteria for impurities should be set, as appropriate. Under certain circumstances (e.g., through process validation), acceptance criteria for selected impurities may not be required.
Potency	A relevant, validated potency assay should be part of the specifications for a biotechnological or biological drug substance and/or drug product. When an appropriate potency assay is used for the drug product, an alternative method (physicochemical and/or biological) may suffice for quantitative assessment at the drug substance stage. In some cases, the measurement of specific activity may provide additional useful information.
Quantity	The quantity of the drug substance, usually based on protein content (mass), should be determined using an appropriate assay. The quantity determination may be independent of a reference standard or material. In cases where product manufacture is based upon potency, there may be no need for an alternate determination of quantity.

Source: From Ref. 1.

Table 3 Test Procedures and Acceptance Criteria for Biotechnological/Biological Drug Product Specifications

Pharmacopeial requirements apply to the relevant dosage forms. Typical tests found in the pharmacopoeia include, but are not limited to, sterility, endotoxin, microbial limits, volume in container, particulate matter, uniformity of dosage units, and moisture content for lyophilized drug products. If appropriate, testing for uniformity of dosage units may be performed as in-process controls and corresponding acceptance criteria set

Appearance and description	A qualitative statement describing the physical state (e.g., solid, liquid), color, and clarity of the drug product should be provided.
Identity	The identity test(s) should be highly specific for the drug product and should be based on the unique aspects of its molecular structure and for other specific properties. The identity test(s) can be qualitative in nature. While it is recognized that in most cases, a single test is adequate, more than one test (physico-chemical, biological, and/or immunochemical) may be necessary to establish the identity for some products.
Purity and impurities	Impurities may be generated or increased during manufacture and/or storage of the drug product. These may be either the same as those occurring in the drug substance itself, process-related, or degradation products that form specifically in the drug product during formulation or during storage. If impurities are qualitatively and quantitatively (i.e., relative amounts and/or concentrations) the same as in the drug substance, testing is not necessary. If impurities are known to be introduced or formed during the production and/or storage of the drug product, the levels of these impurities should be determined and acceptance criteria established.
Potency	A relevant, validated potency assay should be part of the specifications for a biotechnological or biological drug substance and/or drug product. When an appropriate potency assay is used for the drug product, an alternative method (physicochemical and/or biological) may suffice for quantitative assessment at the drug substance stage. In some cases, the measurement of specific activity may provide additional useful information.
Quantity	The quantity of the drug substance in the drug product, usually based on protein content (mass), should be determined using an appropriate assay. In cases where product manufacture is based upon potency, there may be no need for an alternate determination of quantity.
General tests	Physical description and the measurement of other quality attributes is often important for the evaluation of the drug product functions. Examples of such tests include pH and osmolarity.

Source: From Ref. 1.

when exceeded, would trigger proper quality investigations prior to disposition of the material.

As discussed previously, the correlation between a product attribute and safety/efficacy is not an easy one to make. During product development, it is rare that multiple lots are tested in the clinic that cover the range of product variability within manufacturing capability. In addition, the earliest nonclinical and clinical studies where most safety testing occurs happen at a time when most analytical techniques are at the early stage of development, if present at all. Therefore, there is a need to keep retains of all batches of material manufactured during development to maximize data using the final, validated (or most highly product specific) assays.

For biological products in particular, the most difficult assessment that can be made regarding product quality attributes and the link to safety or efficacy is one of the impact on immunogenicity (7,8). There is little understanding of exactly how immunogenicity of marketed products was induced, although there is limited evidence that aggregation, some posttranslational modifications (such as deamidation), and the presence of specific B- or T-cell epitopes may impact immunogenicity rates. Therefore, in the absence of direct data correlations, performing a risk-based assessment when considering the immunogenicity implications in setting specifications should be undertaken. This involves the assessment of factors such as the patient population, disease type, dosing regimen, and the consequences of mounting an immune response, etc.

Setting Acceptance Criteria for Specification Tests

It is clear from previous sections that setting acceptance criteria for a specification test is a multivariate process that occurs once appropriate tests have been selected. Statistical analysis may not always be able to assist in the setting of specification-acceptance criteria, especially if only a few lots of material went into clinical studies. Statistical approaches such as using tolerance intervals can provide some detail for the prediction of variation due to process capability, but cannot provide the assurance that levels of variants falling outside that exposed to patients in the clinic do not impact safety or efficacy. Therefore, the acceptance criteria are often based on a combination of all the information available about product characterization, especially the effect of product variants on safety or efficacy, how the product is affected in vivo, the variability in the manufacturing process, exposure levels in patients, and some level of standard deviation around these ranges.

The failure to meet a specification results in an out of specification (OOS) test result, leading to rejection of the lot if the OOS is confirmed. It is possible in the United States to provide a Prior Approval Supplement to the Food and Drug Administration (FDA), requesting release of the lot if a previously nonapproved reprocessing step is required or if an investigation proves that the OOS does not impact product safety and efficacy for some reason (which may lead to redevelopment of either the specification test or its acceptance criteria).

IPCs versus Specifications

Assessing product quality is not just required at drug-substance or drug-product release. Testing the quality of the product in this way is not deemed appropriate in the absence of a well-characterized and controlled manufacturing process. An understanding of critical process parameters (i.e., those process parameters that can affect product quality or the efficiency of the process) is as crucial to assuring product quality as is an understanding of the product itself.

Therefore, following process characterization, a set of tests and associated criteria in the form of alerts, actions, and, more rarely, rejection limits are created for each step in the manufacturing process—namely in-process communication. The role of limits versus specifications is described in the section titled Limits versus Specifications.

If the process has been characterized and validated to a stage where clearance of product variants or process-related substances has been assured, the need for tests for these materials at lot release may no longer be necessary (e.g., host cell proteins or DNA).

Raw Material and Pharmacopeial Specifications

The assurance of product quality relies not only on a well-characterized and controlled manufacturing and lot-release process, but also on the control of the raw materials that are used in the manufacturing process itself. This includes materials used in fermentation (e.g., sera, medium, antifoam), purification (e.g., buffers, column resins, filters), and formulation/filling (e.g., buffers, excipients, container closures).

A thorough characterization of the critical quality attributes of each raw material is required; i.e., what characteristics of the raw material are important for the assurance of product quality. Methods required to characterize raw materials may well be different from those required for product, and therefore raw material–specific methods may have to be developed. As for the product itself, the design of stability studies on raw materials is critical as an understanding of the degradation pathways of raw materials and their impact on product quality is important. A specification for each raw material should be developed from the data derived from characterization, stability, and small-scale studies as well as from large-scale manufacturing runs.

When setting specifications for raw materials or products, consideration must also be given to the requirements contained in regional pharmacopoeias. Pharmacopoeias contain monographs defining requirements (which are often legally required depending on the regulations of a region) for analytical procedures and acceptance criteria (e.g., sterility, endotoxins, microbial limits, volume in container, uniformity of dosage units, and particulate matter) or quality standards for raw materials (e.g., chemicals such as sodium chloride) or even the product itself (e.g., erythropoietin in the EP). When such a monograph exists, it is likely that the tests must be included in the specification, and acceptance requirements must, at a minimum, be met within the specification.

Limits versus Specifications

Following product and process characterization, both product quality attributes and process parameters can be classified as critical, key, or non-key. As discussed, any quality attribute that impacts safety or efficacy, is a direct measure of product consistency, or is a process parameter that affects product quality, requires a specification test and acceptance criteria. However, there may be attributes or process parameters that are not directly critical to product quality, yet may be an indicator of the control over either the process or the product.

Such attributes and parameters should be controlled via limits, whereby excursion from those limits causes either an "alert," requiring further monitoring but no investigation, or an "action," whereby an investigation is automatically triggered. Alerts are not often linked to lot disposition, but actions (and the subsequent investigations) are usually tied to a lot via a nonconformance—the investigation of which has to result in the assessment of no impact on product quality before a lot can be released. The criteria for an alert versus an action are usually dictated by the limits set around an expected median, with alerts being triggered by excursions from limits tighter than those that then trigger an action. The process for setting limits is similar to that for specifications, with analysis of data derived from the various sources already discussed.

Acceptance criteria for specifications can also have alert and action limits associated within them. It is encouraged that a company derive such limits within a specification. It is generally the case that specification acceptance criteria are often set wider than the clinical experience at the time of licensure, to allow for future process variability, so tighter alert and action limits, more reflective of clinical exposure, encourage process monitoring to ensure no drift or unusual outliers occur in the manufacturing process. There is a distinct advantage to setting alert and action limits within specification acceptance criteria. Moving limits within the specification allows for a change to either the product or the process can be detected and investigated before product quality is adversely affected and/or a lot requires rejection by failing a test with an OOS.

EXPIRATION DATING AND STABILITY ASSESSMENT

Common elements guide the development of specifications and the estimation of expiration dating periods. Requirements for both of these activities are linked through method development, product characterization, product understanding (i.e., impurity and stability profiles of clinical batches), and the capabilities of the manufacturing process. Information is gathered for both specification and expiration dating development throughout the product development life cycle, starting with toxicology testing and continuing throughout product commercialization and marketing.

In an ideal world, specification and expiration dating development would rely on information gathered from a large dataset of batches, simple and precise test methods, straightforward manufacturing processes, and homogeneous, well-defined products. Unfortunately, none of these favorable attributes describe biotechnological products and processes. Those working in biotechnological product development and commercialization must instead deal with small numbers of batches available for statistical analysis, complex and multicomponent products (and the more difficult, less precise analytical methods that measure attributes of these products), and manufacturing processes so complex that they must be considered as part of product definition.

Stability programs and associated product testing provide evidence on how the quality of a product varies with time under the influence of environmental factors such as temperature, humidity, and light. As described above, data collected from stability studies are included in product evaluations for specification development and also permit the establishment of recommended storage conditions (RSCs), retest periods, and shelf lives. A wealth of information is available through regulatory guidance for implementation and management of stability programs. General guidance is applicable to all types of product programs and is quite prescriptive (9). Specific guidance for biotechnological products is of necessity more general (10), due to the complex nature of proteins and peptides, whose integrity may be maintained by noncovalent as well as covalent interactions and whose sensitivity to environmental factors necessitates maintenance of well-defined storage conditions. Because of the complex nature of proteins and peptides, even guidance for biological/biotechnological products only applies to well-characterized molecules (10).

Stability Assessment is Phase Specific

Any stability program must provide sufficient information to ensure identity, strength, quality, purity, and potency of the product. However, the amount of information available for product evaluation will naturally vary with the product developmental phase; therefore, regulatory expectations increase as product commercialization progresses.

Throughout product development, an understanding of comparability between the stability of the preclinical and clinical batches and all subsequent batches must be maintained as the manufacturing process and analytical methods evolve. Results from stability studies, especially stress or accelerated studies, often can elucidate subtle changes in the product during development that are not apparent with release and characterization testing (11). Of course, adequate documentation of the product history during development and consequent testing results is a requirement for success.

Preclinical Stability Program

The purpose of stability studies during preclinical development is to ensure the stability of representative batches of drug substance and drug product for the duration of the preclinical program. Because of the lack of well-defined analytical methods, limited characterization of product, and small-scale manufacturing of batches whose formulation may not represent that of the product at later developmental stages, it is acceptable at this stage to use a limited, templated stability protocol, utilizing general assays that have not yet been tailored to the attributes of the particular molecule under study (12).

Early Clinical Phase Stability Program

The main concern at this phase of development is safety. As with the preclinical product, product formulation may not represent the final, commercial formulation. However, it is expected that studies are performed on drug substance and drug product with assays that are able to confirm that the product remains within established safety parameters for the duration of the clinical studies and that a defined, appropriate stability protocol is used (12).

Late-Stage Clinical Phase Stability Program

Patient safety is still a primary concern, so evaluation of the representative drug substance and drug product for acceptability of material throughout the clinical study duration is a key driver for stability studies. However, at this stage, it is important as well to understand product degradation pathways, have analytical methods in hand that are shown to be stability indicating, and follow a defined stability protocol utilizing these methods. A stability protocol, listing test methods, sampling timepoints, expected duration of study, and container interaction assessment should be developed at this time. By Phase III trials, complete, formal stability protocols should be used for drug substance and drug product in the final formulation and primary packaging material (13).

Commercial Stability Program

A protocol must be defined that meets regulatory expectations for assays, timepoints, and storage conditions. At least one lot must be added to the stability program during each year in which product is manufactured.

Postapproval Changes

Batches representing any changes to process, methods, containers, etc., that have the potential to impact product stability must be added to the commercial stability program.

Development of Stability-Indicating Methods

From the above, it is clear that requirements for stability programs (i.e., batch selection, study duration) are well defined in regulatory guidance for all stages of product development. The value of these stability programs, however, is completely dependent on the development of appropriate stability-indicating assays.

It is expected that methods used in stability programs cover features susceptible to change during storage that could affect quality, safety, or efficacy (10). Methods shown to be stability indicating are used in stability programs to assess the quality of product batches over time under specified conditions of storage and to establish the retest or expiration dating of products. Stability-indicating assays are defined as procedures that are used "to assess the presence/absence of degradants in a product…capable of accurately measuring changes in the product that can occur under conditions of physical or chemical stress" (14).

The complex nature of proteins and peptides requires multiple, complex analytical methods to monitor the appropriate physicochemical, biochemical, and immunochemical properties of the product. These methods should provide quantitative detection of degradation whenever possible to allow for development of specifications and analysis of product trends. Good documentation of product characterization studies and method history is essential both for establishing product comparability and for demonstrating the utility of the chosen analytical methods included in the stability program. The existence of development reports, documenting product characterization and method development progress, is not only best practice but is becoming a regulatory expectation.

Documentation should also be available to demonstrate that analytical methods used in stability studies are stability indicating, and methods should be qualified or validated for stability-indicating capability as appropriate to the product phase of development. There is little regulatory guidance elucidating the different expectations for method qualification and validation; however, the reader is directed to a review of FDA and industry perspectives on the differences between method qualification and validation (15).

Characterizing Degradation Pathways

For the purposes of stability testing, process-related impurities seldom demonstrate increases in concentration over time and are not typically monitored in stability programs. Assays are generally developed to monitor those product-related substances and impurities that do increase over time. Product characterization provides information on types of product-related substances and impurities that should be monitored in a stability program. Product characterization requirements, in alignment with requirements for development of stability programs and stability-indicating assays, are phase dependent. Forced degradation studies on drug substance and formulation-screening studies on drug product are typically conducted during Phase II to determine potential degradants that may increase over shelf-life. By Phase III, such product-related substances and impurities should be characterized and quantified, when possible, in clinical batches. The importance of monitoring product changes is highlighted by the possible adverse consequences of protein degradation, which include such concerns as changes in biological activity and increases in immunogenicity.

Developing appropriate stability-indicating assays, however, requires an understanding of the molecule's potential modes of degradation. Stress testing is typically performed on the drug substance and includes studies of effects of increasing and decreasing temperature (and cycling between temperatures), humidity (when applicable), photolysis, oxidation, pH variation, and agitation.

Such studies provide material for product characterization and method validation, may guide development, and demonstrate suitability of analytical methods. However, for proteins and peptides, such studies may produce degradants not seen when the product is held at RSCs. In such cases, it may not be necessary to characterize such degradants or provide assays for their detection in stability protocols (9).

The information from stress studies, while perhaps not useful for determination of degradation at the RSC, should be used as appropriate to support the validation of analytical procedures.

Expiration Dating Requirements

Expiration dating of clinical trial materials. Accelerated-temperature stability studies may generally be used to support tentative expiration dates for small-molecule products (*CFR Sec. 211.166*), as long as RSC studies are conducted until the tentative expiration date is verified.

However, protein degradation pathways, unlike those of small-molecule products, are complex and must generally be monitored using multiple assays. Because of this complexity, associated with reliance on assay results that are not as well defined as those for small-molecule analyses, and the intrinsic heterogeneity of biological products, it is generally necessary to use data derived from RSC studies to support expiration dating for proteins. Unlike small-molecule products, extrapolation of dating period estimates beyond the observed stability timepoints or reliance on results from accelerated conditions are not considered acceptable for protein products.

This constraint creates a dilemma for stability dating for biotechnological clinical trial materials. In all jurisdictions, RSC studies may be performed in parallel with clinical trials. For clinical trials conducted in the United States, expiration dating is not required on labeling of clinical trial materials. It is expected that products will be stored at RSCs, under strict control, at clinical sites. If a batch should fail a timepoint at its RSC, clinical trial material is removed and supplies are restocked.

The situation in the European Union (EU), however, has changed with implementation of the EU Clinical Trial Directive (2002/20/EC), which requires that each product unit be labeled with an expiration date, which again, can only be obtained from RSC stability data.

Manufacturers of clinical trial materials for trials in the EU have two options: (*i*) manufacture a batch to be used only for expiration dating prior to manufacture of clinical lots, allowing stability data to be collected to support expiration date labeling of product; or (*ii*) label product at risk. The second option, when used, relies on International Conference on Harmonisation (ICH) guidance on extension of expiration dating (16) to statistically extend stability data beyond that collected at RSCs.[a]

---

[a] Note: International Committee for Harmonisation guidances are available on www.ich.org. FDA guidances are available on www.fda.gov.

Both options have drawbacks. The first option generally requires frequent product relabeling at the clinical sites as expiration dating is extended; the second option requires the assumption of risk that the product will meet its acceptance criteria when tested at RSC until the labeled expiration date.

No matter which option is followed, expiration dating of clinical trial materials is subject to the phase-dependent issues previously cited: product characterization, development of process and analytical methods, and qualification/validation of stability-indicating methods concurrent with collection of stability data and conduction of clinical trials. Ensuring that work being done in parallel contributes to overall product commercialization requires teamwork across functions with respect to timing and prioritization of each aspect of development.

Extrapolation of Data

Extrapolation is defined as the practice of using a known set of data to infer information about future datasets. For stability data, the underlying assumption of this practice is that observed change and rate of change in product attributes will continue to apply as future data is collected. Extrapolation therefore relies on a comprehensive understanding of the product through characterization studies and the manufacturing process as well as reliable, specific stability-indicating test methods. These are all in short supply in early product development.

As stated above, extrapolation of dating period estimates beyond the observed stability timepoints is not generally acceptable for biotechnology products. However, manufacturers can utilize regulatory guidance [ICH Q1E: Extrapolation of Data (16)] to justify this approach when requirements are met. Even if extrapolation is used, estimated shelf-life must be limited to the *shortest* of the extrapolated variables.

Considerations for extrapolation of data as described in ICH Q1E include data variability, for both the product and the test method, as well as product stability. For processes/products with low variability and high stability, it is considered appropriate to extrapolate data. For processes/products with high variability and low stability (most protein products), extrapolation is less useful as a predictive tool, even when much product, process, and analytical knowledge is at hand.

Stability Requirements to Support License Applications

Regulatory requirements for drug substance and drug product stability programs to support license applications are well described in regulatory guidance (9,10). The purpose of primary stability studies is to establish, based on testing of a limited number of batches of drug substance and drug product, appropriate retest or expiration dating periods and label storage conditions applicable to all future batches of the drug substance or drug product manufactured and packaged under similar circumstances. This approach assumes that inferences drawn from this small group of tested batches extend to all future batches. Therefore, tested batches should be representative in all respects (e.g., formulation, batch size, container closure system, and manufacturing process) of the population of all batches and conform to all quality test attributes of the drug substance or drug product.

Expiration dating for drug substance to support license application: The number of batches required to support licensure is dictated by relevant guidances—"An adequate number of batches of each drug product shall be tested to determine an appropriate expiration date and a record of such data shall be maintained." (17). "The purpose of a stability study is to establish, based on a *minimum of three batches* of drug substance or drug product, a retest period or shelf-life and label storage instructions *applicable to all future batches* manufactured and packaged under similar circumstances." (9).

A proposed expiration dating period is derived by the applicant from results on three batches of drug substance, manufactured to a minimum of pilot scale and using a method of manufacture and with quality attributes representative of the process to be used for commercial batches. For protein products a minimum of six months of RSC stability data should be available at the time of filing. If data from pilot scale batches are used in the submission, a commitment must be made to place the first three commercial-scale batches into the stability program.

Expiration dating for drug product to support license application: A proposed expiration dating period is derived from analyses of stability results on at least three batches of drug product, of the same formulation and packaged in the same container-closure system as proposed for marketing. As with drug substance, the method of manufacture and the product quality attributes should be representative of the process to be used for commercial batches. For proteins, a minimum of six months of stability data should be provided at the time of filing, with a commitment to place the first three manufacturing-scale batches into the stability program if data from pilot-scale or bench-scale batches were filed in the application. In addition to the real-time and accelerated-stability studies to support expiration dating for drug product, special studies should be performed to determine the impact of handling, shipping, and distribution of the product worldwide. These studies should be initiated prior to license submission and should typically include cycling studies at different temperatures and physical agitation conditions on a representative drug product followed by real-time monitoring under the established stability protocol.

Expiration dating for in-process materials: Stability data should be provided to support hold times and RSC for in-process materials when needed. In-process materials held long-term (generally greater than 30 days) are generally assigned an expiration date on the basis of stability studies. For hold times of less than 30 days, process-validation studies are often used to determine the suitability of held material.

For both drug substance and drug product, stability study results other than those described above, such as data on small-scale batches, can be filed to support expiration dating.

Proposing Alternatives for Commercial Expiration Dating

It is sometimes appropriate, when the drug substance shows little change with time, to propose a retest date rather than an expiration date. Both are defined with respect to the period of time during which the drug substance is expected to remain within acceptable limits. The difference is that drug substance with a retest date can be retested and, on meeting acceptance criteria, can be immediately used. In order to use drug substance with an expiration date for further processing after that date, an expiration extension must first be granted by the appropriate regulatory authorities. Because most protein products are labile, expiration dates rather than retest dates are generally employed (9).

When a product attribute shows marked degradation over time, it may be advantageous to propose release as well as shelf-life acceptance criteria. Requiring tighter control of product at release may prevent a need to shorten expiration dating of batches after product is released. This concept is defined in regulatory guidances (1):

> The concept of release limits versus shelf-life limits may be applied where justified. This concept pertains to the establishment of limits which are tighter for release than for the shelf-life of the drug substance or drug product. Examples where this may be applicable include potency and degradation products.

However, it is considered that a comprehensive understanding of product characteristics coupled with a robust and well-defined manufacturing process can generally provide adequate assurance that a product acceptable at release will remain within specification throughout shelf-life, obviating the need for separate limits.

Extending Commercial Expiry Dating

Requirements for extension of expiration periods include stability data to support the proposed extension from a minimum of three production lots, using the approved stability protocol (18).

Statistical analysis of stability data is not always warranted or possible. When statistical analysis is planned or used, the method should be described and results of the analysis should be provided in the application.

CONCLUSIONS

Development of biotechnology product specifications and expiration periods depends on several parallel product development activities and requires a well-coordinated product commercialization plan. Of great importance is timely development of appropriate product characterization and stability-indicating assays. Tests must be based on known product characteristics and have the ability to detect and measure, if appropriate, all possible product variants, including degradation products.

A thorough product understanding is essential to allow definition of the appropriate tests required to set specifications and define stability studies. Therefore product characterization information must be collected and available as assays are developed, qualified, and validated. A well-defined and well-controlled manufacturing process must be in place to allow representative material to be included into the product characterization studies as well as the stability program to monitor the effects of time, temperature, and other variables on the product quality attributes.

In all commercialization activities, more information is obtained as product development progresses. For this reason, early phase requirements for product and process understanding are less stringent; however, patient safety requires that product quality, including shelf-life, be justified to the greatest extent possible with the limited data available. For commercialization, it is expected that methods are validated for both lot-release and stability-indicating capability and that they are shown to be capable of measuring the true product variant profile, including degradation pathways of the product to allow for lot-release and expiration periods to be set, which reflect the time at which product remains safe and efficacious.

REFERENCES

1. International Conference on Harmonisation: ICH Topic Q6b. Specifications: test procedures and acceptance criteria for biotechnological/biological products www.ich.org.
2. FDA Code of Federal Regulations, 21CFR601.12.
3. Brown F, Mire-Sluis AR. Analytical characterisation and assay of cytokines and growth factors. Dev Biol Stand 1999; 97.
4. Pharmaceutical cGMPs for the 21st century—a risk-based approach. Final Report—Fall 2004. Department of Health and Human Services, U.S. Food and Drug Administration, September 2004 (available from www.fda.gov/cder.)
5. Mire-Sluis AR, Gaines-Das R, Gerrard T, Padilla A, Thorpe R. Biological assays: their role in the development and quality control of recombinant biological medicinal products. Biologicals 1996; 24:351–362.
6. Thorpe R, Wadhwa M, Mire-Sluis AR. The use of bioassays for the characterisation and control of biological therapeutic products produced by biotechnology. Dev Biol Stand 1997; 91:79–88.
7. Koren E, Zuckerman L, Mire-Sluis A. Predicting immune responses to human recombinant biopharmaceuticals. Curr Pharm Biotechnol 2002; 3:349–360.
8. Shankar G, Shores E, Wagner C, Mire-Sluis A. Scientific and regulatory considerations on the immunogenicity of biologics. Trends Biotechnol 2006; 24:274–279.
9. International Conference on Harmonisation. ICH Topic Q 1A. Stability testing of new drug substances and products.
10. International Conference on Harmonisation. ICH Topic Q5C: Stability testing of biotechnology/biological products.
11. International Conference on Harmonisation. ICH Topic Q5E: Comparability of biotechnological/biological products subject to changes in their manufacturing processes.

12. FDA guidance for industry: content and format of INDs for Phase I studies of Drugs, Incl. Well-Char. Therapeutic Biotech-Derived Products.
13. FDA guidance for industry: INDs for Phase 2 and 3 studies; chemistry, manufacturing and controls information.
14. FDA guidance for industry: analytical procedures and method validation—chemistry, manufacturing and controls documentation, CDER/CBER August 2000 [Draft].
15. Method qualification vs validation—what is the difference? CMC Strategy Forum July 2003, Ritter, Simmerman, Advant, McEntire, Hennessey, Mire-Sluis, Joneckis BioProcess Int 2004; 2:32–47.
16. International Conference on Harmonisation: ICH Topic Q1E: evaluation of stability data.
17. Code of Federal regulations 21CFR211. Sub part I. Laboratory controls section 211–166. Stability testing. 2006.
18. FDA guidance for industry: format and content for the CMC section of an annual report.

14

Development of Drug Products: Similarities and Differences Between Protein Biologics and Small Synthetic Molecules

Eugene J. McNally

Gala Biotech, a Catalent Pharma Solutions Company, Middleton, Wisconsin, U.S.A.

Jayne E. Hastedt

ALZA Corporation, Mountain View, California, U.S.A.

INTRODUCTION

The breakthrough in recombinant DNA techniques in the 1970s that allowed for the overexpression of proteins in cell lines and the creation of the first stable hybridoma for producing monoclonal antibodies by Kohler and Milstein in 1975, have collectively brought about entire new classes of therapeutic entities that have begun to enter the market as approved drugs. The first monoclonal antibody to appear on the market occurred 20 years ago with the launch of Johnson and Johnson's Orthoclone product for treating kidney transplant rejection episodes. This was followed by multiple classes of protein therapeutic agents: enzymes, interleukins, cytokines, hormones, etc., via three major pathways for the production of biologics, recombinant expression in cell lines, hybridomas, and transgenic animals. It is estimated that there are more than 400 biologics currently in various stages of drug development; approximately 40% of these are monoclonal antibodies (1).

This is in contrast to the production and sale of small-molecule drugs, which has been occurring for hundreds of years. However, the rational development of small-molecule drugs in the United States is a more recent occurrence that can be traced back approximately 60 years to the passage of several key Food and Drug Administration regulations. The first of these regulations required that all drugs be stable throughout their shelf-life, which forced manufacturers to take a different approach to formulation development to maintain the stability of their products (2). At about the same time, the Guidance on Current Good Manufacturing Practices was published, requiring that manufacturers document their stability programs and perform statistical analysis on the results. This followed from the introduction of mass-produced spectrophotometers and high-performance liquid chromatographs, analytical tools that were capable of detecting and quantitating small amounts of impurities and degradents in drug products (2). This expectation for stability along with the more recent regulatory requirements to justify a chosen formulation has driven the drug product formulation process, on the part of many firms, into a systematic, rational approach to design and optimization. The regulatory expectations for a rational development approach are outlined in the International Conference on Harmonization guidance on pharmaceutical development (3).

Although the differences in the physical and chemical properties between synthetic (small) molecules and protein biologics (peptides, antibodies, and recombinant proteins) are significant, the basic approach taken by the pharmaceutical scientist in formulating and delivery has become remarkably similar. The major differences between small-molecule synthetic drugs and biologics are their degree of heterogeneity and complexity and the higher-order structure of biologics, which gives rise to their functional activity that must be preserved throughout the production, formulation, and delivery process. This higher-order structure, in addition to large molecular size, has dictated specific delivery pathways for proteins. Another mark of time is the recent debate over the production of biologic generics. Patents protecting some of the early protein biological products are expiring and much attention and debate is occurring on the pathway for approving "generic" versions of these drugs and whether the concept of a generic biologic is even possible. The distinction between small-molecule drugs and protein biologics is at the core of this debate. This chapter will provide an overview of the similarities and differences between small and large molecules to provide an introduction to those formulators already skilled in the art of small-molecule formulation and also to give some background to those interested in the debate surrounding biogenerics.

INHERENT DIFFERENCES OF SMALL VERSUS LARGE MOLECULE COMPLEXITY AND SITE OF ACTION

It is interesting to consider the differences between the mechanism and the site of action of small-molecule drugs and their protein counterparts in eliciting a biological response. A recent comparison of the size and complexity of a small-molecule drug versus a large biologic, such as an antibody, used the analogy of comparing a 20-pound bicycle with a 25,000-pound F16 jet fighter airplane (4). Many of the protein biologics being developed as therapeutics are naturally occurring or are at least nearly identical to their human analogs and are being developed as replacement or supplemental therapies. These naturally occurring proteins possess complex structures and often act within the blood stream or on the surfaces of cells to cause their therapeutic effect. In contrast, the vast majority of small-molecule drugs are produced synthetically and possess a small-enough molecular size to be capable of passing through cell membranes to exert their therapeutic benefit by interacting with receptors located inside the cell.

EFFECT OF PROCESSING CONDITIONS ON API PROPERTIES

For both classes of drugs, the processing conditions that the active pharmacentical ingredient is exposed to can have significant effects on the ultimate stability of a batch of drugs. For small molecules, the end stages of synthesis where the molecule is isolated as a solid, often a crystallization process, can have dramatic consequences on the physical properties of the drug, such as solubility, stability, etc. Product development scientists have become accustomed to monitoring and tracking the changes in the crystallization process and monitoring their effects on the formulation and processing of a small molecule. Likewise, for the biologics, a similar exercise must be performed. The conditions under which a protein drug is grown in culture and isolated during downstream processing can have significant effects on the chemical and physical profile of the product. Changes in these conditions can cause a shift in the heterogeneity profile of the biomolecule. Again, little initial thought is given to these changes early in the development process, so they are often poorly understood. Unlike small molecules, the biological API is often isolated in the liquid state. Although little thought is given to the ultimate route of delivery, an initial formulation often emerges during early purification recovery efforts as the bulk API needs to be stored and held, often in a buffered salt solution. This is often the starting point of stability assessment, and by default becomes the beginning of formulation development, often without recognition. The starting point for drug product formulation is the same for small molecules and biologics; it is at the point of synthesis where the chemical entity is isolated, or during fermentation/purification, where the growth and isolation conditions affect the heterogeneity of the isolated protein respectively. The point is that regardless of the class of drug, the formulation group needs to be cognizant of the effects made by changes on the part of the API group on their formulation/delivery efforts.

ANALYSIS

For small synthetic molecules, the complete elucidation of chemical structure is a regulatory expectation for drug approval. This is in stark contrast to many biologics, where a complete proof of structure exercise is not possible due to the complex heterogeneity of the protein. This lack of ability to completely demonstrate the structure of the protein is what complicates the notion of producing a generic version of a biological drug. The premise under which small-molecule drugs are granted abbreviated approval as generics is that by demonstrating that the chemical entity is identical in structure to the innovator product, repeating safety and clinical efficacy studies is not necessary. The concept of a "biogeneric" for the less-complicated protein products that are nonglycosylated—molecules such as insulin and growth hormones—has become a realization. The debate on how to approve those compounds that possess large degrees of heterogeneity is ongoing and will continue for some time, as these molecules are difficult to characterize. The recognition that complete characterization of the highly glycosylated, large proteins is likely to be extremely difficult has given rise to the concept of "biosimilars," where the debate has shifted to one of how much similarity to the innovator must be demonstrated and how much safety and efficacy testing must be undertaken to receive product approval. Another striking difference between the two classes of compounds is the number of analytical methods that need to be deployed. For synthetic molecules, only a few methods are necessary to characterize the drug in the dosage form, while proteins require several complementary methods to provide a profile of the molecule to characterize its heterogeneity, and the result is often a qualitative assessment versus a quantitative one that is often standard with synthetic drugs. In addition to process impurities, consideration also needs to be given to those impurities that are related to the cell line itself and also adventitious impurities that can be introduced during processing. Analytical methods quantitating the amount of host cell proteins, cellular DNA, and other viral and bacterial contaminants need to be incorporated into the final analysis of a biologic product. Another introduction on the biologics front is the recent call for more detailed analyses of extractables and leachables in biological products, similar to the expectations for small molecules delivered via the parenteral and pulmonary routes. One important note is the report of extractables being linked to immunogenicity events, which emphasizes the profound safety implications for this type of testing in biological products (5).

FORMULATION AND DELIVERY

Many of the formulation and product development strategies used for biologics and small molecules are similar. Often, what differs are the development and business strategies companies use to progress them through clinical trials. These strategies often dictate the amount of work and effort expended at a particular point in the development cycle. Phase I formulation and delivery approaches are similar for small molecules and biologics in that many companies do not want to expend significant efforts developing and testing formulations for early first in human clinical trials intended as single dose or limited multidose studies. Instead, small-molecule developers will choose a "powder in bottle" or "drug in a capsule" approach using neat API for support of these quick first human exposure studies. Biologics producers will provide bulk drug in a refrigerated or frozen state directly to the clinical setting for delivery by injection or nebulization for the pulmonary route. These are considered proof of concept studies where the intent is to get some human pharmacokinetic or surrogate endpoint information before investing in full drug development.

Many biologics have been and continue to be delivered by injection versus the large majority of small molecules that are delivered orally. Sixty-four percent of small-molecule prescription drugs are delivered orally (1); while there are no approved protein products delivered orally in the United States, there are a number of companies pursuing this delivery route for proteins, one of which is discussed in Chapter 13. Over the last five years, there has started to be a shift in the delivery of proteins from solely an intravenous injection mode to one that is more compatible with outpatient therapy such as intramuscular and subcutaneous injection and inhalation. In addition to a change in delivery modes, there has also been a slow shift toward formulating proteins under conditions that minimize the requirements for cold storage shipment toward the gold standard for small-molecule storage, which is room temperature stability. All of these changes in formulation and delivery are being driven by the expanding world marketplace, which is demanding greater consumer convenience and room temperature storage products to treat infectious diseases regardless of global location. For the pharmaceutical industry, the evolving product requirements translate into less-invasive means of delivery in outpatient settings and products that will withstand higher temperature fluctuations at low consumer risk and costs while maintaining high benefit to the patient.

When the time comes to formulate either a biologic or a small molecule, the age-old rule of developing the simplest formulation possible should prevail. Excipients used should be justified with a reason for their inclusion. Excipient usage and function vary widely between small molecules and biologics, and it is difficult to make generalizations on the purpose and use of excipients between these two classes of compounds. For many small-molecule formulations, a number of excipients are included as processing aids to ensure adequate flow and ultimate disintegration of the tablet. Some stabilizers for minimizing oxidation can be included but should be done rationally, after studying the degradation process. In protein formulations, excipients are used to buffer a formulation to an optimal pH for delivery or for stabilization purposes. Excipients are added to the lyophilization process to yield pharmaceutically elegant finished products and to stabilize the protein during the freezing process; these have all been discussed in detail in Chapter 9. In addition, for biologics, one has to consider the need for protecting the protein during processing with respect to temperature, agitation, and adsorption to surfaces, and from exposure to oxidative environments, shear forces, pressure, etc. When either a small molecule or a biologic is formulated to provide a solid dosage form, one of the prerequisites is that the bulk drug have good flow properties. The aim of much of small-molecule formulation efforts is to improve on a material's bulk density and hence flow by techniques such as wet or dry granulation. For biologics, it is a significant challenge to isolate the protein in a freely flowing state from a lyophilization or spray drying process; in fact this has been the subject of much development effort in the production of dry powder delivery systems for the pulmonary insulin products (Chapter 10). New processing equipment may need to be developed for isolating biologics in the solid state due to the difficulty in designing both stability and flow properties into the manufacturing process. For formulations that are marketed as liquids, there is often the need to incorporate some means of preservation, and this requirement holds for both small molecules and biologics. The requirement for microbial preservation is universal and is dictated by the route of delivery. For biologics, an additional consideration is that regardless of the route of delivery, preservation must be addressed, as protein systems serve as excellent growth media for bacteria and, when left unprotected, are often degraded by microbial contaminants. One such example is a protein intended for oral delivery. Processing will be conducted under aseptic conditions, even though the oral route of delivery does not require production of a sterile dosage form.

New delivery methods that are more convenient and less invasive to the patient are needed for biologics. Several companies are pursuing new means of needleless injection, which are applicable to both small molecules and biologics. One needleless system of delivery was discussed in Chapter 11. At the time of the writing of this chapter, several versions of inhaled insulin are being either marketed or developed. One means of delivery is via pulmonary deposition and absorption. Another means is by delivery to the back of the pharynx via a simple delivery device (6). Some biologics producers are developing new means for stabilizing protein drugs so that they can then be incorporated into conventional dosage forms using conventional delivery devices. One such approach discussed in a previous chapter using unique stabilizer molecules to enhance stability and the ability to transport the protein across biological membranes (Chapter 13).

THE PRODUCT

One last point that is true for both classes of compounds is that the route of drug delivery and the need for a delivery device must be considered prior to undertaking any formulation development activities. If a delivery device will be used with a product, the physicochemical form of the drug, the product contact material construction, compatibility, and physical effects on the drug must be considered at the onset of formulation development, or else much time and effort can be wasted in having to repeat studies once it is realized that a delivery device must be incorporated into the drug product. While the need for drug delivery is present for small synthetic molecules, it is particularly critical for biologics due to their large complex nature, which often requires parenteral administration.

STABILITY AND EXPIRATION DATING

Stability requirements for biologics are more restrictive than for small molecules. A large majority of proteins require refrigeration, while the large majority of small-molecule drugs are stable at room temperature. These storage requirements for biologics become important clinically when considering shipment of materials, the need for home preparation by the patient, and the consequences of inadvertent freezing and thawing/heating cycles. So there are remarkable differences between the storage requirements for biologics and small molecules. The same can be said for the assessment of stability. When considering small-molecule stability, the ability to perform a mass balance assessment is often possible and serves as a valuable exercise in aiding the understanding of chemical degradation mechanisms. For large molecules, the stability of a single amino acid can be dependent on the flanking order of amino acids in the primary amino acid sequence, and thus not all amino acids of the same chemical structure are of equal stability. This stability is also dependent on the proximity to other amino acids in the tertiary structure, and it can prove extremely challenging to identify which particular amino acid in the primary sequence is degrading. This unique nuance of protein molecules makes their stability assessment and prediction challenging. While expiration dating for small molecules lies heavily on accelerated testing at elevated temperature and humidity, such extrapolations for proteins are difficult to make as some of the higher-order physical structural changes are not accurately predicted by Arrhenius kinetics. This pushes much of the stability assessment into real-time storage conditions, which often limits the shelf-life assigned due to the long waiting periods necessary to collect this data. However, regardless of how complicated the stability assessment might seem, much can still be done to determine protein degradation kinetics and isolate decomposition products. In contrast to the unwritten lore in the literature about protein pharmaceuticals, Arrhenius kinetics can be applied selectively, and formulations containing stabilizing excipients can have a major impact on producing biological drug products with acceptable shelf-lives (7).

CONCLUSIONS

The similarities between formulation of drugs for small molecules and for biologics are many. While the technical intricacies of the chemistry and analysis are vastly different, the considerations for producing a stable dosage form capable of being accurately delivered are remarkably similar. The history of the development of biologics closely follows that of the small synthetic molecules. One such similarity is the emergence of technologies that allow the oral dosing of proteins to mimic the most favored route of delivery enjoyed by the large majority of small molecules. In fact, the two fields are converging in terms of how consumers and regulators think of biologics simply as another category of drugs. The most recent example of this is the attempt to demonstrate that approval of "generic" or "biosimilar" versions of these complicated molecules is possible. In just 20 years, beginning with the approval of the first biological drug, OKT3, markets around the world have begun to see the introduction of "generic" versions of some of these highly successful and important therapeutic products.

REFERENCES

1. Van Arnum P, Greb E, McCormick D. Pharmaceutical Technology 2005 Manufacturer's Rankings, July 2006.

2. Carstensen JT. Drug stability: principles and practices. Vol. 43. Drugs and the Pharmaceutical Sciences. New York: Marcel Dekker, 1990:2–4.
3. ICH Harmonized Tripartite Guideline Q8—Pharmaceutical Development, Nov 10, 2005.
4. http://www.gene.com/gene/about/views/followon-biologics.jsp.
5. Markovic I. Challenges associated with extractable and/or leachable substances in therapeutic biologic protein products. Am Pharm Rev 2006; 9:20–27.
6. Modi P, Mihic M, Lewin A. The evolving role of oral insulin in the treatment of diabetes using a novel RapidMist System. Diabetes Metab Res Rev 2002; 18 (suppl 1): S38–S42.
7. Pearlman R, Nyugen T. Pharmaceutics of protein drugs. J Pharm Pharmacol 1992; 44:178–185.

Index